Sex Determination
in Fish

Sex Determination in Fish

T.J. PANDIAN

Visiting Professor
CAS in Marine Biology
Annamalai University
Parangipettai 608 502
Tamilnadu, India

CRC Press
Taylor & Francis Group
Boca Raton London New York

CRC Press is an imprint of the
Taylor & Francis Group, an **informa** business

A SCIENCE PUBLISHERS BOOK

CRC Press
Taylor & Francis Group
6000 Broken Sound Parkway NW, Suite 300
Boca Raton, FL 33487-2742

First issued in paperback 2017

ISBN 13: 978-1-138-11199-8 (pbk)
ISBN 13: 978-1-57808-748-8 (hbk)

Cover illustration reproduced from **Liu et al. 2007, Genetics, 176: 1023–1034 Figures 5, 7 & 8** by permission from the Genetics Society of America. © Genetics Society of America.

Library of Congress Cataloging-in-Publication Data
Pandian, T. J.
  Sex determination in fish / T.J. Pandian. -- 1st ed.
    p. cm.
  Includes bibliographical references and index.
  ISBN 978-1-57808-748-8 (hardback)
  1. Fishes--Sexing. 2. Fishes--Genetics. 3. Fishes--Sex
    ratio. 4. Fishes--Reproduction. I. Title.
  QL638.99.P36 2011
  571.88217--dc23

                                          2011013330

Preface

Fishes are indeed the most versatile and fascinating group of animals; in them sex, sexuality, simple and multiple sex chromosome systems and sex determining genes remain an elusive evolutionary enigma. In fact no other group of animals offers as much scope as fishes do for the study and research on sex determination. With most divergent expression of sex including many morphotypes within a sex, sexuality ranges from gono-chorism to unisexualism and to an array of almost unbelievable patterns of functional hermaphroditism and reproductive modes from sexual to gyno-genesis and to hybridogenesis, each involving oviparity to viviparity.

The title of the book is a 'hot area' of research. Not surprisingly, there are many reviews and books on this topic. However, these are more concerned with sex differentiation than sex determination; they have not considered unisexualism in the context of sex determination in fishes. In an attempt to find clues to resolve the riddle of sex determination in fishes, this comprehensive book explores it from cytogenetics through hybrids, gynogenics, androgenics, ploidies, allogenics/xenogenics to sexonomics of gonochores, hermaphrodites and unisexuals. About 77 and 50% of references cited here are dated after 1991 and 2001, respectively; they were collected from widely scattered 375 sources of journals, book proceedings, theses and so on. As the book is a continuum of the earlier book 'Sexuality in Fishes', there are a few unavoidable but obligatorily required duplications to keep each chapter complete and independent, besides pointing out areas of research requiring critical inputs. None of the earlier reviews/books have ever considered allogenics/xenogenics; this book is the first to report that researches in this frontier area have conclusively shown that fishes have retained bisexual potency even after sexual maturity and spermiation. The XY genotype found in the unexpected female phenotypes sired by supermales (Y^1Y^2) and androgenic males (Y^2Y^2) points out the need to employ sex specific molecular markers to identify the true genotype of a juvenile, which matures either as a male or female, depending upon the sex of its pair (female or male) and thereby critically assess the environmental role in sex determination. Hopefully, this book bridges the gaps among molecular biologists in search of sex determining gene(s), fishery biologists endeavouring to develop techniques for profitable monosex aquaculture and ecologists interested in conservation of fishes and their genomes.

This book is the second in this series brought out by Science Publishers.

December 2010   TJ Pandian
Madurai 625014

Acknowledgements

It is with great pleasure that I wish to thank Drs. S. Kirankumar (Sheffield) and E. Vivekanandan (Chennai) for critically reading the manuscript of this book and offering valuable suggestions. I am grateful to Drs. S. Arunachalam, R. Jeyabaskaran, G. Kumaresan, P. Murugesan and A.G. Ponniah for useful discussions. Special thanks are due to Prof. T. Balasubramanian for his consistent support and encouragement. I gratefully appreciate my former students. Drs. K. Varadaraj (USA), S. Kavumpurath, T. George, S.Kirankumar (UK), R. Koteeswaran and C.J. David (USA), whose publications from my laboratory indeed form the backbone of this book. The Central Marine Fisheries Research Institute, Kochi provides the best library in India for fishery science and excellent services to visitors. The MS for the book was prepared at Dr. R.D. Michael's Fish Immunology Laboratory, Lady Doak College, Madurai by Ms. S.P. Geetha and I wish to thank them for their patience, co-operation and excellent work.

To reproduce figures and tables from published domain, I need to thank many. Firstly, I wish to gratefully appreciate the open door policy and record my very sincere thanks for issuing permission to The American Society of Ichthyologists and Herpetologists (*Copeia*), The Brazilian College of Animal Reproduction (*Animal Reproduction*), Cambridge University Press, Editor in Chief, *Folia Zoologica*, Executive Editor, *Proceedings of the National Academy of Sciences, USA*, The Fisheries Society of the British Isles/Blackwell-Wiley (*Journal of Fish Biology*), The Genetic Society of America (*Genetics*), The Ichthyological Society of Japan (*Japanese Journal of Ichthyology, Ichthyological Research*), Indian Academy of Science (*Current Science, Journal of Genetics, Proceedings of the Indian Academy of Sciences: Animal Science*), The International Journal of Developmental Biology (Spain), The Japanese Society of Fisheries Science (*Fishery Science*), National Institute of Science Communication and Information Resource (*Indian Journal of Experimental Biology*), The Royal Society, London, UK (*Proceedings of the Royal Society* London, Series B), The Society of Endocrinology and Bioscientifica (*Journal of Molecular Endocrinology*) and The Zoological Society of Japan (*Zoological Science*).

I welcome and gratefully appreciate the new policy of *Evolution, Aquaculture Research* and *Journal of Fish Biology* for choosing to leave the copyright with respective authors; accordingly, I thank very sincerely my fellow scientists Drs. G. Caetano, R. Rosa, L.R. Franca, Judith E. Mank, Vladimir P. Margarido, Orlando Moreiro-Filho, Phil P. Molly, Konrad Ocalewicz and Shuming Zou. Special thanks are also due to Dr. VP. Margarido for facilitating the permission securing process with South American colleagues. I also wish to thank the Society for Biology of Reproduction (*Biology of Reproduction*) for issuing the copyright permission. For advancing science in this area, the rich contributions made by the authors of the publications, whose figures and tables are reproduced in this book, I extend my grateful appreciation.

TJ Pandian

Contents

Preface   v
Acknowledgements   vii

1. Introduction   **1**
   1.1 Gonadal origin, eggs and sperm   3
   1.2 Chromosomes and gamety   7
   1.3 Prevalence of gonochorism   8
   1.4 Escape from Muller's ratchet   8
   1.5 Multiple parentage and genetic diversity 10
   1.6 Prelude   15

2. Cytogenetics and Sex Chromosomes   **18**
   2.1 Chromosomes and genomes   19
   2.2 Cytogenetics   23
   2.3 Heterogamety   26

3. Hybridization and Sex Ratio   **36**
   3.1 Extent of hybridization   37
   3.2 Hybrid identification   38
   3.3 Viability and fertility   40
   3.4 Introgressive hybridization   44
   3.5 Sex ratio   47

4. Gynogenesis and Consequences   **51**
   4.1 Gynogenic types   51
   4.2 Purity of gynogenics   56
   4.3 Survival and gamety   58
   4.4 Fertility and F_1 sex ratio   64
   4.5 Sex ratio and departures   65

5. Androgenesis and Autosomes   **69**
   5.1 Production methods   70
   5.2 Survival   78
   5.3 Fertility   82
   5.4 Overriding autosomes   84

6. Triploidy and Sterility   89
6.1 Types of triploids   90
6.2 Survival and sex ratio   96
6.3 Male's fertilizability   99
6.4 Female sterility   105

7. Tetraploidy and Polyploidy   112
7.1 Survey and history   112
7.2 Scope for polyploidization   115
7.3 Tetraploid types   117
7.4 True induced tetraploids   118
7.5 Polyploidy   122
7.6 Interploid hybridizations   125
7.7 Unreduced "gamete bank"   126

8. Allogenesis and Xenogenesis   130
8.1 Primordial Germ Cells (PGCs)   131
8.2 PGC transplantation system   134
8.3 PGCs in hybrid embryos   139
8.4 PGCs in heterospecific embryos   141
8.5 Spermatogonial Stem Cells (SSCs)   146
8.6 Renewal of SSCs   148
8.7 SSC transplantation system   149
8.8 Relevance of PGCs and SSCs   155
8.9 Surgical gonadectomy   158

9. Sex change and Hermaphroditism   160
9.1 Extant of hermaphroditism   161
9.2 Sex change in gonochores   165
9.3 Sex allocation theory   166
9.4 Sex changing age   167
9.5 Genetic or environment?   171

10. Unisexualism and Reproduction   177
10.1 Mutational meltdown   178
10.2 Exploration of reproductive modes   181
10.3 *Squalius alburnoides*   184
10.4 Males in unisexuals   188

11. Genetic Sex Determination   191
11.1 Molecular markers   193
11.2 Sex chromosomes   195
11.3 Sexonomics   196
11.4 Sex changers   201
11.5 Polygenic sex determination   203
11.6 Concluding remarks   205

References	209
Author index	249
Species index	257
Subject index	265
Colour Plate Section	269

1

Introduction

Sex determination refers to that event which compromises bipotential gonad to develop as an ovary or a testis (Hayes et al., 1998). Fishes are characterized by a level of sexual plasticity that is unrivalled among other vertebrates. They display most divergent expressions of sex. The range of morphotypes encountered within a sex is astounding (Mank and Avise, 2006a); for instance, genetic morphotypes reported within a species are primary and secondary males (all diandric protogynous hermaphrodites, e.g. *Thalassoma bifasciatum*), parental and cuckolder (e.g. blue-gill sunfish, *Lepomis macrochirus*), hooknose and jack (e.g. coho salmon, *Onchorhynchus kisutch*), pairing and haremic males (e.g. the Nigerian cichlid *Pelvicachromis pulcher*) and phenotypic morphotypes are the streaker, satellite, sneaker and piracy males (e.g. shell-brooding cichlid *Telematochromis vittatus*). Whether these morphotypes represent the expression of alleles of the same sex determining gene(s) or different genes operate to determine these male genotypes is not known. In *O. tshawytscha*, sneaker males are known to sire sneaker sons (Heath et al., 2002).

In fishes sexuality ranges from gonochorism to unisexualism, and self-fertilizing hermaphroditism to sequential and serial hermaphroditism (Fig. 1). Teleosts are the only vertebrate group, within which self-fertilizing, sequential and serial hermaphroditisms are found. In the gonochoristic platyfish *Xiphophorus helleri*, hermaphroditism has been reported and few individuals have also been found to switch from female to male (Lodi, 1979). Some fish species like *Cobitis taenia bilineata* have hermaphroditic and gonochoristic sub-populations (Lodi, 1980), while others like *C. taenia* have unisexual and bisexual complex (Bobyrev et al., 2002). Every conceivable form of mating system ranging from monogamy to promiscuity is employed by fishes. Polygamy in both sexes is most common among nest-tending species (e.g. *Ophiodon elongatus*, King and Withler, 2005). Among

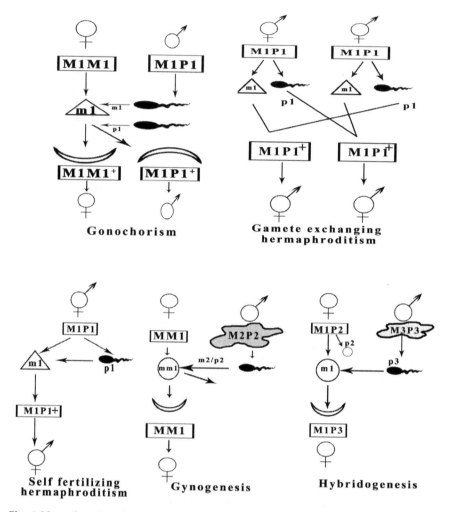

Fig. 1 Natural modes of reproduction in fishes. Boxes with continuous (conspecific) and pseudopodial (heterospecific) lines represent parents. Circles with small letters represent unreduced clonal eggs. Triangles with small letters represent eggs produced following regular meiosis. Fertilized eggs with a + represent products of recombination.

sequential hermaphrodites, the mating system varies even within a species; for instance, within *Sparisoma cretense*, some are haremic, while others are group spawners (deGirolamo et al., 1999); *Halichoeres maculipinna* is haremic in Florida but not in Panama; the reverse is true for *H. garnoti* (Robertson, 1981). Modes of reproduction also ranges from sexual to gynogenesis and to hybridogenesis (Fig. 1) and each of these modes may involve external fertilization and oviparity, or internal fertilization and viviparity, which itself ranges from zygoparity to true viviparity (Pandian, 2010). In fact no other vertebrate group offers as much diversity for the study of sex determination as teleost fishes (Mank and Avise, 2006a).

1.1 Gonadal origin, eggs and sperm

In higher vertebrates, the gonad originates in an undifferentiated state and consists of a cortex, derived from stromal cells and a medulla from the peritoneal wall, which itself is drawn from the mesonephric blastema. During ovarian differentiation, the cortex proceeds through further development, but with simultaneous degeneration of the medulla; in the case of testicular differentiation, the developmental programme is reversed. However, the somatic component of the gonads of fishes is of unitary origin, i.e. the somatic component of ovary and testis originates from the peritoneal wall alone (Nakamura, 1978). Usually bilateral gonads are developed; however, a single (e.g. *Fundulus dispar*, Taylor and Burr, 1997) or fused (e.g. viviparous fishes, Parenti and Grier, 2004) may also be developed in some species; in the mud eel *Monopterus albus*, the left gonad is alone developed (Mei et al., 1993).

Following the arrival of Primordial Germ Cells (PGCs), as guided by chemo-attractants and colonization of the gonadal anlage, the gonad now consists of both germ cells and gonadal supporting somatic cells. The somatic cell lineages then develop into granulose and thecal cells in the ovary and differentiate into Sertoli cells and Leydig cells in the testis. The granulose and the Stertoli cells are known as germ cells supporting cells and directly enclose the germ cells, whereas thecal and Leydic cells function as sex steroid hormone producing cells. To study the interplay between the maternally-derived PGCs and the germ cell supporting somatic cells in sex determination and differentiation in medaka, Kurokawa et al. (2007) generated PGC and *nanos* deficient mutants of medaka through morpholino-mediated (MO) knockdown of *Cxcr4*, a chemo PGC-attractant receptor gene and *nanos*, a gene essential for the maintenance of germ cell identity. These germ cells-deficient XX morphants reversed sex by inhibiting the maturation of granulosa cells and induced transdifferentiation of Sertoli cells. Hence the germ cell supporting somatic cells are inherently predisposed to develop towards the male and the presence of PGCs, germ cells, originally derived from the maternally inherited mRNA (Hashimoto et al., 2004), is essential for the sustenance of sexual dimorphism. This conclusion of Kurokawa et al. (2007) is also consistent with the previous observation of Slanchev et al. (2005); the germ cells-deficient morphants of zebrafish generated by MO-inhibition of *dead end*, a gene involved in development of germ cells, reversed the sex to sterile males. Thus the germ cells and germ cell supporting somatic cells have a complementary role in sex determination.

Among vertebrates, fishes are unique for the lack of acrosome in their spermatozoa (Fig. 2). The sperm enters the egg through micropyle of the egg (Fig. 3). This has facilitated heterologous fertilization in many fishes (Pandian and Kirankumar, 2003) and led to hybridization and polyploidization, which may considerably alter the process of sex determination. Hybridization is now known to occur between more than 300 fish species (Argue and Dunham, 1999; Scribner et al., 2001). Fishes are unique in that polyploidization has occurred extensively, independently and is often repeated in many fish groups (Ohno et al., 1968, Pandian and Koteeswaran, 1998; Leggatt and Iwama, 2003; Otto, 2007). To provide a ready reference, Table 1 lists fish species, in which heterospecific insemination leading to heterologous fertilization/activation, is not an uncommon phenomenon and results in the production of viable progenies. Upto now, experiments on heterospecific insemination have mostly been made in commercially important food and ornamental fishes. From available information, *Cyprinus carpio* may serve as a universal donor, as its sperm is accepted by a dozen species belonging to Cyprinidae, Cichlidae and Gobitidae. Among salmonids, *Oncorhynchus mykiss* is perhaps a universal recipient. Successful reciprocal heterospecific inseminations do occur: e.g. *C. carpio* and *Ctenopharyngodon idella*, *C. carpio* and *Hypophthalmichthys molitrix*. Not surprisingly, hybridizations among the Indian major carps are prevalent and have led to genetic retrogression. However, such reciprocal

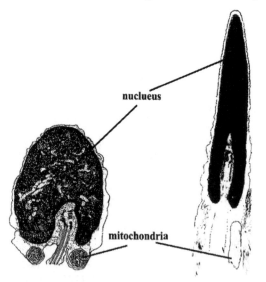

Fig. 2 Longitudinal sections of spermatozoa of oviparous *Oncorhynchus tshawytscha* (left) and viviparous *Poecilia latipinna* (right). Note the mitochondria in *O. tshawytscha* sperm and the large modified ones in *P. latipinna* (from Jamieson, BGM, 1991 *Fish Evolution and Systematic Evidence from Spermatozoa*, Cambridge University Press).

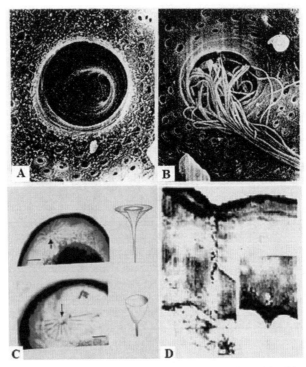

Fig. 3 Electron microscopic views of (A) egg showing micropyle and (B, D) entry of sperm through the micropylar canal in an egg of *Alcichthys alcicornis* (from Munehara et al., 1989). Apical view of unfertilized eggs of (C) *Gymnocorymbus ternetzi* (lower) and *Hemigrammus caudovittatus* (upper) and schematic views of their respective micropylar canal in the eggs (from David, 2004).

heterospecific insemination does not necessarily result in hybridization but may induce gynogenesis, as in *Puntius conchonius* ♀ x *P. tetrazona* ♂, or may also induce paternal (*C.idella* ♀ x *H. nobilis* ♂), or maternal triploidy, as in many cyprinids and salmonids. *Pangasius sutchi* successfully donates sperm to *Clarias macrocephalus*, but does not serve as recipient of the sperm of *C. macrocephalus*. Likewise *P. conchonius* can be a sperm donor to *P. tetrazona*, whose sperm is, however, not acceptable to *P. conchonius* (see Pandian and Kirankumar, 2003).

Incidentally, intra-cytoplasmic sperm injection (ICSI) is an *in vitro* fertilization technique, in which a single spermatozoon is injected directly into an egg. The ICSI has been widely applied for mammals, making useful contributions to both basic and applied aspects of gamete biology. In fishes, the technique has to encounter the presence of micropyle as an advantage but the scattered yolk granules throughout the cytoplasm and the invisible nucleus as disadvantages. Consequently, these disadvantages have limited

Table 1 Heterospecific insemination in fishes (from Pandian and Kirankumar, 2003).

Sperm donor	Sperm recipient
Cyprinus carpio	*Ctenopharyngodon idella*
C. carpio	*Carassius auratus*
C. carpio	*Hypophthalmichthys molitrix*
C. carpio	*Cirrhinus mrigala*
C. carpio	*Misgurnus anguillicaudatus*
C. carpio	*Cobitis biwae*
C. carpio	*Tinca tinca*
C. carpio	*Oreochromis niloticus*
C. carpio	*O. mossambicus*
C. auratus	*C. idella*
C. auratus	*M. anguillicaudatus*
C. auratus	*O. niloticus*
C. idella	*C. carpio*
H. nobilis	*C. idella*
Puntius conchonius	*P. tetrazona*
P. gonionotus	*C. carpio*
C. biwae	*M. anguillicaudatus*
M. anguillicaudatus	*C. biwae*
M. mizolepis	*Paralichthys olivaceus*
Barbus barbus	*C. carpio*
T. tinca	*O. niloticus*
Pangasius schwanenfeldii	*P. gonionotus*
P. sutchi	*Clarias macrocephalus*
Ictalurus furcatus	*I. punctatus*
Gnathopogon elongatus elongatus	*M. anguillicaudatus*
Herichthys cyanoguttatus	*O. mossambicus*
Osteochilus hosselti	*C. carpio*
Acanthopagrus schlegeli	*P. olivaceus*
Pagrus major	*Sparus aurata*
Menidia notata	*Fundulus heteroclitus*
Acipenser ruthenus	*Huso huso*
A. baeri	*A. ruthenus*
Salmo salar	*O. mykiss*
Salmo trutta	*O. mykiss*
Salvelinus fontinalis	*O. mykiss*
O. kisutsch	*O. mykiss*
O. tshawystcha	*O. mykiss*
O. masou	*O. mykiss*
S. trutta	*S. salar*
Thymallus thymallus	*O. mykiss*
Abramis brama	*C. carpio*
S. fontinalis	*S. trutta*
Poecilia velifera	*P. sphenops*
P. sphenops	*P. velifera*
Oreochromis aureus	*O. niloticus*
O. hornorum	*O. niloticus*
O. hornorum	*O. mossambicus*

Table 1 contd....

Table 1 contd....

Sperm donor	Sperm recipient
O. macrochir	*O. niloticus*
O. veriabilis	*O. niloticus*
O. hornorum	*O. aureus*
O. vulcani	*O. aureus*
O. niloticus	*O. leucostictus*
O. niloticus	*O. spilurus niger*
O. niloticus	*O. mossambicus*
*O. aureus hornorum**	*O. niloticus*
O. mossambicus	*O. spilurus niger*
Prinotus paralatus	*P. alatus*
Semotilus atromaculatus	*Phoxinus oreas*
Gila eremica	*G. ditaenia*
Micropterus dolomieui	*M. salmoides*

*Hybrid between *O. aureus* and *O. hornorum*

the hatching success to 1.6% in *Danio rerio* (Poleo et al., 2001) and 8.5% in *Oreochromis niloticus* (Poleo et al., 2005). Injecting a single sperm of the medaka *Oryzias latipes* into the cytoplasm through micropyle 10–15 seconds following the onset of egg activation induced by piezo-actuated vibration, comparable to the site and timing of penetration in normal fertilization, Otani et al. (2009) improved the hatching success to 13.4% .

1.2 Chromosomes and gamety

In most fishes, sex chromosomes remain cytologically indistinguishable, as they are perhaps at a low level of differentiation. According to Arkhipchuk (1995), karyotype of more than 1,700 species has been described. Of them, 176 species (10.4%) alone are reported to have 'heteromorphic sex chromosomes'; however, all these claims are yet to be confirmed (Haaf and Schmid, 1984; Sandra and Norma, 2009). Hence the mechanism of sex determination involving sex chromosomes are based more on genetic rather than cytological evidence. Yet, the early classical genetic experiments by Winge (1930) recorded the presence of XX males and XY females in a male heterogametic guppy *Poecilia reticulata*, which perhaps paved the way for the polygenic sex determination hypothesis (Price, 1984). In the European sea bass *Dicentrarchus labrax*, sex determination by polygenic system has been hypothesized (Vanderputte et al., 2007).

Unlike in other vertebrates, there is a range of diverse sex determination mechanisms (SMD) in fishes; as many as seven mechanisms have been so far reported (Chourrout, 1989), but male heterogamety (XX/XY) and female heterogamety (WW/WZ) are the basic SMDs and others are variants of them (Sandra and Norma, 2009). Unlike in mammals and birds, within a genus like *Oreochromis*, males are heterogametic in *O. mossambicus* (Varadaraj and Pandian, 1989a) but females are heterogametic in *O. aureus* (Mair et al., 1991b).

1.3 Prevalence of gonochorism

According to the estimates of Fish Base (2010, www.fishbase.org), there are 30,000 fish species (Table 2). A vast majority of fishes are gonochores; i.e. individuals develop and remain as either female or male throughout their life. Not surprisingly, more than 98% of fishes are gonochores. Just 1.9% of fishes alone are non-gonochores. Yet, they make fishes as the most fascinating group among vertebrates, especially from the point of sex determination. Besides, sex changing sparids, serranids, scarids and lethrinids sustain valuable fisheries both in temperate and tropical seas (Chopelet et al., 2009).

With regard to differences in the gonadal differentiation process, gonochores may be classified into three groups: 1. In *Primary gonochores*, gonadal development proceeds from an undifferentiated gonad directly to ovary or testis (e.g. *Cyprinus carpio*, Komen et al., 1992), 2. In *Secondary gonochores*, gonads develop first into the ovary with oocytes. Later oocytes degenerate by apoptosis in half of the embryos and the gonads then undergo masculinization (e.g. *Gambusia affinis*, Koya et al., 2003) and 3. In the third group, juveniles initially possess a bipotential intersexual gonad, which then develops directly into either an ovary or a testis (e.g. *Anguilla anguilla*, Beulbens et al., 1997).

1.4 Escape from Muller's ratchet

In most organisms, deleterious mutation occurs at the rate of one per diploid per generation and causes 1% reduction in fitness per generation. With almost total absence of segregation and recombination in unisexual and self-fertilizing fishes (Fig. 1), the loss of fitness due to the 'mutational meltdown' may invariably lead to reduction in population size and eventually drive the population into the evolutionary dead end, 'the Muller's ratchet' (Lynch et al., 1993).

Unisexuals: Barring *Squalius* (= *Rutilus*) *alburnoides*, all the other unisexuals produce unreduced diploid or triploid eggs and thereby these eggs have missed genomic recombination. They reproduce clonally by gynogenesis or hemiclonally by hybridogenesis (Fig. 1). Yet, they have thus far escaped from the Muller's ratchet (Loewe and Lamatsch, 2008), by inheriting (i) leaked paternal sub-genomic fractions (Rasch and Balsano, 1989; Jia et al., 2008) and (ii) host-species derived microchromosomes (Nanda et al., 2007; Lamatsch et al., 2004). Being hybrid themselves, the unisexuals are

Table 2 Sexuality, mating systems and modes of reproduction in teleostean fishes.

Sexuality	Species (no)	Mating system	Mode of reproduction	Reference
Gonochorism	29, 431	Monogamy→Promiscuous	Sexual	Pandian (2010)
Haremic gonochorism[1]	135	Polygynous	Sexual	Pandian (2010)
Unisexualism (♀♀ only)	8	Polygynous	Gynogenesis, hybridogenesis	Pandian (2010)
Self fertilizing hermaphroditism	2	Androdioceous	Selfing	Tatarenkov et al. (2009)
Potential selfing hermaphroditism	3		Selfing	Fishelson (1992)
Simultaneous hermaphroditism	14	Gametic exchange	Sexual	Sadovy and Liu (2008)
Protogynous hermaphroditism	323	Polygynous	Sexual	Nakazona and Kuwamura (1987)
Protandrous hermaphroditism	27	Polygynous	Sexual	Pandian (2010)
Serial hermaphroditism[2]	17		Sexual	Nakazona and Kuwamura (1987)
Total number of species	30, 000			Fish Base (2010)

1. belonging to Acanthuridae, Cichlidae, Labridae, Ostraciidae, Pomacanthidae
2. including Marian and Okinawan hermaphroditism

heterozygous; their heterozygosity is further increased by genome addition. Triploids occur in all the seven of the eight unisexuals, but the presence of tetraploids is limited to *Carassius* and *Cobitis* complexes, and *S. alburnoides*. Recently, the occurrence of tetraploid has been reported in the Amazon molly *Poecilia formosa* (Lampert et al., 2008).

Self-fertilizing hermaphrodites: As selfing hermaphrodite *Kryptolebias* (= *Rivulus*) *marmoratus* may suffer from (i) intense inbreeding and (ii) inherent physiological and hormonal 'conflict' by simultaneous production of eggs and sperm within an individual but has the benefit of assured fertilization without the need for a mate. Microsatellite analyses have shown that the androdiocious (populations consisting of hermaphrodites and males) *K. marmoratus*, and *K. ocellatus*, another selfing species, exhibit very high self fertilization frequencies of 98 and 97%, respectively (Tatarenkov et al., 2009). Using DNA fingerprinting with an array of microsatellite and minisatellite probes, Turner et al. (1992) have found very high clonal diversity in the Florida population of *K. marmoratus*. As many as 42 clones have been found among 58 individuals (1.4 individuals per clone). Such an unprecedented high clonal diversity may provide selective advantage to perpetuate *K. marmoratus* in a particular environment (see also Martin, 2007). However, studies on relocating *K. marmoratus* to other habitats are yet to be undertaken.

Despite their selfing frequency of ≈ 98% , the available genetic data imply a considerable antiquity for the populations of *K. mamoratus* and *K. ocellatus*. The conventional mtDNA clock calibration for vertebrate mtDNA (1% sequence change per million years) indicates that *marmoratus* and *ocellatus* were separated as independent species probably 2 million years ago (MYa). Even if mtDNA of *Kryptolebias* evolves 10 times faster than the vertebrate norm, then *ocellatus* and *marmoratus* were phylogenetically separated over 200,000 years ago (Tatarenkov et al., 2009). These data arguably provide evidence that despite selfing *K. marmoratus* and *K. ocellatus* have escaped from Muller's ratchet for not less than 200, 000 years.

1.5 Multiple parentage and genetic diversity

By facilitating recombination of maternal and paternal genomes in a fertilized egg, sexual reproduction increases genetic diversity among progenies. Hence recombination is regarded as the major source driving evolution. For instance, the gonochoric sailfin molly *Poecilia latipinna* has about 1.5 and 2.6 times more alleles for (major histocompatibility complex) MHC class I and II genes than those in the gynogenic molly *P. formosa* (Schaschl et al., 2008). Consequently, gonochores display high levels of diversity in their genetic make-up. But monogamous and haremic strategies may considerably reduce multiple parentage of their progenies. Yet they adopt a wide range of reproductive strategies to elaborate the genetic diversity (Mank and Avise, 2006b). Among vertebrates, fishes seem to have maximally exploited the alternate mating strategy (AMS) to achieve multiple paternity or maternity within a brood to increase genetic diversity and to ensure 100% fertilization success (FS).

Fishes display a staggering array of AMS. Genotypic morphotypes namely primary males among protogynous hermaphrodites and cuckolders among gonochores compete for fertilization by streaking and sneaking. Likewise phenotypic morphotypes, the streakers, satellites, sneakers and piracy males compete against the territorial male by streaking and sneaking. The AMS is so widely prevalent among the teleosts that in 1998 Taborsky found evidence for the utilization of AMS among 140 oviparous species belonging to 28 families. Since then publications on this topic have indeed been numerous (e.g. Jones et al., 2001; Immler et al., 2004, Mobley et al., 2009). The AMS is now classified into nine groups (Pandian, 2010). Some representative examples are provided to show the high levels of multiple paternity/maternity achieved by fishes and thereby increasing genetic diversity, using a variety of AMS.

Monogamy: As a mating system, monogamy may lead to inbreeding and decrease in genetic diversity. However, the teleostean monogamy seems to be a continuum from monogamy to polygyny/polyandry (Wickler and Seibt, 1983, Kuwamura, 1997). Limitation of space and resources may impose monogamy. Exclusive monogamy, in which a pair of female and male confines at least themselves to one reproductive cycle (see Kuwamura, 1997), is more common among fishes (Pandian, 2010). Experimental evidence suggests that among fishes monogamy lasts mostly for a single reproductive cycle; for instance, 56% monogamous pairs get separated after the first reproductive cycle in the Japanese goby *Valenciennea longipinnis* (Takegaki and Nakazona, 1999) and St Peters tilapia *Saratherodon galilaeus* (Fishelson and Hilzerman, 2002).

Harems: Extremely skewed sex ratios, prevalent in harems of gonochores and sequential hermaphrodites, may drive harem dwellers to the polygynic or polyandric mode of the mating system (Pandian, 2010). However, some sequential hermaphroditic harems lack stable dominance and a permeable harem is maintained by more than two males (e.g. *Anthias squamipinnis*, Shapiro, 1981). Even in non-permeable harems each with a single territorial male, incidences of extra-haremic spawnings by female members of the harem are not uncommon (e.g. *Labroides dimidiatus*, Sakai et al., 2001). Hence exclusive polygyny among these harems is diluted to an extent to allow a little more genetic diversity.

Pelagic spawners: Among pelagic-spawning coral fishes, there are two types of males: pair spawners and group spawners. Typically, group spawning may include 5–20 primary males and the males jointly release 50–80 times more sperm than that by a pair-spawning male (Petersen et al., 1992; Shapiro et al., 1994). Hence the pair-spawners economize sperm but the group spawners ensure multiple paternity of fertilized eggs. Using the microsatellite technique, Hutchings et al. (1999) have shown that the most dominant males fertilized 85, 11 and 4% of eggs of the pelagic broadcast spawning *Gadus morhua*, although the spawning female was encircled by as many as 17 males. It appears that a minimum of three males sire the progenies of pelagic spawners.

The occurrence of one or more morphotypes like sneaker, satellite, cuckolder (female mimic), piracy male and so on are not uncommon among both gonochores and hermaphrodites. These morphotypes seem to have been produced through differential expression of reproductive hormones (Knapp, 2004); for instance, a higher titre of 11-ketotestosterone seems to produce piracy males (Brantley et al., 1993). Figure 4 describes the possible evolutionary pathways, through which these morphotypes may have arisen (Mank and Avise, 2006a). When the hormonal adaptation becomes heritable, a sneaker father may sire sneaker sons (Heath et al., 2002)

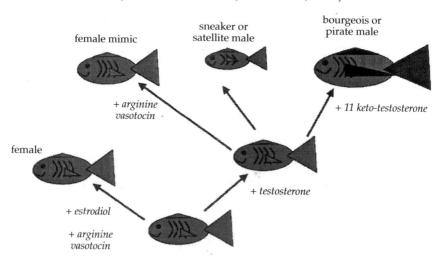

Fig. 4 A generalized model for hormonal adaptation resulting in the production of different morphotypes within a sex in a fish species (from Mank and Avise, 2006a).

Demersal spawners: Due to limited number of eggs in a clutch, demersal spawners achieve 100% FS. Kanoh's (2000) observations provide a representative example for multiple paternity among demersal spawners and relevant information from his observations is compiled in Table 3.

Accordingly, pair spawning in the rose bitterling *Rhodeus ocellatus* involves one female and one male ($1♀ + 1♂$); on arrival of a female at the mating site of the territorial male, namely the mussel *Anadonta woodiana*, courtship activity is commenced. The female repeatedly oviposits eggs in 4–5 pulses and the male ejaculates following each pulse (for more details see Pandian, 2010). In group spawning ($1♀ + 5♂$), this alternate sequence of oviposition and ejaculation is not followed by the males, who ejaculate continuously one after another. Pair spawning accompanied by sneakers also alters oviposition-ejaculation sequence so much that inappropriate timing of

Table 3 Estimated sex ratio, mating and reproductive success of the Japanese rose bitterling *Rhodeus ocellatus* (compiled from Kanoh, 2000).

Parameter	Pair spawning	Group spawning	Pair spawning + sneakers
Operational sex ratio	$1♀ : 1.1♂$	$1♀ : 4.9♂$	$1♀ : 3.1♂$
Mating success (%)	36	18	46*
Reproductive success (%)	60	10	30

*18, 11, 10, 5 and 1% pair spawnings were accompanied by 2, 3, 4, 5 and 7 sneakers, respectively

ejaculations occurs. Despite 64% mating success achieved jointly by group spawners and 1–7 sneakers interrupting pair spawners, their reproductive success—as estimated by allozyme analysis—is limited to 40%. This limitation is mainly due to inappropriate timing of sperm ejaculation. Yet 40% progenies are sired by (average) eight males belonging to the group spawners and sneakers, which achieve a significant level of multiple paternity of the rose bitterling.

Cuckoldry: An AMS successfully adopted by a large number of nest-holding fishes is the cuckoldry. A territory holder is the parental male and precociously fast grown cuckolders compete for mating against the parentals. For instance, the females of blue-gill sunfish *Lepomis macrochirus* do not discriminate cuckolders from parental and spawn in the presence of parental and cuckolders. In fact they may release three times more number of eggs than when they spawn in the presence of parental alone (Fu et al., 2001). Cuckolders are excellent sperm competitors and fertilize more than 80% of the eggs that a female spawns (Neff, 2004). Hence cuckolders facilitate multiple paternities in a clutch of eggs.

Nest-holders: Nest–holding oviparous fishes provide excellent opportunity to study multiple paternity/maternity using one or other genetic marker. Jones et al. (2001) reported that cuckolders of the sand goby *Pomatochistus minutus* fertilize eggs in 50% of the assayed nests. In nest holders, 5–30% embryos in a nest are not sired by the nest holders (polyandry, Table 4). Each nest also contains embryos, which have been generated by 2–6 mothers,

Table 4 Multiple paternity/maternity among nest-holding and other fishes.

Species	Incidence (%)	Progenies not sired by nest holder (%)	Minimum mothers (no/nest or ♂)	Reference
		Nest holders		
Lepomis punctatus		5	4.4	DeWoody and Avise (2001)
Micropterus salmoides		7	1.2	DeWoody and Avise (2001)
L. auritus		12	3.6	DeWoody and Avise (2001)
Etheostoma olmstedi		14	2.6	DeWoody and Avise (2001)
Gasterosteus aculeatus		14	6.0	Rico et al. (1992)
Pomatochistus minutus		15	–	Jones et al. (2001)
Spinachia spinachia		15	–	Jones et al. (1998)
Gobiusculus flavescens	100	–	4.3	Mobley et al. (2009)
Salmo salar			5.0	Martinez et al. (2000)
		Pelagic spawners		
Gadus morhua	76		3.0	Hutchings et al. (1999) Bekkevold et al. (2002)
		Mouth brooders		
Pseudotropheus zebra	86			Parker and Kornfield (1996)
		Pouch brooders		
Syngnathus floridae			1.9	Jones et al. (2001)
S. typhle			3.1	Jones et al. (2001)
		Live bearers		
Sebastes alutus	71			van Doornik et al. (2008)
S. altrovirens	100			Sogard et al. (2007)
S. schlegeli	89			Yoshida et al. (2001)
Ditrema temmincki	50		2.0	Takagi et al. (2008)
Neoditrema ransonneti	33		2.0	Takagi et al. (2008)

who have visited the nest and mated with the nest-holder (polygyny). Hence, in both sexes polygamy is rampant in the nests, and the nests render both sexes to be promiscuous (King and Withler, 2005). Thus nest-tenders maximize multiple paternity and maternity and hence genetic diversity within populations.

Live bearers: Most surprisingly, 71–100% females in populations of marine viviparous fishes are reported to gestate embryos sired by more than one male (Table 4). Equally high values have also been reported for marine oviparous cod, pouch-brooding seahorses and mouth brooding freshwater cichlids. The females of viviparous fishes can store sperm for periods of 2 (Takagi et al., 2008) to 4 months (Love et al., 2002) and may selectively fertilize their eggs using the stored sperm from more than one male. In a sense polyandry seems to ensure multiple paternity in these viviparous fishes. Thus, irrespective of differences in breeding sites and reproductive modes, fishes seem to increase multiple paternity and thereby genetic diversity among progenies.

1.6 Prelude

With most divergent expression of sex including many genotypes within a sex and with sexuality ranging from gonochorism to unisexualism and to self-fertilizing hermaphroditism and from simultaneous to sequential and to serial hermaphroditism, fishes are indeed the most versatile and fascinating group of animals, in which sex, sexuality and sex determination remain an elusive evolutionary enigma. Cytological studies have laid the foundation for the basic understanding of the chromosomal mechanism of sex determination and male/female heterogamety. Owing to the rapid ongoing evolution of the sex determining system, the presence of sex chromosomes is realized in fishes more from genetic rather than cytological evidence. Hybridization brings together an amalgamation of two or three distinct genomes. Due to human interventions, its frequency has been accelerated by several orders of magnitude. Though observed in fishes more than any other vertebrates groups, very few hybridizations have occurred and have been made to cross (i) male heterogametic species with female heterogametic and (ii) protandrous hermaphroditic species with protogynous hermaphroditic, the results of which could have given a greater insight into the vagaries of departures from the expected sex ratio of 0.5 ♀ : 0.5 ♂.

With no paternal genomic contribution, gynogenics offer a unique opportunity to study the independent role played by females on sex determination. The reverse is true of androgenics. Induction of gynogenesis and androgenesis has given an idea on the dimensions of departures from the expected sex ratio and broadly indicates the ubiquitous but diverse overriding role of one or more minor genes located on autosome(s). Unisexuals are of hybrid origin but share no recombination with the genomes of their parental species. With no male ever recorded, *Poecilia, Poeciliopsis* and *Menidia* alone are true unisexuals and reproduce by gynogenesis/hybridogenesis. In *Carassius* and *Phoxinus* males, when present, have either no role or a limited role in fertilizing females. Maintenance of males in a population is a real luxury for a species to sustain itself but its critical need to ensure recombination is displayed by *Squalius*.

Surprisingly, hybridization results in male sterility but triploidization in female sterility. Triploidization has been much on search to deliver the commercially valuable 'broiler' fishes and as a widely practicable means to contain exotics and transgenics. The easiest means to mass produce triploids is to induce interploid hybridization between tetraploids and diploids. However, despite repeated attempts, fertile bisexual tetraploids have been generated only in five species, and the credit goes to the Chinese for having produced three of them. Departures from the expected sex ratio in triploids and polyploids seem more due to sex dependent mortality suffered throughout the life span, suggesting the inability of females to tolerate genome addition beyond diploidy.

In the context global warming, conservation of genome has become a 'hot area' of research. Due to mesolecithality, eggs and zygotes of fishes are not amenable for preservation. However, Primordial Germ Cells (PGCs) and Spermatogonial Stem Cells (SSCs) are readily cryopreserved. On transplantation of the PGCs or SSCs to a homospecific (allogenics) or heterospecific (xenogenesis) surrogate, a fish species can be restored. Of them, the xenogenics may revolutionize genome conservation and aquaculture by realizing the ideas like mass production of tuna from surrogate mackerel. From the point of sex determination, these novel experiments have clearly shown that bisexual potential is retained by the fishes even after sexual maturity. In this context, it may not be difficult to reconcile with the existence sex changing fishes among 367 species belonging to 34 families characterized by sequencial or serial hermaphroditism (Sadovy and Liu, 2008) as an intriguing phenomenon. Changing sex seems to enhance the life time reproductive success, especially at the savings of 20% investment on gonads. Experiments undertaken to show the social pressure induced sexual maturation, sex differentiation and sex change do require evidence from genomic DNA analysis before the total elimination of the role of genetic sex determination in the sex changing fishes.

To escape from Muller's ratchet and/or mutational meltdown, the unisexuals have adopted one or more of the following strategies: Incorporation of (i) subgenomic amounts of DNA (ii) B chromosomes and (iii) genome of the sperm donor. Thus *Poecilia* and *Poeciliopsis* have perpetuated for more than 100,000 generations over a period of 270, 000 years. Amazingly, *P. formosa* has retained the structural genes of males as functional but not expressed for 280,000 years. Thus the bisexual potency has been retained for 280,000 years.

In fishes, very little is known about the genetic and molecular mechanisms that control sex determination. Molecular markers have been developed to identify genetic sex of a couple of barbs and tetras, sex linked molecular markers to identify sex chromosomes and genome mapping to identify the locus SEX in sex chromosome of a few salmonids. The 280 bp length *Dmrt1bY*, a duplicated copy of autosomal *DMRT* gene, is the only sex determining gene known to control sex in *Oryzias latipes* and *O. curvinotus*; the *vas-s* and *vas*, *Dmrta2* and *Amh* are the other proposed genes associated with sex determination in other medaka and cichlid. Thus there is no common or universal sex chromosome and sex determination gene controlling the sex in fishes.

2

Cytogenetics and Sex Chromosomes

More than 98% fishes display gonochorism; the others exhibit either any one pattern of hermaphroditism [e.g. simultaneous, sequential (protogynous: monandric, diandric; protandrous: monogynic, digynic) or serial (including Marian, Okinawan, cyclical)], or unisexualism, suggesting the presence and operation of diverse sex determination mechanisms. Expectedly, sex determination in fishes ranges from strictly genetic to primarily genetic with environmental modulation perhaps at the downstream levels of genetic cascade of sex differentiation. Male heterogamety (XX/XY), and female heterogamety (WZ/ZZ) with or without autosomal influence on sex determination have been reported (Devlin and Nagahama, 2002). Sex chromosomes are also known to display variable levels of molecular differentiation (e.g. Libertini et al., 2008). Different genetic sex determination mechanisms have been reported to be in operation for species belonging to a single genus *Poecilia* (e.g. *P.reticulata*, Nanda et al., 1990a, *P.sphenops*, George and Pandian, 1995) and *Oreochromis* (e.g. *O. mossambicus*, Varadaraj and Pandian, 1989a, *O. aureus*, Mair et al., 1991b), as well as for different populations of the same species (*Gambusia affinis*, Black and Howell, 1979) and *Xiphophorus maculatus* (Kallman, 1984). Briefly, sex determination in fishes appears to involve repeated emergence of new sex chromosomes from autosomes and thus bringing the sex determination cascade under new master regulators (Volff et al., 2007).

2.1 Chromosomes and genomes

With the presence of evolutionarily labile sex chromosome systems and relatively undifferentiated chromosomes, each measuring 2–5 μm (Gold, 1979), the light microscopic resolution aided by differential staining techniques has not yielded adequate information on sex chromosomes. Hence karylogical analysis of fishes is of limited value to cytogenetics, which is concerned with the study of chromosomes and cytological mechanisms of hereditary transmission by chromosomes as carriers of genes. Methods and choice of tissue for karyotype preparation, differential banding and other molecular techniques used to understand the cytogenetics of sex chromosomes of fishes have been summarized from time to time by Gold (1979), Manna (1984), Rishi (1989) and Pandian and Koteeswaran (1998).

The diploid number of chromosomes in Actinoptergiian fishes ranges from 12 in *Gonostoma bathyphillum* to 500 in *Acipenser mikadoi*. Summarizing available information drawn from over 1,000 publications on karyology of more than 3, 000 fish species, Klinkhardt (1998) recorded that our present knowledge on chromosome structures of fishes is largely based from six thoroughly examined families. Of them Cyprinidae alone includes the highest number of 2,420 species belonging to 250 genera; of these cytogenetic studies have been made on 1,410 species belonging to 152 genera. In them, the diploid (2n) chromosome number ranges from 34 to 446 and averages to 50 (Mani et al., 2009), which is not very different from the modal number 48 (Klinkhardt, 1998). Listing cytogenetic characterization of 58 species belonging to Pleuronectiformes, Azevedo et al. (2007) have noted (i) a reduction in 2n number in several species due to centric fusions and (ii) an increase in chromosome arm number.

Incidentally, the Neotropical fishes call for special attention. The number of fish fauna described and catalogued from the Neotropical freshwater is 4,475 species, belonging to 71 families (Reis et al., 2003). According to Schaefer (1998), about 8,000 freshwater species may be present in the Neotropics and may represent nearly 25% of all fish species of the world. In 1978, information on chromosome number was recorded for 252 species belonging to 88 genera and 23 families (Almeida-Toledo, 1978). Ten years later, the information was available for 433 species belonging to 145 genera and 33 families (Oliveira et al., 1988). Thanks to the sustained endeavour of the Brazilian ichthyologists, today a strong database has been established to provide information on fundamental chromosome number, karyotype formulae, number of nuclear organizing regions, chromosome banding, B chromosomes, sex chromosomes and DNA content of some 1,156 Neotropical fish species. Of them, the presence of sex chromosomes has been claimed only for 5.9% species (Oliveira et al., 2009).

According to Ohno (1970), the DNA content of the ancestral fish genome is only 20% of that of mammals. A significant positive correlation between chromosome number and genome size has been found for fishes (Hinegardner and Rosen, 1972). Using 1,764 estimates of genome size (C value = DNA content in pg per haploid nucleus) available for 1,270 fish species, Smith and Gregory (2009) noted the following: 1. Genome size of fishes ranges from 0.4 pg in the pufferfish *Tetraodon fluviatalis* to 7.2 pg in green sturgeon *Acipenser mediarostris* and rarely to 120.7 pg in the lungfish *Lepidosiren paradoxa* and to 142 pg in *Polypterus aethiopicus* (Pedersen, 1971). 2. For marine, anadromous and freshwater fishes, the DNA content averages to 0.95, 1.35 and 1.70 pg, respectively. 3. Fishes with fastest population doubling time of <1.3 years have the smallest genomes (1.15 pg), followed in order by species with doubling times of 1.4–4.5 years (1.2 pg), 4.5–14 years (1.3 pg) and those with longer than 14 years have the largest genome size of 1.65 pg. Using fishes belonging to 22 freshwater families and 18 marine families, Hardie and Hebert (2004) have found the mean genome size as 0.975 pg for live-bearers including mouth brooders, 1.0 pg for the embryo-guarders and 1.44 pg for the non-guarding egg broadcasters.

In general, sex chromosomes in fishes are rich in Transposable Elements (TEs) and other types of repetitive elements (Nanda et al., 2002, Kondo et al., 2006, Wang et al., 2009), possibly due to suppression of recombination (Steinemann and Steinemann, 2005). Among diploid fishes, the differential accumulation of non-coding DNA is the major contributor to the diversity in genome size and TEs constitute the most significant component of it (Kidwell, 2002). The estimated number of TE copies is only 4, 000 in the pufferfish *T. nigroviridis* against more than a million in *Alu* elements alone in human, but in terms of diversity, there are 73 different types in TEs in the pufferfish, as compared to about 20 in human and mice (Jaillon et al., 2004). Moreover, 40 TE families are active in *Takifugu rubripes* against just six in human (Aparicio et al., 2002).

The differential accumulation of TEs, which are relatively more diverse and active in fishes, might play a role in the differentiation of sex chromosomes (Volff et al., 2007). Suppression of recombination in the sex determining region, an early step towards the differentiation of sex chromosomes, has been reported in several fishes (Charlesworth et al., 2005). Depending on the nature of the genetic factor that has gained dominance in the sex determination process, either male specific Y chromosome or female specific W chromosome emerges. The isolation of such sex chromosomes in one sex leads to the remarkable cytogenetic transformation, i.e. the establishment of sex chromosome, and heterogamety and homogamety. Thus in these monogenic systems, sex is determined by a gene located on a certain chromosome and the genes on other chromosomes have little effect (Devlin and Nagahama, 2002).

"Protected by the constant presence of the X (or Z) chromosome, which is evolutionarily constrained by homozygosity in one sex, and restricted in ability to recombine with it, the Y (or W) chromosome may degenerate from the condition of initial homomorphism with the X (or Z) to a small heterochromatic element with few functional loci" (Green, 1990). Barring those belonging to *Triportheus*, the W chromosome is larger in size due to hetrochromatization (e.g. *Parodon hilarii*, Fig. 7c), in contrast to the Y chromosome in male heterogametic system. Among *Triportheus* spp, however, the W chromosome is smaller than the Z chromosome and reduction in size of the W chromosome ranges from 25% in *T.elongatus* to 50% in *T.guentheri* and to 65% in *T.albus* (Artoni et al., 2001). Figure 5 is a schematic representation of the transformation undergone by W chromosome through reduction in its size in *Triportheus* spp and increasing its size in all others. However, the average length of sex chromosome in fishes is estimated as 5% of the total karyotype and it is larger than the average chromosome size (Arkhipchuk, 1995). This observation suggests that degeneration of sex chromosomes in fishes has not resulted in the loss of sex chromosome fragment(s) or drastic genetic damage, where Y or W is lost. In fact, of 186 fish species recognized to have heteromorphically distinguishable sex chromosomes, only 7.5% of fishes seem to have lost Y (XX/XO) and W

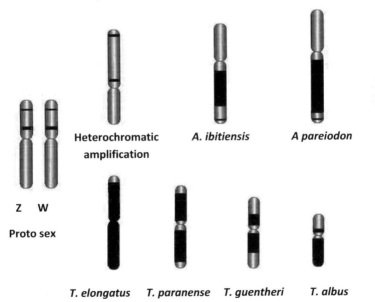

Fig. 5 Schematic illustration of the transformation of W chromosome in female heterochromatic fishes. The upper row shows the increasing size of W chromosome in most female gametics, e.g. *Apareiodon* and the lower row shows the rare decreasing size of W chromosome in *Triportheus* spp (compiled from Artoni et al., 2001, Bellafronte et al., 2009).

(ZO/ZZ) chromosomes. In humans, over 90% of X-chromosomal genes do not have any counterpart on the Y chromosome (Graves, 2006). That the YY supermales and ZZ superfemales are viable and fertile in many fish species (Table 5) indicates that gene contents of the Y and W chromosomes are almost similar to their respective counterpart X and Z chromosomes, and still share a similar repertoire of functional genes (Woram et al., 2003). In fact the evolution of sex chromosomes in teleost fishes clearly includes repeated emergence of male heterogametic systems in different subtaxa (see Fig. 1 of Volff et al., 2007). Due to the rapid turnover of sex chromosomes, mainly by the transposition of an existing sex-determination gene, the appearance of a new sex determination gene on an autosome, and the fusions between sex chromosomes and autosomes (Van Doorn and Kirkpatrick, 2007, Ross et al., 2009) of fishes, new master regulators have been repeatedly formed just 10 MYa or later (Peichel, 2005). Consequently the sex chromosomes are

Table 5 Survival and fertility of YY and ZZ superfemale fishes.

Family & Species	Survival (%)	Fertility	Reference
Poeciliidae			
Poecilia reticulata		+	Winge (1934)
P. reticulata	+, 12	+	Kavumpurath and Pandian (1992a)
*P. sphenops**	+ 8	–	George and Pandian (1995)
Adrianichthyidae			
Oryzias latipes	+ 3–5	+	Yamamoto (1955)
Ictaluridae			
Ictalurus punctatus	+	+	Dunham and Smitherman (1987)
Salmonidae			
Oncorhynchus mykiss	+, 1–3	+	Scheerer et al. (1991)
O. tshawytscha	+	..	Devlin et al. (2001)
Cichlidae			
Oreochromis niloticus	+ 1–4	+	Tuan et al. (1999)
O. mossambicus	+ 2–3	+	Varadaraj and Pandian (1989a)
Cobitidae			
Misgurnus anguillicaudatus	+, 8	+	Arai et al. (1995)
Cyprinidae			
Danio rerio	+, 22	..	Coreley-Smith et al. (1996)
Cyprinus carpio	+	..	Bongers et al. (1999)
Carassius auratus	+	..	Yamamoto (1975)
Puntius tetrazona	+, 15	+	Kirankumar and Pandian (2003)
P. conchonius	+, 7–15	F_1, F_2, F_3	Kirankumar and Pandian (2004b)
Characidae			
Hemigrammus caudovittatus	+, 11	+	David and Pandian (2006b)
Gymnocorymbus ternetzi	+, 10	+	David (2004)
Anabantidae			
Betta splendens	–	–	George et al. (1994)

+ = yes; – = No; .. = no data available, * = ZZ superfemale

at the early stage of differentiation (Volff et al., 2007). In contrast, birds have maintained a single sex chromosome system (female ZW heterogamety) across their 150-million years of existence (Fridolfsson et al., 1998) and mammals have retained another sole mechanism of sex determination (male XY heterogamety) since their origin approximately 250 MYa (Foster and Graves, 1994, see also Mank and Avise, 2006a).

2.2 Cytogenetics

More than 1, 700 fish species have been cytogenetically characterized, of which 186 species belonging to 86 families, i.e. 11% fish species are reported to have cytogenetically distinguishable heteromorphic sex chromosome(s) (see the list of Devlin and Nagahama, 2002). Of 13,000 marine fish species, so far fewer than 2% of them have been studied cytogenetically (Brum, 1996). Of 400 and odd species displaying one or other form of hermaphroditism, sex chromosomes have been recognized only in half a dozen species. A reason for the identification of heteromorphic sex chromosome only in such low percentage of the cytogenetically characterized fishes may be traced to the fact that sex chromosomes of most fishes are in an early stage of evolution (Matsuda et al., 1998). However, many fishes are known to utilize genetic determination systems that are primarily associated with sex chromosomes (e.g. Phillips and Rab, 2001). Genes involved in sex determination may be distributed in genome or located on single chromosomes or restricted to a single gene locus. This single locus control restricts sex to be determined by a single sex chromosome namely Y or W (Devlin and Nagahama, 2002). Hence investigations employing more sensitive Synaptonemal complex (SC), Fluorescent *in situ* hybridization (FISH), and Nucleolar Organization Region (NOR) techniques may detect heteromorphic sex chromosomes.

Synaptonemal complex (SC): Examination of chromosome pairing in SC during the pachytene stage of spermatocyte and oocyte can reveal the presence of sex chromosome regions, which are otherwise not visible. Meiotic recombination between the differentiated regions of gonosomal pair in the heterogametic sex (XY or ZW) does not usually occur but occurs between the homogametic sex (XX or ZZ) of the same species. SC analysis of three sexual genotypes of *Oreochromis niloticus* (XX, XY, YY) identified the unpaired segment with its features (unpaired axial elements, self-folding etc) characteristic of sex chromosome in the longest bivalent in normal male (XY) but the same was not present either in XX neomale or in YY supermale (see Fig. 1, 2 of Carrasco et al., 1999, see also Foresti et al., 1993). A similar SC analysis of the heterogametic females (ZW) and neomale (ZW) of *O. aureus* revealed the presence of two separate regions of unpairing. The first region present in the subterminal region of the longest bivalent, is similar to that found in the closely related species *O. niloticus*, in which this region has been associated with the heterogametic XY male genotype. Intriguingly, the same chromosome is associated with both XX/XY (e.g. *O. niloticus*) and ZW/ZZ (e.g. *O. aureus*) determination system in such closely related species suggesting how remarkably labile heteromorphic sex chromosomes systems in fishes can be (Mank and Avise, 2006a). The second region of unpairing is present in a small bivalent. However, its role in sex determination is not clearly understood (Campos Ramos et al., 2001). Incidentally, having found that the SC is completely paired in both spermatocyte and oocyte of the zebrafish *Danio rerio* and since their SC karyotype resembles the somatic karyotype, Wallace and Wallace (2003) questioned the very existence of sex chromosome in the zebrafish.

Fluorescent in situ hybridization (FISH): Flow sorting and microdissection are two major techniques used to isolate sex chromosomes and to generate chromosome specific DNA sequences, which are used as a probe in FISH to identify specific chromosomes (e.g. *O. niloticus*, Harvey et al., 2003). Laser microdissection (LMD) is a newly developed powerful tool for isolation of specific chromosomes (Rab et al., 2008). Using LMD system, Diniz et al. (2009) have obtained a probe for identification of Z chromosome from *Triportheus* and used it for whole chromosome painting and analysis of ZZ/ZW chromosome system. Sex chromosomes of *Xiphophorus maculatus*, which are hardly distinguishable in metaphase spread, were detected by FISH using identical repeat sequences or Bacterial Artificial Chromosomes (BACs) as probes (Nanda et al., 2000, Schultheis et al., 2006, 2009). The sex determining region of *X. maculatus* was located in the subtelomeric region of the sex chromosomes. FISH studies have also detected sex chromosomes in the catfishes *Liobagrus marginatus* and *L. styani* (Chen et al., 2008)

In an attempt to identify the sex chromosomes of the tongue sole *Cynoglossus semilaevis*, Wang et al. (2009) have used more sophisticated and reliable techniques to characterize WZ chromosomes. With Giemsa staining, W chromosome is readily recognizable by virtue of its large size (see Fig. 1, 3 of Wang et al., 2009). After blocking *Cot-1* DNA, the use of DOP-PCR products of laser dissected WZ chromosomes as probes, FISH produces clear signals to detect W and Z chromosomes. W chromosome library and its sequence analysis have shown that it is rich with 54.3% AT and 45.7% GC. With the DNA size of 29.5 MB, the W chromosome abounds with simple repeats LTR elements, TEs and low complexity. The accumulation of repetitive DNA is typical of sex chromosomes and their differentiation. Such repetitive sequences, described for *O. niloticus* (Harvey et al., 2002, 2003), *Oryzias latipes* (Kondo et al., 2006), *Hoplias malabaricus* (Ferreira et al., 2007) and *Mastacembelus aculeatus* (Zhao et al., 2008), provide insights into the initial process of sex chromosome evolution. Briefly it is tempting to state that by isolation of sex chromosome specific painting probes and construction of W chromosome library and so on, Wang et al. (2009) have almost successfully identified the existence of sex chromosome in the tongue sole. Yet even with the use of sensitive FISH and other techniques, Sola et al. (1997a) noted that the cytogenetic features of fishes like the carangid greater amberjack *Seriola dumerili* rule out the presence of morphologically differentiated sex chromosomes.

Heterochromatization is also known to play a fundamental role in the evolution of sex chromosomes. The level of heterochromatization of Y chromosome of the lake trout *Salvelinus namaycush* (Phillips and Ihssen, 1985) and Z chromosome of the black molly *Poecilia sphenops* var *melanosticta* (Haaf and Schmid, 1984) represents early stages of sex differentiation. In the European bitterling *Rhodeus amarus* too, heterochromatin is limited in quantity, compared to the larger blocks detected in *Tanakia koreensis* and *T. signifier* (Libertini et al., 2008). In many *Triportheus* spp, the morphological differentiation of the W chromosome is accomplished by a heterochromatization process together with reduction in size (Fig. 5). Yet the rDNA accumulation possibly influences the initial step of the sex chromosome differentiation in *Triportheus* (Artoni and Bertollo, 2002). The W chromosome has undergone a process of heterochromatization and accumulation of repetitive sequences in the differentiation process in relation to the Z chromosome. In this process, the long arm of the W sex chromosome of the parontids like *Apareiodon ibitiens* is entirely heterochromatic (Bellafronte et al., 2009). Thus the accumulation of repetitive DNA in the W chromosome constitutes a decisive step in the differentiation of this chromosome followed by structural changes in the Z and W chromosomes (see Fig.5). In *Characidium lanei*, the major rDNA sites are located on the sex chromosomes (Noleto et al., 2009), whereas these sites in *C. gomesi* and *C. alipionis* are on independent chromosomes (Centofante et al., 2001, 2003; Vicari et al., 2008). Likewise other classes of repetitive DNA are present in sex chromosomes of *C. lanei* (Vicari et al., 2010). Thus information available on the classes of repetitive DNA and their accumulation sites on sex or other chromosomes seems species specific and is not in the unidirectionary process of heterochromatization of sex chromosomes alone.

Nucleolar organization regions (NOR) with ribosomal gene sequences have also been used to detect sex chromosome (e.g. Moran et al., 1996; Khuda-Bukhsh and Datta, 1997). However, a large volume of available publications on the use of NORs is related to cytotaxonomy rather than cytogenetics. NOR, identified by Ag-NOR staining, FISH and Double FISH techniques with 18S rDNA and 5S rDNA probes, has been shown to successfully detect the presence of rDNA site, rich in GC on the telomeric region of the short and long arms of the Z and W chromosomes of *Characidium lanei* (Noleto et al., 2009). 5S sequences have been found to be linked to the short arm of the Y chromosome in chinook salmon *Oncorhynchus tshawytscha* (Stein et al., 2002). But FISH has not detected a sex-specific organization of the NORs in Atlantic salmon *Salmo salar* (Pendas et al., 1994) and Arctic charr *Salvelinus alpinus* (Reed and Phillips, 1997). Again available information on heterochromatization and NORs remains diverse and inconclusive.

On the whole, relevant cytogenetic information available on identification of sex chromosomes in fishes using simple staining to highly sensitive molecular techniques do not permit in resolving that cytologically distinguishable heteromorphic sex chromosomes are present in all fishes and that they can be detected by using a specific method independently or in combination with one or another technique. Yet, the findings of Traut and Winking (2001) seem to provide an answer. They undertook a microarray comparative genomic hybridization (CGH) on mitotic and meiotic chromosomes of the zebrafish *Danio rerio*, the platyfish *X. maculatus* and the guppy *P. reticulata* and found that these fishes appear to represent three basic levels of sex chromosome differentiation. Accordingly, 1. The zebrafish has an all-autosome karyotype. 2. The platyfish has genetically defined sex chromosomes but no differentiation between X and Y is visible in the SC or with CGH in meiotic and mitotic chromosomes and 3. The guppy has genetically and cytogenetically differentiated sex chromosomes. The acrocentric Y chromosome of the guppy consists of a proximal homologous and a distal differential segment. The proximal segment pairs in early pachytene with respective X chromosome segment. The differential segment is unpaired in the early pachytene but synapses later in equalization. This segment is postulated by Traut and Winking as the sex determining region. Its CGH consists of a large block of predominantly male-specific repetitive DNA.

2.3 Heterogamety

A heterogametic individual produces genetically different gametes, one of which may determine the sex of its progeny generated by its fusion with the gamete of the opposite sex. The three cytological criteria, by which heterogamety is recognized, are: 1. Regular occurrence of a heteromorphic chromosome pair in mitotic cells in one sex but not in other. 2. Atypical behaviour of a single bivalent at meiosis I, and 3. The presence of two different haploid karyotypes at meiosis II, each possessing one of the heteromorphic chromosome pair (Ebeling and Chen, 1970). However, in view of difficulties encountered in cytological detection of sex chromomes, heterogamety in fishes is adduced from genetic rather than cytological evidence. The existence of sex chromosome(s) has been realized in many fishes by observations with sex linked colour genes (Ohno, 1967), and experimental crossing with sex reversed fishes (e.g. male heterogamety:

O. mossambicus, Varadaraj and Pandian, 1989a; female heterogamety: *P.sphenops* (George and Pandian, 1995). Uyeno and Miller (1971) were the first to report multiple sex chromosome system (X_1X_2/Y); since then at least another three multiple sex chromosome systems have been reported (Table 6).

On the basis of reliable evidence from sex chromosome heteromorphism, 188 fish species belonging to 86 families are recognized as heterogametic. Of them, 135 species belonging to 61 families are male heterogametic but female heterogamety is limited to 53 species belonging to 25 families (Devlin and Nagahama, 2002), i.e. only 28% of fish species alone are female heterogametic. Since the publication of this valuable document by Devlin and Nagahama in 2002, new reports on heterogamety in fishes have appeared. An attempt has been made to summarize the available information on heterogamety in fishes (Table 7). Thanks to the major contributions by Brazilian ichthyologists, new records for the presence of 40 species belonging to 10 families characterized by female heterogamety have been added (Table 8). Indeed the Neotropics appears to be the 'gold mine' for the female heterogametic fishes. From the non-neotropic regions, female heterogamety has been reported for another 15 species using cytogenetic or gynogenetic evidence (Table 9). From the available reliable records, 176 male heterogenetic fishes versus 88 female heterogametics are now known (Table 7). Briefly, 33.3% of the fishes are heterogametic and the updated data still confirm the conclusion of Devlin and Nagahama (2002) that only a third of fishes are female heterogametic. Clearly, male heterogamety has been the choice of natural selection. It is, however, not clear whether the male heterogamety has any relevance to the size advantage model (see Kraak and de Looze, 1993). Secondly, it is also difficult to comprehend that despite males being carriers of a greater mutational load than females , why selection is in favour of male heterogamety?

From Table 7, the following may also be noted: 1. Of 264 species, in which the sex determining Y and W chromosome has been identified, only in 15 species, i.e. 6% of fishes carrying Y or W chromosome is considered to have been lost. 2. Unusually, the X or Z chromosome has not been detected in three species, the causes for which are not known. Incidentally, the Y chromosome is fused to an autosome and thus, chromosome number is also different in the two sexes (Bertollo et al., 1997; Almeida-Toledo et al., 2000). Table 10 lists the occurrence of multiple sex chromosome system in the Neotropical and marine, especially from the deep sea and the Antarctic Sea. 3. Simple sex chromosome system is in operation in ($156\ X_1X_2/X_1Y_2$ + $83\ Z_1W_2/Z_1Z_2$) 239 species, i.e in 91% fishes, in comparison to the presence of multiple sex chromosome system in 9% fishes only. Incidentally, the Antarctic may prove a 'gold-mine' for male heterogametic multiple sex chromosome system; of 13 endemic species thus far studied, five belong to

Table 6 Described sex chromosome systems in fishes (compiled from Moreira-Filho et al., 1980, 1993). 'no' in column 2 indicates the number.

Sex chromosome system	Diploid number	Karyotypic feature	Meiotic feature	Example
		I Male heterogamety		
XX : XY	♂ 2n : ♀ 2n	heteromorphic ♂	atypical bivalent in ♂	Betta splendens
XX : XO	♀ 2n : even no ♂ 2n–1 : odd no	a single homologue in given pair in ♂	a univalent in ♂	Tricanthus brevirostris
$X_1X_2 X1X_2 : X_1X_2Y$	♀ 2n even no ♂ 2n–1 : odd no	3 chromosomes without homologues in ♂	a trivalent in ♂	Hoplias malabaricus
$XX : X_1Y_1Y_2$	♀ 2n : even no ♂ 2n–1: odd no	3 chromosomes without homologues in ♂	a trivalent in ♂	Coregonus sardinella
		II Female heterogamety		
ZW : ZZ	♀ 2n = ♂ 2n	heteromorphic ♀	atypical bivalent in ♀	Poecilia sphenops
ZO : ZZ	♂ 2n : even no ♀ 2n–1 : odd no	a single homologue in given pair in ♀	a univalent in ♀	Colisa lalius
$ZW_1W_2 : ZZ$	♂ 2n : even no ♀ 2n + 1 : odd no	3 chromosomes without homologues in ♀	a trivalent in ♀	Apareiodon affinis
		III Male and female heterogamety		
XX / XW / WW / WY ♀ : XY / YY ♂	♂ 2n = ♀ 2n			Xiphophorus maculatus

Table 7 Gamety in fishes including the contributions from the Neotropics and marine* systems.

Sex chromosome system	Described in Devlin and Nagahama (2002) (No)	Neotropics (No)	Others (No)	Total (No)
Male heterogamety				
X_1X_2/X_1Y_2	135 +	9	+	144
X_1X_2/X_1O	11	1	–	12
X_1X_2/Y	1	9	8*	18
X/Y_1Y_2	1	1	–	2
Total	148	20	8	176
Female heterogamety				
Z_1W_2/Z_1Z_2	25	40	15	80
Z_1O/Z_1Z_2	3	–	–	3
$Z_1Z_1Z_2Z_2/Z_1Z_2W_1W_2$	2	2	–	4
$Z_1Z_1Z_2Z_2/Z_1W_1W_2$	–	1	–	1
Total	30	43	15	88

*Species reported from marine, especially the Antarctic Sea

Table 8 Female heterogametic system (ZZ/ZW) reported with cytogenetic evidences for the Neotropical fishes.

Family/Species	References
Anostomidae	
Leporinus reinhardti	Galetti, Jr and Foresti (1986, 1987)
L. macrocephalus	Galetti, Jr and Foresti (1986, 1987)
L. conirostris	Moreiro-Filho et al. (1993)
L. trifasciatus	Moreiro-Filho et al. (1993)
L. obtusidens	Oliveira et al. (2009)
L. aff. Brunneus	Oliveira et al. (2009)
Characidae	
Triportheus albus	Falco (1988)
T. angulatus	Oliveira et al. (2009)
T. elongatus	Oliveira et al. (2009)
T. flavus	Falco (1988)
T. guentheri	Olivera et al. (2009)
T. paranensis	Artoni et al. (2001)
T. signatus	Artoni et al. (2001)
Odontostible heterodon	Oliveira et al. (2009)
O. microcephala	Oliveira et al. (2009)
O. paranaensis	Oliveira et al. (2009)
Serrapinus notomelas	Oliveira et al. (2009)

Table 8 contd....

Table 8 contd....

Family/Species	References
Chrenuchidae	
Characidium alipionis	Oliveira et al. (2009)
C. fasciatum	Oliveira et al. (2009)
C. gomesi	Vicari et al. (2010)
C. lanei	Vicari et al. (2010)
Doraididae	
Opsodoras sp	Oliveira et al. (2009)
Gasteropelecidae	
Thoracocharax stellatus	Oliveira et al. (2009)
Heptapteridae	
Imparfinis mirini	Oliveira et al. (2009)
Laricariidae	
Microlepidogaster leucofrenatus	Andreata et al. (1993)
Hisonotus leucofrenatus	Oliveira et al. (2009)
Hisonotus sp	Oliveira et al. (2009)
Otocinclus vestitus	Oliveira et al. (2009)
Hemiancistrus spilomma	de Oliveira et al. (2009)
Hypostomus sp G	Oliveira et al. (2009)
Paradontidae	
Paradon ibitiensis	Bellafronte et al. (2009)
P. vladii	Rosa et al. (2006)
P. moreirai	Centofante et al. (2002)
Apareiodon sp	Vicari et al. (2006)
A. ibitiensis	Bellafronte et al. (2009)
Prochilodontidae	
Semaprochilodus taeniatus	Oliveira et al. (2009)
Sternopygidae	
Eigenmannia virescens	Bellafronte et al. (2010)
Ancistrus dubius	Mariotto and Myiazawa (2006)
A. ranunculus	de Oliveira et al. (2009)
Ancistrus sp Piagacu	de Oliveira et al. (2009)

the X_1X_2Y group (Morescalchi et al., 1992). Galetti et al. (2000) have also listed another three from marine fishes characterized by multiple sex chromosome system. 4. Notable is the occurrence of different cytotypes in sympatric populations within a species. The existence of two or more cytotypes is not uncommon. For instance, cytotypes with ♀ 52 : ♂ 51 chromosomes and ♀ 54 : ♂ 53 chromosomes are found in *Erythrinus erythrinus* but this variation does not alter the sex chromosome system (X_1X_2Y) (Bertollo et al., 2004). However, such variations in cytotype change the sex chromosome system in *Hoplias*

Table 9 Female heterogametic (ZZ/ZW) system reported for the non-Neotropic fishes using cytogenetic (1) or gynogenetic (2) evidence.

Species	Reference
Acipenseridae	
Acipenser baeri[2]	Fopp-Bayat (2008)
A. brevirostrum[2]	Flynn et al. (2006)
A. transmontanus[2]	Van Eenennaam et al. (1999)*
Adrianichthyidae	
Oryzias hubbsi	Kondo et al. (2009)
O. javanicus	Kondo et al. (2009)
Anabantidae	
Marcropodus opercularis	Gervai and Csanyi (1984)*
Congridae	
Conger myriaster[1]	Ojima and Uedo (1982)*
Cyprinidae	
Barbus barbus[2]	Castelli (1994)*
Paramisgurnus dabryanus[2]	You et al. (2008)
Rhodeus ocellatus ocellatus[2]	Kawamura (1998)
Danio rerio	Uchida et al. (2002)
Cynoglossidae	
Cynoglossus semilaevis[1]	Wang et al. (2009)
Scophthalmidae	
Scophthalmus maximus	
Esocidae	
Esox masquinongy[2]	Rinchard et al. (2002)
Poeciliidae	
Xiphophorus alvarezi	Schultheis et al. (2009)

*Left out by Delvin and Nagahama (2002)

Table 10 Multiple sex chromosome systems reported for the Neotropical and marine fishes.

Family, species	chromosome (no)	Sex chromosome system	Reference
Neotropical fishes			
Erythrinidae			
Erythrinus erythrinus	52, 51 or 54, 53	X_1X_2/Y	Bertollo et al. (2004)
Hoplias malabaricus	40 or 42 40, 39	$X_1X_2/X_1X_2Y_2$ X_1X_2/Y	Rosa et al. (2009)
Gobitidae			
Awaous strigatus	♀ 46	X_1X_2/Y	Oliveira et al. (2009)
Hypopomidae			

Table 10 contd....

Table 10 contd....

Family, species	chromosome (no)	Sex chromosome system	Reference
Brachyhypopomus pinnicaudatus		X_1X_2/Y	Oliveira et al. (2009)
Hypopomus sp	42, 41	X_1X_2/Y	Oliveira et al. (2009)
Loricariidae			
Harttia carvalhoi	52, 53	X_1X_2/Y	Oliveira et al. (2009)
Ancistrus sp Balbina	38, 39	X/Y_1Y_2	de Oliveira et al. (2009)
Ancistrus sp [2]	39, 40	X_1X_2/XO	Alves et al. (2006)
Sternopygidae			
Eigenmannia sp	32, 31	X_1X_2/Y	Almeida-Toledo et al. (2000)
Anostomidae			
Leporinus elongatus		$Z_1Z_1Z_2Z_2/$ $Z_1Z_2W_1W_2$	Parise-Maltempi et al. (2007)
Laricariidae			
Ancistrus sp Barcelos	52	$Z_1Z_1Z_2Z_2/$ $Z_1Z_2W_1W_2$	de Oliveira et al. (2009)
Parontidae			
Apareiodon affinis	54, 55	$Z_1Z_1Z_2Z_2/$ $Z_1W_1W_2$	Jesus and Moreira-Filho (2000)
Marine fishes			
Balistidae			
Stephanolepis cirrhifer		$X_1X_1X_2X_2/X_1X_2Y$	Galetti et al. (2000)
Benniidae			
Blennius tentacularis		$X_1X_1X_2X_2/X_1X_2Y$	Galetti et al. (2000)
Clupeidae			
Brevoortia aurea		$X_1X_1X_2X_2/X_1X_2Y$	Galetti et al. (2000)
Channichthyidae			
Chaenodraco wilsoni		$X_1X_1X_2X_2/X_1X_2Y$	Morescalchi et al. (1992)
C. hamatus		$X_1X_1X_2X_2/X_1X_2Y$	Morescalchi et al. (1992)
C. myersi		$X_1X_1X_2X_2/X_1X_2Y$	Morescalchi et al. (1992)
Chinobathyscus dewitti		$X_1X_1X_2X_2/X_1X_2Y$	Morescalchi et al. (1992)
Pagetopsis macropterus		$X_1X_1X_2X_2/X_1X_2Y$	Morescalchi et al. (1992)
Salmonidae			
Coregonus sardinella		$X_1X_1X_2X_2/X_1Y_1Y_2$	Frovlov (1990)

malabaricus from (40 chromosomes) X_1X_2/X_1Y_2 in a male cytotypes to X_1X_2/Y in male cytotypes carrying 39 chromosomes (Table 10; Fig. 6).

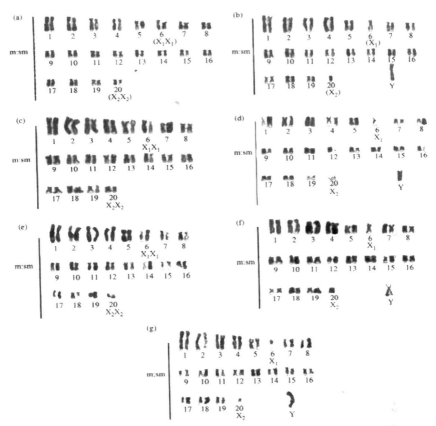

Fig. 6 Karyotypes of *Hoplias malabaricus* with $X_1X_1X_2X_2/X_1X_2Y$ sex determining system: (a) ♀ from Vermeho River, (b) ♂ from Vermeho River, (c) ♀ from Rancho Alegre, (d) ♂ from Rancho Alegre, (e) ♀ from Tres Bocas Stream, (f) ♂ from Tres Bocas Stream, and (g) ♂ from Paranapanema River m : sm = metacentric : sub-metacentric. Note the deletions in 6X1 and 20X2 in the male (from Rosa et al., 2009).

Structural changes in chromosome may facilitate readily recognizable changes in the shape and number of sex chromosomes: for instance, additions or deletions of heterochromatic blocks (e.g. *Hoplias malabaricus*, Rosa et al., 2009, Fig. 6), reduction in chromosome size (e.g. *Triportheus guentheri*, Bertollo and Cavallaro, 1992), increase in chromosome size (e.g. *Parodon hilarii*, Moreira-Filho et al., 1993, Fig. 7c) and chromosome rearrangements (e.g. *Zeus faber*, Vitturi et al., 1991). In the most common XX/XY system, X and Y chromosomes are genetically distinguishable but the 2n number between the sexes remains equal (e.g. *Betta splendens*, Kavumpurath and Pandian, 1994). But in an XX/XO system, the Y chromosome has been lost (e.g. *Tricanthus brevirostris*, Chaudhury et al., 1982). Hence males possess one chromosome less than females. In an $X_1X_2 X_1X_2/X_1X_2Y$ system, the Y

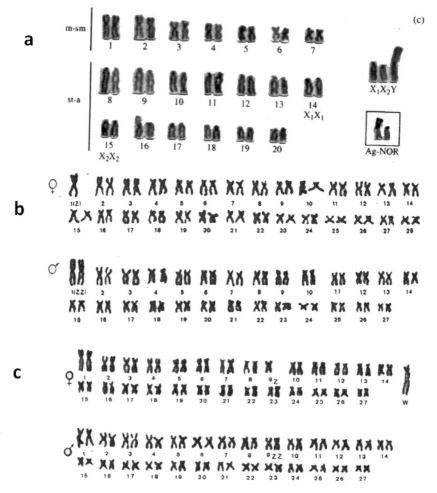

Fig. 7 Giemsa staining karyotypes of (a) *Gymnotus pantanal* female. Male X_1X_2 / Y sex chromosomes are detailed. Ag-NOR-bearing chromosome pairs are boxed (from Margarido et al., 2007). (b) *Apareiodon affinis* (ZO/ZZ). Note the Z chromosome in female and ZZ chromosomes in male (from Moreira-Filho et al., 1980) and (c) *Parodon hilarii* : Note the easily distinguishable W chromosome in female (from Moreira-Filho et al., 1993).

chromosome is attached to an autosome by translocation or Robertsonian fusion; hence males and females have a different chromosome number (e.g. *Gymnotus pantanal,* Margarido et al., 2007, Fig. 7a). An acrocentric X chromosome is fused with an autosome in the XX/XY_1Y_2 system (e.g. *Coregonus sardinella,* Frolov, 1990). In female heterogametic species, the structural changes observed in sex chromosome systems are exactly the opposite of those in male heterogametic species (see also Table 6). Figure 7 b and c shows selected karyotypes of sex chromosome systems in ZO/ZZ and ZW_1W_2/ZZ female heterogametic species.

Gordon (1952) has noted that the Mexican race of the platyfish *Xiphophorus maculatus* is male heterogametic and that of Honduras is female heterogametic. Surveying many races of the platyfish, Kallman (1984) has resolved that *X. maculatus* is polymorphic for three sex chromosomes: W, X and Y. Four (XX, XW, WW, XY) of the six possible mating combinations result in brood sex ratio of 0.5 females. However, the two other mating combinations (WY, YY) results in biased sex ratios; matings between WX females and XY males produces female-biased broods (0.75 ♀ : 0.25 ♂) and matings between XX females and YY males produce all male broods. This is the one of the few systems, in which sex chromosomes allelic constitution affects brood sex ratios such that predictable deviation from 0.5 results (see Basolo, 2001). Incidentally, a cross between *X. helleri* (Sarabia strain from south Mexico, WY/YY) and *X.maculatus* (XX/XY) also produces all male progenies (Walter et al., 2006).

Equally interesting is the presence of male heterogamety in *Gambusia affinis holbrooki* but female heterogamety in *G. affinis affinis* (Black and Howell, 1979). The poeciliids *Xiphophorus* and *Gambusia* are relatively younger teleosts and the diversification of different genera and origin of the extant species have occurred in a time frame of 5–30 MYa. It appears that the same master sex determining gene operates differently in males and females within the species (Volff et al., 2007).

The very existence of male heterogametic (X^1X^2/X^1Y^2) in about 67% of fishes and female heterogametic (Z^1Z^2/Z^1W^2) in about 33% of fishes, and simple sex chromosome system in 91% of fishes and multiple sex chromosome system in the rest clearly indicate the absence of common, universal sex chromosome system in fishes. Secondly, the presence of male and female heterogamety within a genus (e.g. *Oreochromis, Poecilia*) and within a species (e.g. *Gambusia, Xiphophorus*) indicates that not a single sex determining chromosome system governs sex determining mechanism in fishes.

Hybridization and Sex Ratio

Hybrids carry amalagamations of divergent genomes. In a broad sense hybridization is considered as a 'heterospecific insemination' (Chevassus, 1983); it occurs widely across taxonomically diverse array of fish species (Schwartz, 1981; Argue and Dunham, 1999) and is observed in fishes more than other vertebrate groups like amphibians. Some 200 and odd hybrid records are published every year. Of more than 16,050 records on hybrids, fishes have a lion share of 21% (Schwenk et al., 2008). The number of hybrid records available is 0.12/species for fishes against 0.0065/species for amphibians. This wide difference in the number hybrid records may be due to the fact that taxa like the economically important fishes have received far greater attention from taxonomists and population geneticists than most other taxa.

By compressing the time scale within one generation, hybridizations expose novel genomes to natural selection and provide a key to unravel the ultimate causes for adaptation and speciation (Schwenk et al., 2008). Hence examples of hybridizations have been described as 'natural laboratories for evolutionary process' (Hewitt, 1988), 'windows on evolutionary process' (Harrison, 1990) or 'as an ecologically dependent behaviourable phenomenon with genetic consequences' (Grant and Grant, 2008). Over evolutionary time scale, hybridization has been a source of diversification of species. For instance, hybridization has been the major source for the origin of unisexuals like the *Poecilia* complex, *Poeciliopsis* complex and *Cobitis* complex. Examples for bisexual diploid taxa of hypothesized hybrid origin include *Catasomus discobolus*, *Chasmistes brevirostris* and *Luxilus albeslus* (Scribner et al., 2001). On the other hand, introgressive hybridizations are considered as contributing factors leading to extinction of *Cyprinodon nevadensis calidae, Gambusia amistadensis* and *Coregonus alpenae* (McMillan and Wilcove, 1994). These extreme examples suggest that no single evolutionary factor is universally applicable to fishes. Hence predictions of the outcome of a hybridization event are difficult to generalize across species.

3.1 Extent of hybridization

Incidences of natural hybrid range from 2% in clupeids *Drosoma capedianum* x *D. petenense* to 37% in cyprinids *Rutilus rutilus x Scardinius erythrophthalmus* and to 39% in salmonids *Salvelinus alpinus x S. fontnalis* (Argue and Dunham, 1999). Summarizing available information on hybrid fertility, Argue and Dunham (1999) listed a total of F_1 fertile 130 natural hybrids and 150 artificially produced F_1 hybrid fishes. The presence of fertile F_1 hybrids in 47 intergeneric crosses may pose a big question on the validity of current classification. A number of factors contribute to the high incidence of hybridization among related taxa of fishes: external fertilization, lack of acrosome in spermatozoa, weak behavioural isolating mechanisms, unequal abundance of the two parental species, competitions for limited spawning habitat, and decreasing habitat complexity are some of them. A hybrid zone is typically a narrow region, where genetically distinct species meet, mate and produce hybrids (Barton and Hewitt, 1985) that are viable and at least partially fertile (Arnold, 2006). Scribner et al. (2001) have identified four major contributing factors as likely causes of hybridization (i) habitat loss/ habitat alteration, (ii) aquaculture, (iii) range expansion and (iv) introduction of new species. Drawing information from 158 publications covering 168 species and 139 species pairs, of which 47 are intergeneric hybrids from 19 freshwater families of fishes, Scribner et al. (2001) concluded that anthropogenic influences have historically been an important contributor to hybridization. Of 163 hybridization events, 81 involved human influences. For instance, aquaculture alone is the contributor to 32 events, introduction of new species to 22 events, habitat loss/alteration to 21 events and range expansion to three events. Hybridization events in 10 of the 19 families surveyed by Scribner et al. are attributed to the introduction of new species. Hybrids have been purposely created to enhance productivity of aquaculture strain (e.g. Ictalurids, Lutz, 1997) or to enhance recreational angling opportunities.

Scribner et al. (2001) have also found that species within the most speciose families Cyprinidae (40%) and Centrarchidae (20%) hybridize more frequently and account for 60% of all the recorded hybridizations. Within cyprinids, 68 species are involved in 56 hybridization events and 18 Centrarchid species have produced 28 hybrid crosses. Salmonidae accounts for another 8% of the total hybridization events. Hybridization within Percidae, Catostomidae, Poeciliidae accounts for 4% each,

Cichlidae, Clariidae and Esocidae for 3% each, Cyprinidontidae for 2 % and Acipenseridae, Anguillidae, Athrerinidae, Clupeidae and Cottidae for 1% each. Interestingly, among the cyprinids, *Campostoma anomalum* and *Alburnus alburnus* are involved in 31% of the non-aquacultural hybridization events. Among the Centrarchids, *Lepomis* species hybridize extensively. Of the surveyed hybridization events, *L. macrochirus* is involved in 8 (29%), *L. cyanellus* 7 (25%), *L. gibbosus* 6 (21%), *L. gulosus* 5 (18%), *L. auritus* and *L. humulis* 4 (14%) each and *L. punctatus* 1 (4%) (Scribner et al., 2001). Table 11 provides a representative list for the extent of hybridizations among 4 cyprinids, 3 adrianichthyids and 9 apolocheilids.

Table 11 Representative examples for the extent of successful hybridization and backcrossing in fishes.

Cyprinidae (Nikoljukin, 1972)
(Rutilus rutilus x Abramis brama) x Scardinius erythrophthalmus
(R. rutilus x A. brama) x Blicca bjoerkna
(R. rutilus x A. brama) x R. rutilus
(R. rutilus x B. bjoerkna) x S. erythrophthalmus
(R. rutilus x A. bjoerkna) x (R. rutilus x B. bjoerkna)
(S. erythrophthalmus x B. bjoerkna) x A. brama
(S. erythrophthalmus x B. bjoerkna) x B. bjoerkna
(S. erythrophthalmus x B. bjoerkna) x R. rutilus
(S. erythrophthalmus x B. bjoerkna) x S. erythrophthalmus
S. erythrophthalmus x (S. erythrophthalmus x B. bjoerkna)

Adrianichthyidae (Kurita et al. (1992)
Oryzias luzonensis x O. curvinotus
(O. curvinotus x O. latipes) x O. latipes
(O. curvinotus x O. latipes) x O. curvinotus
(O. latipes x O. curvinotus) x O. latipes
(O. latipes x O. curvinotus) x O. curvinotus
(O. latipes x O. curvinotus) x O. luzonensis

Apolocheilidae (Schwartz, 1981)
Aphyosemion australe   x (A. gardneri x   A. australe)
(A. australe x A. gardneri) x A. australe
(A. cognatum x A. schoutedeni) x   A. schoutedeni
(A. cognatum x A. schoutedeni) x (A. christyi x A. cognatum)
(A. fasciolatus x   A. albrecti) x (A. albrecti x A. fasciolatus)
A. gardneri x (A. gardneri x A. scheeli)
A. scheeli x (A. scheeli x A. cinnamomeum)

3.2 Hybrid identification

The F_1 hybrid may assume an intermediary body form and colour of the parental species. However some hybrids like those of carp × goldfish are not always easy to identify due to blending of parental characters (Taylor and Mahon, 1977). The F_1 hybrids of the cross between channel catfish *Ictalurus punctatus* ♀ × blue catfish *I. furcatus* ♂ can not easily be distinguished from the channel catfish offsprings (Waldbieser and Bowsworth, 2008). Among the crosses between the poeciliids *Poecilia sphenops* and *P. velifera*, F₁ hybrids assume two different body colours (Fig. 8). Among hybrids resulting from crosses between brook trout *Salvelinus fontinalis* and bull trout *S. confluentis*, some display the bull trout features, while others those of the brook trout, indicating that the hybrid crosses have taken place in both reciprocal directions (Leary et al., 1993). In a cross between silver carp *Hypophthalmichthys molitrix* × *Cyprinus carpio*, two distinct hatching periods

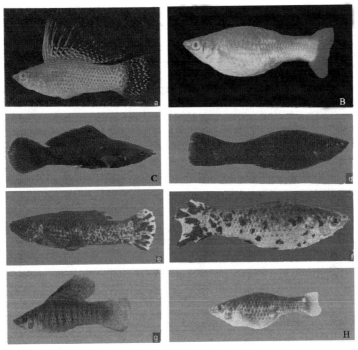

Fig. 8 External appearance of *Poecilia velifera* (a) male, (b) female, *P. sphenops* (c) male, (d) female, and their mottled hybrid (e) male and (f) female and striped hybrid (g) male and (h) female (from George and Pandian, 1997).

Colour image of this figure appears in the color plate section at the end of the book.

have been observed, the early hatchlings are gynogenics and the latter are hybrids. Similar results have also been published for the cross between *H. molitrix* × *Aristichthys nobilis* with incidences of not only gynogenics but also triploid hybrids (Chevassus, 1983; Mia et al., 2005). These examples emphasize the need for the use of protein and genetic markers and novel methods of statistical analysis to infer the extent, rate, direction and likely causes for hybridization (Scribner et al., 2001). Not surprisingly, multiple markers are often used concurrently. For example, Dowling et al. (1989)

have used phenotype, allozymes and mt DNA in the analysis of *Notropis chrysocephalus* and *N.cornutus*. Using diagnostic allozyme markers, Wallis and Beardmore (1980) have identified F_1 and F_2 (produced by crossing $F_1 \times F_1$) hybrids in the goby *Pomatoschistus*. Yet, phenotypic markers are the most commom method (45%) employed for hybrid identification; the other markers used less frequently are allozymes (35%), mt DNA (12%); n DNA (4%) and karyology (2%).

To elucidate the usefulness of employing multiple markers in identification of true hybrids, a couple of representative examples are described. Using diagnostic markers for identification of each parental species, an allozyme locus and two nuclear PCR–based DNA markers (ITS-1 and Ca M), Imai et al. (2009) identified the hybrid genotype of the cross between *Nematolosa japonica* and *N. come* and found that the hybrid frequency ranged from 1 to 67% and a backcross level of 2% in three Okinawan populations. You et al. (2007) used karyotype and a diagnostic microsatellite marker to identify the hybrids between the large scale loach *Paramisgurmus dabryanus* (2n = 48) and pond loach *Misgurnus anguillicaudatus* (2n = 100). The analyses confirmed that the progenies were true hybrids with 2n = 74 chromosomes. However, only one of 20 individuals, which were suspected from phenotypic features as hybrids, was found to be a true hybrid from the satellite marker analysis. This observation reveals the rarity of hybrids between these loaches. Incidentally, the hybrid frequency between Atlantic salmon *Salmo salar* (2n = 58) and brown trout *S.trutta* (2n = 80) is also reported to be low (Garcia-Vazquez et al., 2001). Using multilocus-nuclear and mtDNA analyses, none of the F_1 hybrids among 277 surveyed events within *Lepomis macrochirus*, *L. auritus*, *L. gulosus*, *L. cyanellus* and *L. microlophus* is found to have a successful backcross or F_2 progenies (Avise and Saunders, 1984), despite the claim by Childers (1967) that *Lepomis* F_1 hybrids are partially fertile.

3.3 Viability and fertility

Reciprocal crosses between *Danio rario* and *D. frankei* produced about 80% viable hybrids that were phenotypically a blending between the parental species (Kavumpurath and Pandian, 1992c). Survival of intergenic hybrids is generally low; for instance, almost all combinations of the crosses between the clariid catfishes *Clarias gariepinus* and *Heterobranchus longifilis* resulted in 30–72% fertilization success but due to high zygotic mortality, the fry survival is limited to 6–56% only (Table 12). But the hybrids suffer progressive mortality almost throughout the life span; for instance, survival of the hybrid between *Heteropneustes microps* ♀ x *H. fossilis* ♂ is 90% at hatching but is further reduced to 70, 52 and 48% following 3, 25 and 60 days after hatching, respectively (Sridhar and Haniffa, 1999). Similarly, the intergeneric cross between the characids namely the tambaqu *Colossoma macroponum* and the pacu *Piaractus mesopotomicus* produced 80% fertilized eggs but 37% tambacu progenies alone survived (Toledo-Filho et al., 1994). Interfamily level cross between *Clarias macrocephalus* (Clariidae) and *Pangasius sutchi* (Pangasiidae) produced only 1.8% hybrids (NaNakorn et al., 1993a). Fertile F_1 hybrids were produced from the cross between *Oryzias luzonensis* ♀ x *O. curvinotus* ♂ but the reciprocal cross yielded sterile hybrids only (Sakaizumi et al., 1992). A cross between cyprinids *Richardsonius balteatus* and *Mylocheilus caurinus* generated fertile female but sterile male hybrids (Aspinwall and McPhail, 1995). The cross *Barbus meridionalis* x *B.barbus* too produced sterile males and fertile females (Phillippart and Berrebi, 1990). However, the female hybrids are less fecund; for instance the relative fecundity of hybrid between *C. gariepinus* and *H. longifilis* was just 12 eggs/g body weight, against 70–106 eggs/g body weight of the parental species (Legendre et al., 1992). On the other hand, female hybrids of the cross between *O. latipes* x *O. curvinotus* produced unreduced diploid eggs (Sakaizumi et al., 1993). As a culmination, no viable hybrids were produced from the cross between the dusky grouper *Epinephelus marginatus* and white grouper *E.aeneus* (Glamuzina et al., 1999). None of the intergenic hybrids survived to attain feeding stage among the reciprocal crosses between the tench *Tinca tinca* and bream *Abramis brama* or *Cyprinus carpio* (Mamcarz et al., 2006).

Gametic sterility: It is not known why hybrid males suffer gametic sterility, but the female does not. The testes of *Barbus barbus* x *B.meridionalis*

Table 12 Ferilization, hatching success and survival of hybrids from crossings and backcrossings of the clariid catfishes *Clarias gariepinus* and *Heterobranchus longifilis* (from Legendre et al. (1992)** and Nwadukwe (1995)*.

Crosses and backcrosses	Fertilization success (%)	Hatching success (%)	Survival (%)
H. longifilis ♀ x *C. gariepinus* ♂**	65	58	52
C. gariepinus ♀ x *H. longifilis* ♂**	72	64	56
(*C. gariepinus* x *H. longifilis*) ♀ x (*C. gariepinus* x *H. longifilis*) ♂*	29	20	15
(*C. gariepinus* x *H. longifilis*) ♀ x *C. gariepinus* ♂*	33	19	6
(*C. gariepinus* x *H. longifilis*) ♀ x *H. longifilis* ♂*	30	16	6
(*C. gariepinus* x *H. longifilis*) ♀ x (*H. longifilis* x *C. gariepinus*) ♂*	–	12	–
H. longifilis ♀ x (*H. longifilis* x *C. gariepinus*) ♂*	–	47	–
(*H. longifilis* x *C. gariepinus*) ♀ x (*H. longifilis* x *C. gariepinus*) ♂*	–	45	–
(*H. longifilis* x *C. gariepinus*) ♀ x *H. longfilis* ♂*	–	80	–
(*H. longifilis* x *C. gariepinus*) ♀ x *C. gariepinus* ♂*	–	73	–

hybrids were often irregular with absence of a lobe, marked longitudinal asymmetry and contained no liquid milt (Phillippart and Berrebi, 1990). Intratesticular semen production by male hybrid between *Heterobranchus longifilis* and *Clarias gariepinus* was two orders of magnitude lower (1.5 –9.8 x 10^7/ml) than in parental species (2.9–4.0 x 10^9/ml) (Legendre et al., 1992). Microscopic examination of seminal fluid of *Lepomis gibbosus* x *L. cyanellus* hybrid showed 'small number of deformed sperm often with multiple tails' (Dawley, 1987). Though in small numbers and that too with deformed sperm, spermiogenesis seemed to take place in centrarchid hybrid males but it failed to proceed beyond the spermatocyte stage in bitterlings (Kawamura and Hosoya, 2000). Apparently, gametic sterility among male hybrids may fail at different stages of spermiogenesis. More research is required in this area.

Causes for zygotic mortality: Causes adduced for the zygotic/embryonic mortality of hybrids are: (i) Differential yolk content (ii) maternal karyotypic incompatibility and (iii) parental chromosome cytoplasmic incompatibility. For instance, Vrijenhoek and Schultz (1974) found that the eggs of *Poeciliopsis viriosa* contain inadequate yolk to produce viable hybrid, when crossed with *P. monacha*, whose embryo is nourished by its large ova of 2.2 mm. Secondly between closely (e.g. *Poecilia sphenops* (2n = 46) x *P. velifera* (2n = 46), (George and Pandian, 1997) or distantly [e.g. *Catla catla* (2n = 50) x *Labeo rohita* (2n = 50), Desai and Rao, 1970] related species but with similar karotypes hybridize readily with parental species. With parental species possessing unequal number of chromosomes, true hybrids are also produced (e.g. *Paramisgurnus dabryanus* (2n = 48) x *Misgurnus anguillicaudatus* (2n = 100, You et al., 2007). A cross between *O. gorbuscha* (2n = 52) x *O. keta* (2n = 71) produces F_1 hybrid with 2n = 63 but F_2 hybrid has about 62 (56–69) only, indicating that the F_2 begins to eliminate some of its chromosomes. In fact, *O. mykiss* (2n = 60) is amenable to diploid and triploid hybridizations with *Salmo salar* (2n = 60), *Salvelinus fontinalis* (2n = 42), *S. trutta* (2n = 40), *O. keta* (2n = 74), *O. kisutch* (2n = 60) and *O. masou* (2n = 66); however, its compatibility with the genome based on chromosome number, structure or DNA content of any one of the hybridizing species is not easily reconcilable (see Pandian and Koteeswaran, 1998).

An escape route to neutralize chromosomal incompatibility is gynogenesis and/or triploidization. The cross between *Clarias macrocephalus* ♀ (2n = 54) and *Pangasius sutchi* ♂ (2n = 60) results in the production of diploids (2n = 57) and triploids (3n = 54+30 = 84, NaNakorn et al., 1993a, b). In many such events, the hybrid females produce unreduced diploid eggs through premeiotic endomitosis (see Pandian and Koteeswaran, 1998). The crosses between *S. salar* (2n = 60) and *S. trutta* (2n = 60, Johnson and Wright, 1986) and *Oryzias latipes* (2n = 48) x *O. curvinotus* (2n = 48; Sakaizumi et al., 1992) are good examples.

The third reason for zygotic inviability has been traced to the incompatibility between maternal cytoplasm of *Oncorhynchus masou* (Ms) and parental chromosome of *O.mykiss* (Rb) (Fujiwara et al., 1997). Nearly half of the haploid Rb chromosomes are eliminated during the period from just after fertilization to late blastula. However such uniparental elimination of parental chromosomes is seldom or never observed with the reciprocal crossing between Rb ♀ x Ms ♂.

Fertility: Interestingly, the F_1 hybrid males and females of the reciprocal crosses between the roach *Rutilus rutilus* (R) x silver bream *Blicca bjoerkna* are viable and fertile (Matondo et al., 2008). F_1 hybrid female resulting from all possible crosses are equally fecund ($\approx 7.5 \times 10^3$ eggs) and their fertilization success is also equally high, despite the sperm density of all F_1 hybrid males being less than 10% of their respective parental male and a significant decrease in egg size of F_1 female hybrids of the cross between *Rutilis rutilus* ♀ and *B.bjoerkna* ♂. However, measured by the number of progenies in viviparous fishes, fecundity of the hybrids arising from reciprocal crosses between *Poecilia sphenops* x *P.velifera* dramatically decreases after the second impregnation by F_1 male hybrids (Fig. 9). This decrease results in 40% reduction in fecundity (George and Pandian, 1997). It is likely that the milt of F_1 male hybrids contains inadequate viable sperm. Further, the hatching success of F_2 progenies resulting from all possible eight crosses of *R.rutilus* x *B.bjoerkna* is low, especially those arising from the crosses between F_1 hybrid males and females (Matondo et al., 2008). Examples for low viability and ferility of F_2 hybrid fishes belonging to 10 families compiled from Argue and Dunham (1999) and others are listed in Table 13.

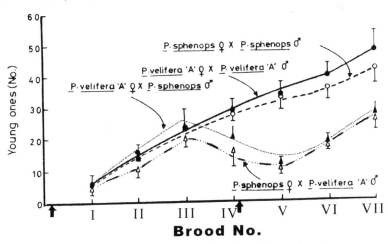

Fig. 9 Fecundity of conspecifics and hybrids of *Poecilia velifera* and *P. sphenops* as function of successive broods. Arrows indicate the first and second impregnations (from George and Pandian, 1997).

Table 13 Information on viability of F_2 generation of hybrid fishes compiled from Legendre et al. (1992), McElroy and Kornfield (1993), Argue (1996), Argue and Dunham (1999), Matondo et al. (2008) and others.

Family	Cross	Survival of F_2 hybrid (%)
Acipenseridae	*Huso huso* × *Acipenser ruthenus*	15
Cyprinidae	*Rutilus rutilus* × *Blicca bjoerkna*	<2
	Richardsonius balteatus x *Mylocheilus caurinus*	11
Ictaluridae	*Icatalurus punctatus* × *I. furcatus*	Low
	Clarias gariepinus x *Heterobranchus longifilis*	6
Coregonidae	*Coregonus fera* × *C. pallasi*	Reduced
Salmonidae	*Salvelinus namaycush* × *Salmo trutta*	Reduced[1]
	S. fontinalis x *Oncorhynchus masou*	0
Aplocheilidae	*Aphyosemion australe* × *A. gardneri*	Fertile[2]
Poeciliidae	*Xiphophorus maculatus* × *X. helleri*	Delayed maturity
Meronidae	*Merone saxatilis* × *M. chrysops*	50
Centrarchidae	*Lepomis auritus* × *L. macrochirus*	Low[3]
Cichlidae	*Cichlasoma cyanoguttatum* × *C. nigrifasciatus*	Low
	Haplochromis burtoni × *H. nubilus*	Less fecund
	Oreochromis mossambicus × *Sarotherodon galilaeus*	5–8

1. Hybrids called splake were produced upto F_8 generation, but F_7 ♀♀ were themselves rarely fertile

2. Many generations and many hybrids of ornamental fishes, but techniques kept as trade secret

3. F_1 produced 2 % ♀♀; F_2 0.6% ♀♀ only

Despite survival and partial and complete fertility in some F_1 hybrids, hatching success of F_2 hybrid progenies perceptibly declines but no author has so far traced cytogenetic causes for the ubiquitous low reproductive performance of these hybrids. That very few progenies produced by F_3 hybrids of the cross between *Ictalurus punctatus* and *I. furcatus* clearly indicates the break down of reproduction in the hybrids (Argue, 1996). There are also reports on fertility of intergeneric F_2 hybrids: an example is *Labeo rohita* and *Catla catla* (Desai and Rao, 1970). However, the phenotype-based obervations recorded in Table 11 of Argue and Dunham (1999) for the cyprinids and that of Desai and Rao require confirmation using protein and/or molecular markers. For, the use of such molecular markers has revealed that the conclusions arrived purely based on phenotypic markers, as made by Childers (1967, 1971), are not reliable (Avise and Saunders, 1984). However, it must also be stated that at least a few hybridization events have led to the production of fertile F_1 to F_{11} hybrids. For instance, the hybrid of *S. namaycush* x *S. fontinalis*, called splake proceeds upto F_7 generation and breaks down at F_8 generation (see Argue and Dunham, 1999). The cross between *Carassius auratus* (red var., 2n = 100) x *C.carpio* (2n = 100) produces viable diploid F_1–F_2 male and female hybrids; F_3 produces fertile 2n sperm and eggs, which fertilize to produce F_4 allotetraploids: these allotetraploids are maintained up to F_{16} as fertile hybrids (see Liu et al., 2001).

3.4 Introgressive hybridization

It is defined as the incorporation of foreign genes in parental species through backcrossing (see Gante et al., 2004). Concurrent use of mitochondrial and nuclear markers has more specifically defined paternal contribution to hybrid genome. Besides, advances in the field of genomics have magnified the said definition to an unprecedental level of details (Arnold, 2006). With these developments, introgression appears to occur much more frequently than previously considered. For instance, the frequency ranges from 2–4% among the hybrids of European and American anguillids (Avise et al., 1990) to nearly 100% among some salmonids. Bematchez et al. (1995) reported that all the individuals in a population of *Salvelinus fontinalis* possessed the mitochondrial genome of *S. alpinus*. The mtDNA of the cardinal shiner *Luxilus cardinalis* was present in the dusky stripe shiner *L. pilsbryi*, as a result of introgressive hybridization. With the introduction of five species of *Cichla* between 1980s and 1990s, 15 different morphotypes of *Cichla* are now recorded in the Brazilian freshwater bodies suggesting the fast rate of hybridization (Oliveira et al., 2006).

Instead of inducing such 'genetic pollution', some backcrosses induce triploidy. Backcrossed to *Salmo salar*, the F_1 hybrid of *S. trutta* x *S. salar* produced 40% diploid and 60% triploid F_2 progenies (Johnson and Wright, 1986). Involving a third species, the backcrosses generate trigenomic triploid progenies. The F_1 hybrids of *S.trutta—S.salar*, crossed with *S. fontinalis* generated triploids with haploid genomes of all the three species (Wilkins et al., 1993). Similarly, a cross of *Oryzias celebensis* with F_1 hybrid of *O. latipes* x *O. curvinotus* produced trigenomic triploids (Kurita et al., 1992). The introgression among *O. apache*, *O.mykiss* and *O. clarki* was so extensive that all the hybrids carried alleles of the three species (Wilkins et al., 1993).

Backcrosses of *S. fontinalis* with either (*S. trutta* x *S. fontinalis*) or (*Oncorhynchus mykiss* x *S. fontinalis*) produced only non-viable eggs. Similarly, that of *S. confluentis* x *S. fontinalis* yielded so few fry that their backcrosses may not pose a threat to the parental species. However, the 'genetic pollution' of one or another parental species by backcrosses may lead to progressive reduction of parental species. For instance, backcrossed to *Mylocheilus caurinus*, *Richardsonius balteatus* suffered 20% reduction in survival and high proportion (8.4%) deformed fry (Aspinwall and McPhail, 1995). Backcrosses of *Merone saxatilis* x *M.chrysops* produced 50% deformed fry (Bishop, 1967). A similar negative effect of genetic introgression reduced 60% population of the pupfish *Cyprinodon pocosensis* (Wilde and Echella,

1992). Using mtDNA and multinuclear DNA markers in a series of field and laboratory studies, Scribner et al. (1993, 1994a, b) showed a directional introgression of *Gambusia holbrooki* and 'swamping' of *G. affinis*. Their experiments showed that each population experienced an initial flush of hybridization followed by backcrossing and a rapid decline in the frequencies of nuclear and cytoplasmic alleles arising from *G. affinis*. After a period of two years, there was a dramatic increase in *G. holbrooki* mtDNA and nuclear gene frequency as well as that of parental *G. holbrooki*. Thus, a directional introgression of *G.holbrooki* and 'swamping' of *G. affinis* led to the abundance of *G. holbrooki* and hybrids of *G. holbrooki* matrilineal descent and progressive disappearance of *G.affinis* and hybrids of *G. affinis* descent (see also Avise, 2001).

The use of mtDNA and nuclear markers has also led to the discovery of sex based mating asymmetrics in hybridizations. The strict maternal mtDNA inheritance is due to the fact that the low number (ca 100) of paternal mitochondria in sperm is simply overwhelmed by the large number of maternal mitochondrial number (ca 1,00,000) in the egg (see Brown, 2008). Thus Gante et al. (2004) noted that backcross matings among female parent and female hybrid of the Iberian *Chondrostoma* are not at random. This kind of sex based mating asymmetry appears to be common among coral fishes. One parent has alone been found to act as mother during hybridization (e.g. *Thalassoma*, Yaakub et al., 2006; *Plectropomus*, van Herwerden et al., 2006) in many coral reef fishes. However, in an investigation using mtDNA and nuclear marker, Marie et al. (2007) found that *Acanthurus nigricans* is the mother in hybrid matings with *A. leucosternon*. Although *leucosternon* may also act as a female partner in hybridization, due to its rarity, *A. leucosternon* may become locally extinct in some islands of the Pacific. These descriptions do emphasize the need for concurrent use of multiple markers to identify a true hybrid and its parentage, as well as the level and direction of hybrid introgression.

In endemic species, hybridization can lead to an influx of novel and beneficial genetic variations but high rates of introgression may cause genetic swamping of the endemic species and may have detrimental effects on the survival potential. Investigations on the endemic *Barbus carpathicus* and potential for hybridization with its widespread congener *B.barbus* seem to suggest that such hybridization is limited to gain novel genes but not to allow introgression. Interestingly, *B. barbus* is female-heterogametic (Castelli, 1994) and tetraploidy is not also uncommon in it (Berrebi et al., 1993). Using six diagnostic allozyme loci, a microsatellite locus and mtDNA, Lajbner et al. (2009) found no hybrid carrying the mtDNA of *B. carpathicus*; cytonuclear linkage disequilibria also showed a significant positive association between hybrid genotype and *B. barbus* mtDNA, suggesting unidirectionality in the interspecific mating with disproportionate contribution by *B. barbus* mothers. This may be due to the reduced viability and fertility of the hybrid female (Chenuil et al., 2004). Consequently the introgression seems to affect *B. barbus* and not *B. carpathicus* and *B. meridionalis*, another closely related species inhabiting the same river system (Berrebi et al., 1993).

In a dimensionally opposite situation, in which hybridizations and backcrossings are free between all the three cyprinids, the bream *Abramis brama*, roach *Rutilus rutilus* and rudd *Scardinius erythrophthalmus*, Wyatt et al. (2006) investigated these events using both nuclear internal transcribed spacer region 1 (ITS 1) and mitochondrial (cytochrome B) markers. Analysis of the hybrid populations showed that hybridization occurs in both directions. F_1 roach-bream and rudd-bream hybrids are fertile and the hybrids are capable of backcrossing with either of the parental species. With no apparent post-zygotic barriers to reproduction, behavioural mechanisms may alone limit the occurrence of backcrossing. Irrespective of the infrequent backcrossings, the genes from one hybridizing species may become introgressed into the genome of another. Analysis of cloned hybrid nuclear ribosomal DNA revealed evidence for recombination between parental IST1 sequences. Post-F_1 hybridization and introgression may be occurring between roach-bream and rudd-bream, although some unknown barriers appear to suppress backcrosses within the hybrid population.

Incidentally, inbreeding, a distinctly different phenomenon from introgression, also poses serious problems to aquaculture. It is caused by sib- or brother-sister matings leading to homozygosity. The deleterious recessive allele, which is generally masked by the dominant ones in random mating, begins to express leading to death. A population becomes homozygous due to inbreeding. In some south Indian carp hatcheries inbreeding incidence ranges from 2 to 17% (see Padhi and Mandal, 2001).

3.5 Sex ratio

The foregone description clearly reveals that very little information is provided on sex ratio or sex determining mechanism on natural and man-made hybridizations involving the 300 and odd fish species. Yet, bits and pieces of information widely scattered in literature shows that quite a few hybridizations have induced gynogenesis in crosses: e.g. *Cyprinodon alvarezi* ♀ x *Megupsilon aporus* ♂ (Haas, 1979), *Oryzias latipus* ♀ x *O. curvinotus* ♂ (Hamaguchi and Sakaizumi, 1992), *P. nigrofasciata* ♀ x *P. caudofasciata* ♂ (Yamamoto, 1969). Interestingly, hybridizations among sunfishes result in progressive decrease in female ratio: 0.28 ♀♀ in *Lepomis cyanellus* ♀ x *L. microlophus* ♂ (Argue and Dunham, 1999), 0.13 ♀♀ in *L. cyanellus* ♀ x *L. macrochirus* ♂ (Laarman, 1979) and 0.06 ♀♀ in *L.macrochirus* ♀ x *L.microlophus* ♂ (Whitt et al., 1973). Similarly, a cross between *Labeo calbasu* ♀ x *L. rohita* ♂ is recorded to produce reduced female ratio of 0.31 (Bhowmick et al., 1987). Conversely, the female biased sex ratio of 0.6 ♀♀ : 0.4 ♂♂ observed among parental *Poecilia* is maintained by F_1 hybrids of *P.sphenops* and *P. verlifera* (George and Pandian, 1997). Incidentally, *L. cyanellus* is a male heterogametic species (see Devlin and Nagahama, 2002) but *P. sphenops* and *P. velifera* are female heterogametic.

Perhaps, the contribution by Kawamura and Hosoya (2000) is an ideal publication providing desired information on sex ratio and sex determining mechanism among the hybrids of bitterlings from over 100 crosses involving 10 bitterling species belonging to *Tanakia*, *Acheilognathus* and *Rhodeus*. Congeneric *Tanakia* hybrids are fertile with 1♀: 1♂ ratio (see Table 1 of Kawamura and Hosoya, 2000). All other crosses produce mostly sterile males and very few females. In sterile males, spermatogenesis fails even at the spermatocyte stage. In contrast oogenesis is normal in most hybrid females. Sterile intersexes are also produced in 16 crosses, as against the stated five by Kawamura and Hosoya. Gynogenesis occurs at least in one cross: e.g. *T. limbata* ♀ x *A. cyanostigma* ♂. Fecundity of hybrids remains in between those of parental species; for instance, *T. limbata* ♀ x *T. lanceolata* ♂ hybrid produces 14 to 59 eggs, while *T.limbata* and *T. lanceolata* 10 and 70 eggs, respectively.

Taking the female heterogametic sex determining system (ZW), the basic genotype ZZ, drawn from each of the parental species, masculinizes the bitterlings and WZ is a derivative for feminization. When parental species are related, the sex of phenotype of hybrids coincides with genetic sex (see Fig. 7 of Kawamura and Hosoya, 2000). When parental species differ, the sex of phenotype is reversed to become male by abnormal interaction between the Z and X chromosomes. The rare occurrence of females and intersexes among the male-predominant hybrids may be due to complex or partial functional expression of the W chromosome (Kawamura and Hosoya, 2000).

Crosses between protogynous and protandrous hermaphrodites can be a useful model in genetic research to have an insight into the riddle of sex determination in fishes, especially when the crosses are made between male and female heterogametic hermaphrodites. Interestingly, a cross between the protandrous gilthead seabream *Sparus aurata* ♀ and protogynous Mediterranean red porgy *Pagrus pagrus* ♂ was successfully made (Pavlidis et al., 2006). Most hybrids (87%) had either undifferentiated gonads or one differentiated gonad and another undifferentiated gonad. The other 13% hybrids possessed fully differentiated gonads, similar to the ovotestis of the parental species. The ovotestis consisted of a dorso-medial ovarian and ventro-lateral testicular fraction that were separated by a thin layer of connective tissue. Two types of ovotestis, one with the dominance of ovarian structure and the other with that of the testis (see Fig. 1 of Pavlidis et al., 2006) were present. With advancing age, a sex ratio of 0.71 ♀ : 0.29 ♂ (12 ♀♀ : 5 ♂♂) was recognized among maturing hybrids. As a fraction of females is known to remain females in the red porgy (see Pandian, 2010) and the sea bream are known to change sex early the predominance of females among hybrids may not be a surprise. However, no vitellogenic oocytes were found. Although spermatozoa were detached from the lobular wall and filled the semniferous tubules, spermatogenic activity was low or rare. In the parental *P. pagrus* too, precocious primary males were with small volume of unreleased spermatozoa (Kokokiris et al., 1999). Briefly, the males were more like the primary males of the red porgy.

To produce all-male progenies for aquaculture farms, experimental crosses made between selected cichlids have produced a considerable body of information on mechanism of sex determination. Sex in tilapias is controlled by genetic determinants, mostly located on sex chromosomes with an XX, XY model (e.g. *O. mossambicus*; Pandian and Varadaraj, 1988b). Table 14 lists the other models so far proposed for the cichlids. That most interspecific crosses in *Oreochromis* resulted in the production of less than

Table 14 Summary of crosses in *Oreochromis* with respective known / proposed genotypes and sex ratios of F_1 progenies from Hickling (1968), Pruginin et al. (1975), Lovshin, (1982) Pandian et al. (Unpublished observations), Muller-Belecke and Horstgen-Schwark (2007).

Female parent	Known/proposed genotype	Male parent	Known/proposed genotype	Male progeny (%)
O. niloticus	XX	*O. hornorum*	XY	98–100
O. hornorum	XX	*O. niloticus*	XY	75
O. niloticus	XX	*O. macrochir*	YY	100
O. macrochir	XY	*O. niloticus*	XY	75
O. niloticus	XX	*O. aureus*	ZZ	50–100
O. aureus	ZW	*O. niloticus*	XY	75
O. niloticus	XX	*O. veriabilis*	XY ?	100
O. niloticus	XX	*O. nigra*	ZZ ?	85
O. niloticus	XX	*O. leucostictus*	XY ?	94
O. niloticus	XX	*O. mossambicus*	XY	70
O. niloticus	YY	*O. niloticus*	YY	100
O. mossambicus	XX	*O. hornorum*	YY	98–100
O. hornorum	XX	*O. mossambicus*	XY	75
O. mossambicus	XX	*O. niloticus*	XY	70
O. nigra	ZW ?	*O. hornorum*	YY	100
O. nigra	ZW ?	*O. niloticus*	XY	43
O. nigra	ZW ?	*O. leucostictus*	XY ?	96
O. vulcani	XX	*O. hornorum*	YY	98.5
O. vulcani	XX	*O. aureus*	ZW	93–98
O. aureus	ZW	*O. hornorum*	XY	90

100% males indicates that autosomal genetic factors modify the primary effect of sex chromosomes. The cross of *O. nigra* ♀ x *O. hornorum* ♂ produced 100% male progenies; however, *O. nigra* ♀ x *O. niloticus* ♂, and *O.niloticus* ♀ x *O. hornorum* ♂ failed to consistently produce 100% male progenies; likewise the cross of *Macropodus opercularis* x *M. chinensis* yielded 50% ♂♂ and 50% ♀♀, but *M. concolor* x *M. chinensis* resulted in the production of 68 to 91% sterile males (Yamamoto, 1969); these again include the modifying effects of autosomal factors. Incidentally, supermales (YY) in *O. mossambicus* (Varadaraj and Pandian, 1989a) and *O. niloticus* (Scott et al., 1989) have been generated. When YY *O. mossambicus* or *O. niloticus* were crossed with their respective normal females, it was noted that 2–12% of females were produced (see Pandian, 1993) indicating again the modifying effect of autosomal factors on the primary effect of sex chromosomes . However, 100% male progenies are generated, when YY male and YY female of *O. niloticus* are crossed (Muller-Belecke and Horstgen-Schwark, 2007).

Interestingly, Desperz et al. (2006) studied the sex ratio of the red tilapia, a hybrid between *O. mossambicus* ♂ and *O. urolepis-honorum* ♀ using 111 progeny groups including 20, 577 fry sexed from 46 females and 12 males. The male ratio varied widely from 5 to 89% . However, an analysis of distribution of number of progenies showed three groups with respective mean ratios of 27, 50 and 70%. More specific analysis of sex ratio of three selected males was made; the first male crossed with seven females produced 61% males, the second mated with four females produced 48% males and the third after mating with seven females, sired 17% males only. All available monofactorial sex chromosome models do not explain the complicated sex ratio observed in the red tilapia. Hence Desperz et al. concluded that the stability of sex ratios in repeated single pair matings and the parental influence on sex ratios obtained are in agreement with a polygenic basis of sex determination in this hybrid and in many female heterogametic fishes. Hybridization experiments on proven male heterogametic vs female heterogametic (e.g. *Poecilia reticulata* x *P. sphenops*) and protogynous vs protandrous fishes may provide some keys to unlock the riddle in sex determination of fishes.

Briefly, hybridization among fishes is more prevalent than presumed, but many events thus far reported may not result in the production of true hybrids. Human intervention rather than nature is responsible for the prevalence of hybridization events. Survival of hybrids is low; fertility of surviving true hybrids is poor and is broken at F_3 in the ictalurid catfishes and F_8 in 'splake', the hybrid salmonid. By and large, the definition of species characterized by randomly mating populations resulting in intraspecific fertility and interspecific sterility may hold good for fishes.

4

Gynogenesis and Consequences

It is an oogenetic process that exclusively facilitates matrilineal inheritance. In most cases, it is achieved by activation of the egg by genome eliminated or inactivated homologous or heterologous sperm followed by diploidization with retention of the polar body. It may prove useful to (i) identify the existence of homo- (e.g. *Betta splendens*, Kavumpurath and Pandian, 1994) or heterogamety (*Paramisgurnus dabryanus*, You et al., 2008) in female fishes (ii) rapidly produce isogenic or clonal lines for improvement of brood stock (e.g. Liu et al., 2004c, 2009) and (iii) produce all female progenies in candidate species like the acipenserids to obtain the luxurious and expensive caviar (e.g. Saber et al., 2008). The progenies arising from gynogenesis are called gynogenics, instead of gynogens (Liu et al., 1992), gynes (Nam et al., 2000), gynogenetics (Muller-Belecke and Horstgen-Schwark, 1995); this term is used in line with androgenics, allogenics, xenogenics and transgenics. The credit for perfecting the technique of inducing gynogenesis goes to the Russian scientists, Romashov and Beleyaeva (1964). The ease with which the second polar body can be retained for diploidization has resulted in the production of a large number publications and these publications have been reviewed by Cherfas (1981), Pandian and Koteeswaran (1998), Komen and Thorgaard (2007).

4.1 Gynogenic types

Gynogenics include three types: 1. Spontaneous, 2. Meiotic and 3. Mitotic. Spontaneous gynogenics comprise four groups, which are diverse in their mode of origin, ploidy status and existence (Table 15). 1a. Natural unisexual gynogenics like *Poecilia formosa* and *Poeciliopsis* complex produce

Table 15 Natural occurrence of spontaneous gynogeneics.

Species	Reference
(i) Natural unisexual gynogenics	
Poecilia formosa	Hubbs and Hubbs (1932)
(ii) Natural bisexual gynogenics	
Carassius auratus gibelio (3n)	Cherfas (1966)
Carassius auratus langsdorfi (3n)	Umino et al. (1997)
(iii) Induced bisexual gynogenics	
Cobitis biwae	Kusunoki et al. (1994a)
Misgurnus anguillicaudatus	Itono et al. (2007)
(iv) Gynogenic products	
Tinca tinca	Linhart et al. (1995)
Esox masquinongy	Dabrowski et al. (2000)
Cyprinus carpio	Gomelsky (2003)
Oreochromis niloticus	Ezaz et al. (2004b)
Acipenser ruthenus	Fopp-Bayat et al. (2007)
A. baeri	Fopp-Bayat (2008)

unreduced diploid eggs through premeiotic endomitosis (Fig. 1) and are described elsewhere. 1b. Natural gynogenics like the triploids *Carassius auratus gibelio* and *C.a.langsdorfi*, which belong to bisexual species. 1c. The loach *Misgurnus anguillicaudatus* occurs in diploid, triploid (including diploid —triploid mossaics), tetraploid, pentaploid and hexaploid morphs (Fig. 10). Almost all these morphs are amenable for induction of gynogenesis. Some crosses in each morph produce bisexual progenies, while others gynogenics; barring the haploid eggs, production of all other gynogenics involves neither the retention of the second polar body nor the suppression of first mitotic cleavage for duplication of chromosomes. Thanks to the excellent contributions of Arai and his team, the unprecedented level of plasticity displayed by the loach *M. anguillicaudatus* in its gynogenic and sexual modes of reproduction involving haploids to hexaploids has been brought to light through a series of publications (Suzuki et al., 1985; Arai and Mukaino, 1997, 1998; Arai, 2000; Arai et al., 2000; Itono et al., 2006, 2007). Table 16 shows the similarities exhibited by bisexual gynogenics of the loach with one or other unisexual species. More than the unisexuals, the loach can simultaneously produces haploid and diploid eggs (Arai and Mukaino, 1997; Arai et al., 2000) and can also produce tetraploid gynogenics (Arai and Mukaino, 1997). Similarly, the spinous loach *Cobitis biwae* exists as diploid and tetraploid morphs; interploid hybridization produces no surviving progeny. However, diploid eggs of tetraploids are amenable for induction of gynogenesis without involving chromosome duplication (Kusunoki et al., 1994a). 1d. The fourth group includes *Acipenser ruthenus, A. baeri, Tinca*

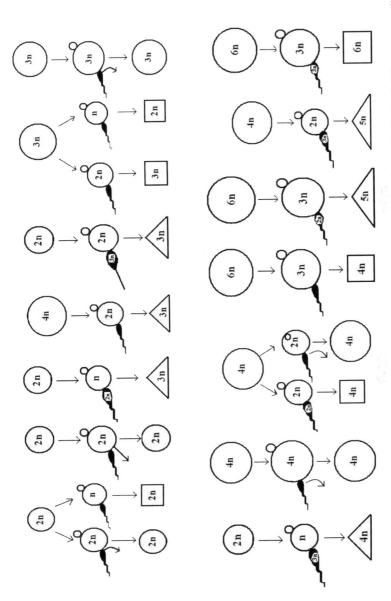

Fig. 10 A compilation of bisexual (□) gynogenic (○) and interploid hybridization (△) modes of reproduction in the loach *Misgurnus anguillicaudatus* involving natural diploid, triploid, tetraploid (bold circles) and induced hexaploid morphs. Note the sizes of eggs and sperm; also note the role played by sperm to activate or fertilize an egg. Polar body is shown by straight flagellum attached to the head.

tinca and others, in which 0.1 to 5.5% diploid gynogenics are spontaneously produced, when gynogenesis is being induced, i.e these diploid gynogenics spontaneously occur, even without undergoing the shock required for diploidization. Heritable nature of such a trait of the biproduct gynogenesis has been confirmed in fancy carp *C. carpio* (Cherfas et al., 1995).

On being activated by irradiated homologous or heterologous sperm, the haploid embryo may at the best survive upto the hatching stage, when all of them succumb due to haploid syndrome. Very rarely gynogenic haploids do survive, as in *Oreochromis mossambicus* (Varadaraj, 1993), but they too die at puberty. From a survey of the Indian catfish *Heteropneustes fossilis* from Tamilnadu and Kerala states, Pandian and Koteeswaran (1999) found that the haploids occur at 1.7% frequency with the sex ratio of 1 ♀ : 1 ♂. The haploid (n = 30) male produced euploid sperm with 30 chromosomes, and initiated development in eggs of normal diploid catfish. However most developing embryos suffered from failure of low jaw formation (Fig. 11a), eye formation (Fig. 11b), and the formation of tumour (Fig. 11c). From Japan, Yamaki et al. (1999) reported the incidence of haploid-diploid mosaic charr *Salvelinus leucomaensis*, in which haploid organs were sustained in the haploid and diploid mosaic system.

Incidentally, with reference to the production of gynogenics using unreduced diploid eggs, requiring no need for diploidization, gynogenic hybrids have been generated in many other fishes, but involving hybridization. Cherfas et al. (1994a) established a hybrid line by crossing

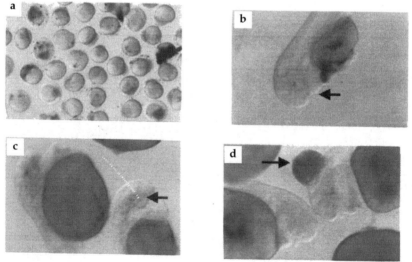

Fig. 11 (a) Haploid eggs of normal diploid *Heteropneustes fossilis* initiated development by haploid sperm derived from a haploid male. Note the failure of (b) jaw formation, (c) eye formation, and (d) formation of tumor (from Koteeswaran and Pandian, 2011).
Colour image of this figure appears in the color plate section at the end of the book.

Table 16 Similarities between unisexual gynogenics and bisexual gynogenic *Misgurnus anguillicaudatus* for which references are given in column 3. For unisexuals representative examples are given in column 2 but corresponding references can be obtained from Pandian (2010).

Characteristics	Unisexual gynogenics: examples	Bisexual gynogenics: Reference
Unreduced 2n/3n eggs through premeiotic endomitosis	*Poecilia formosa* *Poeciliopsis*	Matsubara et al. (1995), Zhang and Arai (1996); Zhang et al. (1998)
Reproduce gynogenetically	*P. formosa, Menidia clarkhubbsi*	Arai and Mukaino (1997)
Clonals and aclonals present	*Cobitis, Phoxinus eos-neogaeus*	Arai et al. (2000) Arai and Mukaino (1997)
Triploids by fertilization of 2n egg by haploid sperm of the donor	*Poeciliopsis, P. eos-neogaeus*	
Hybridogenesis occurs	*Poeciliopsis*	Morishima et al. (2008)
Diploid-triploid mosaics present	*P. eos-neogaeus*	Itono et al. (2007)
Males rare; when present, they are sterile	*P. eos-neogaeus, Carassius auratus*	Oshima et al. (2005)
Males present; can also sexually reproduce	*P. eos-neogaeus, Squalius alburnoides*	

C. auratus gibelio ♀ with *C. carpio* ♂; their hybrid eggs, when activated by the irradiated milt of the hybrid, hatched as diploid gynogenics; the proportion of the fertile hybrid gynogenics increased from 22% in F_1 to 83% and to 97% in F_2 and F_3 progenies. Similarly, Liu et al. (2001, 2004c) established a hybrid line by crossing *C. auratus gibelio* ♀ with *C. carpio* ♂. Their F_2 hybrids produced three types of eggs : 14% of haploid eggs each with the diameter of 1.3 mm, 33% diploid eggs (1.7 mm) and 50% eggs each with the diameter of 2 mm. Using molecular methods to identify and expression analysis of genes involved in early ovary development of the gynogenic hybrids, Liu, J et al. (2004c) and Liu, D et al. (2009) have also shown that diploid eggs were produced through premeiotic endomitosis, whereas the tetraploids produced diploid eggs through regular meiosis (see Liu et al., 2001; Sun et al., 2003). The other reports on the production of diploid eggs generated by diploid hybrids are: (i) (2n *Salmo salar* ♀ x 2n *S. trutta* ♂) 2n F_1 ♀ x 2n *S. salar* ♂ (Galbreath and Thorgaard, 1995), and (ii) (2n *S. salar* ♀ x 2n *S. trutta* ♂) 2n F_1 ♀ x 2n *S. trutta* ♂ (Johnson and Wright, 1986).

Ionizing and UV-radiations are strong physical mutagens listed as carcinogenic and clastogenic agents capable of chromosomal and DNA breaks, respectively. The ionizing radiation may damage chromosomal DNA by inducing double strand breaks (DSB), the most lethal form of DNA lesions. UV-radiation triggers lesions on DNA that may cause DSB (Pfeifer et al., 2004). However both of them are useful for inactivation of nuclear DNA in fish gametes to induce gynogenesis and androgenesis, and to provide individuals possessing maternal or paternal nuclear DNA alone.

Following activation of egg through genome elimination by ^{60}Co irradiation (or X-rays) or inactivation by UV-irradiation homologous/ heterologous sperm, diploidization of haploid gynogenics is achieved by shocking the activated eggs. Two types of gynogenics are possible: Meiotic and Mitotic. Shock given at an early (4 minutes activation *Paramisgurnus dabryanus*, You et al., 2008) zygotic stage results in retention of the second polar body and production of meiotic gynogenics. That given at a relatively later zygotic stage (18 minutes post fertilization) leads to the suppression of first cleavage and generation of mitotic gynogenics. Retention of the polar body permits effective diploidization of the genome without interference to crossing over events and hence some level of heterozygosity is ensured. Frequency of crossing over depends on the distance between a given gene and centromere, and the frequency determined for fishes is low (Kallman, 1984). Supposing that the average frequency of crossing over is not more than 0.1%, then one gynogenic generation is equal to 12 generations of sib-mating with respect to the degree of inbreeding (see Nagy et al., 1978). However, 0.6 to 0.95% cross over frequencies have been estimated for fishes (e.g. *Misgurnus anguillicaudatus*, Morishma et al., 2001). But mitotic gynogenics are completely homozygous for all the loci and are absolutely required as mother fish to produce isogenic clones by the second cycle of gynogenesis in the next generation (see Arai, 2001).

4.2 Purity of gynogenics

In irradiated sperm, ^{60}Co or X-rays may eliminate the chromosome by fragmentation. On the other hand, the UV-irradiation inactivates the genome by introducing the enzymatically photoreactivable cyclobutyl rings between adjacent pyrimidines (mostly thymine) (Voet and Voet, 1990). To prevent photoreactivation, the entire process of UV-irradiation is completed in darkness (e.g Van Eenennaam et al., 1996). The estimated paternal genetic contamination ranges from 0.006% for ^{60}Co-irradiated sperm (Thorgaard, 1986) to 0.03% for the UV-irradiated sperm (Quillet, 1994). In the sea bass *Dicentrarchus labrax*, AFLP analysis has shown 10% genetic contamination in one of the three inductions (Felip et al., 2000). An alternative is to choose a donor species, which will not yield hybrid or triploid; for instance, sperm of *O. mykiss* do not produce hybrids or triploids, when crossed with *S.salar* (Quillet and Gaignon, 1990). Castelli (1994) has chosen *C. carpio* as a donor, for it is known that neither hybrid nor triploid is produced, when crossed to *Barbus barbus*.

To avoid paternal genetic contamination, many authors have chosen one or other phenotypic markers like body colour, scale character, protein markers like allozyme, isozyme and others like sex specific molecular marker, multilocous satellite markers, DNA fingerprinting and so on. In fact, a few authors have chosen to use two or three markers concurrently (Table 17). Microsatellite markers are used to detect any paternal contribution to the diploid gynogenic's genome. Using LS 19 and 54 probes, Saber et al. (2008) have detected the presence of a complete maternal genome among the gynogenics of *Acipenser stellatus*, whereas equal parental and maternal genomes among the control diploids. DNA fingerprints of diploid gynogenic *M. anguillicaudatus* are very similar to those of the mother fish (Fig. 12). The similarity (purity) of gynogenics, as measured by Band Sharing Index (BSI), to the mother ranges between 0.933 and 0.978. The BSI among the F_2 mitogynics was closer to 1 (Arai and Ishimoto, 1996).

4.3 Survival and gamety

With reference to gynogenesis, a basic difference between male- and female heterogametics must be recognized. In species with male heterogamety, gynogenesis may produce all female progenies with genotypes of X^1x^1 (X^1 and Z^1 represent contributions from respective egg nuclei proper and x^1 and z^1 the contributions from respective polar bodies) in meiogynogenesis and X^1X^1 (X^1X^1 and Z^1Z^1 represent contributions from the respective nuclei

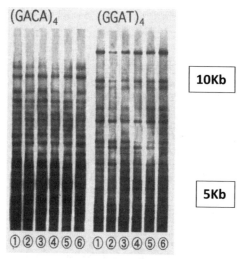

Fig. 12 DNA fingerprints {probed with (GACA) 4 and (GGAT) 4} of gynogenic diploid progeny (lane 1–5 = gynogenics, lane 6 = mother of *Misgurnus anguillicaudatus* fish. Molecular weights are given in kilo base pairs (from Arai et al., 2000).

arising from first mitotic nuclear division) in mitogynogenesis. But in female heterogametic species, about 50% males bearing Z^1z^1 genotype and superfemale with W^1w^1 and/or normal female are produced, depending upon the recombination frequency between the sex determining element and the centromere during meiotic prophase. Secondly, gynogenics, bearing homogametic genotypes X^1x^1 and Z^1z^1, are more viable than the superfemales (W^1w^1).

Survival: Comparison of data on survival of gynogenics is a difficult task. Survival of meiogynogenic of *Cyprinus carpio* varied from 13% in one population to 95% in another (Nagy et al., 1978). In general, pressure shocks seem to ensure higher survival and cold shocks the lower (Pandian and Koteeswaran, 1998). Hatchability of gynogenics of *Cobitis biwae* was 87, 22 and 4%, when its eggs were activated by the donor sperm of *Misgurnus anguillicaudatus, Cyprinus carpio* and *Gnathopogon elongatus elongatus,* respectively. Apparently, differences in head size, motility and sperm count of the irradiated sperm should account for the observed differences (Kusunoki et al., 1994b). Many authors have reported survival of gynogenics at hatching, while others at the feeding stage and yet others at sexual maturity. For instance, following successful induction of meiotic gynogenesis, 48 and 26% of *Oncorhynchus mykiss* survived at hatching and feeding stages (Chourrout and Quillet, 1982); survival of mitogynogenics of *O. mykiss* was reduced to 2% at the age 2+, when sexual maturity was attained (Quillet, 1994).

Table 17 Marker and/or technique used for confirmation of successful induction of gynogenesis.

Species	Marker	Reference
Oncorhynchus mykiss	Karyotype	Chourrout (1984)
O. mykiss	Scale	Quillet (1994)
Cyprinus carpio	Colour	Komen (1990)
Misgurnus anguillicaudatus	Colour	Suzuki et al. (1985)
Oryzias latipes	Colour	Naruse et al. (1985)
Danio rerio	Black spot	Kavumpurath and Pandian (1992c)
D. frankei	Blue stripe	Kavumpurath and Pandian (1992c)
Ictalurus punctatus	Allozyme	Liu et al. (1992)
M. anguillicaudatus	Allozyme	Arai and Mukaino (1998)
Cynoglossus semilaevis	DNA marker	Chen et al. (2009)
Paramisgurnus dabryanus	Microsatellite, Karyotype	You et al. (2008)
Acipenser stellatus	Microsatellite	Saber et al. (2008)
A. baeri	Microsatellite, Karyotype	Fopp-Bayat (2007)
Carassius langsdorfi	Flow cytometry, DNA fingerprinting	Umino et al. (1997)
M. anguillicaudatus	DNA fingerprinting, Microsatellite	Itono et al. (2007)
M. anguillicaudatus	Flow cytometry, Allozyme	Arai and Mukaino (1998)
M. anguillicaudatus	DNA fingerprinting, Erythrocyte size	Arai and Mukaino (1997)
Plecoglossus altivelis	DNA fingerprinting, Allozyme, Tissue grafting	Taniguchi et al. (1994)

Survival of mitotic gynogenics is very low, in comparison to that of meiotics. A few authors have made a comparative study on survival of these two gynogenics in the same species; for instance, 19 and 1% for the meiotic and mitotic gynogenics of the loach *Cobitis biwae* (Kusunoki et al. 1994a) and 4 and 0.2% for the meiotic and mitotic gynogenics of *Ictalurus punctatus* (Goudie et al., 1995). Other reported values are: 19% (Peruzzi et al., 1993) and 2% (Hussain et al., 1993) for the meiotic and mitotic gynogenics of *Oreochromis niloticus* and 29% (Komen et al., 1988) and 1% (Gomelsky et al., 1992) for the meiotic and mitotic gynogenics of *Cyprinus carpio*. A cause for zygotic mortality has been traced to the presence of 40–49 chromosomes only, instead of modal number of 50 in *M. anguillicaudatus* (Suwa et al., 1994).

Gamety: Gynogenesis is one of the powerful methods to examine the genetic sex determination system of bisexual species. Table 18 provides a few representative examples for gamety in fishes using gynogenesis as a method to determine gamety. In this context, the conclusion made by Devlin and Nagahama (2002) that less than a third of fishes are characterized by

Table 18 Gynogenesis as a technique to determine sex determining system in representative fishes.

Species	Reference
Male heterogamety (X^1X^2/X^1Y^2)	
Aristichthys nobilis	Shelton (1986)
Hypopthalmichthys molitrix	Mirza and Shelton (1998)
Limanda yokohamae	Aida and Arai (1998)
Oreochromis mossambicus	Pandian and Varadaraj (1990a)
Betta splendens	Kavumpurath and Pandian (1994)
Puntius gonionotus	Pongthana et al. (1995)
Polyodon spathula	Mims et al. (1997)
Misgurnus anguillicaudatus	Nomura et al. (1998)
Clarias gariepinus	Galbursera et al. (2000)
Silurus asotus	Nam et al. (2001a)
Cyprinus carpio	Gomelsky (2003)
Paralichthys lethostigma	Luckenbach et al. (2004)
Hippoglossus hippoglossus	Tvelt et al. (2006)
Female heterogamety (Z^1W^2/Z^1Z^2)	
Macropodus opercularis	Gervai and Csanyi (1984)
O. aureus	Mair et al. (1991b)
Barbus barbus	Castelli (1994)
Rhodeus ocellatus ocellatus	Kawamura (1998)
Acipenser transmontanus	Van Eenennaam et al. (1999)
A. brevirostrum	Flynn et al. (2006)
Esox masquinongy	Rinchard et al. (2002)
Paramisgurnus dabryanus	You et al. (2008)
Cynoglossus semilaevis	Chen et al. (2009)

female gamety, it is notable that gynogenesis seems to have detected female heterogamety in more number of species.

In male heterogametic species, gynogenics produces 100% founder females (Table 19). Experiments have also confirmed that the crosses of hormonally sex reversed gynogenic neomales (X^1X^2) with normal females (X^1X^2) have produced almost (>98%) all female progenies (e.g *Oreochromis mossambicus*, Varadaraj and Pandian, 1988; *Limanda yokohamae*, Aida and Arai, 1998). On the other hand, in female heterogametic fishes gynogenesis produces ZZ males, ZW females and WW superfemales in various proportions (Table 20), depending upon the cross over between the sex determining factor and the centromere. If a cross over does not take place between these two, no heterozygote is generated. However, if one cross over takes place between them, about 100% heterozygotes are produced. When the sex determining factor is independently assorted of the centromere, about 67% heterozygotes may be produced and the value may go up to

Table 19 Appearance of unexpected males on induction of meio and mitogenesis in male heterogametic fishes.

Species	♂(%)	♀(%)	Remarks	Reference
Betta splendens	0	100	Meio- and mitogynogenics	Kavumpurath and Pandian (1994)
Ictalurus punctatus	0	100	Meio- and mitogynogenics	Goudie et al. (1995)
I. furcatus	0	100	Meio- and mitogynogenics	Goudie et al. (1995)
Clarias gariepinus	8		Males only in two batches of pressure-shocked gynogenics	Volckaert et al. (1994)
Silurus asotus		100	12 meiogynogenic crosses produce gynogenics Neomales produced 100% females	Nam et al. (2001a)
Oncorhynchus mykiss	0	100	21 meiogynogenics	Krisfalusi et al. (2000)
O. rhodurus	0	100	Mitogynogenics by 4 heterozygous ♀♀ Meiogynogenic died prior to maturity	Kobayashi et al. (1994)
O. kisutch	0	100	34% gynogenic ♀ suffer reduced ovaries with areas devoid of oocytes	Piferrer et al. (1994)
Puntius gonionotus	0	100	Meiogynogenics from 10 mothers	Pongthana et al. (1995)
P. gonionotus			8 out of 10 neomales (X¹x¹) sired 100% ♀♀ when crossed with (X¹X²) ♀♀; others sired 12% males	Pongthana et al. (1999)
Limanda yokohamae	2	98	10 gynogenic neomales (X¹x¹) crossed with normal females (X¹X²)	Aida and Arai (1998)
Misgurnus anguillicaudatus	2	98	In 56/57 crosses homogamous orange coloured ♀♀ were present	Suzuki et al. (1985)
Oreochromis mossambicus	0	100	Meiogynogenics from 11 ♀ mothers Gynogenics from 6 ♀ mothers	Varadaraj and Pandian (1989a)
O. niloticus		100	89 Gynogenics from one ♀	Penman et al. (1987)
O. niloticus	10	90	10 heterozygous meiotic crosses	Karayucel et al. (2004)

Table 20 Fertility and sex ratio of gynogenics and their F_1 progenies in female heterogametic fishes.

Species/References	Observations
Barbus barbus Castelli (1994)	54% ♀ meiogynogenics. Hatching success 53% and 55% in F_0 and F_1 gynogenics. 3 crosses involving 3 ($X^1 x^1$) ♀♀ with 19 (Z^1Z^2) males produced 99.8% F_1 female progenies
Rhodeus ocellatus ocellatus Kawamura (1998)	Fertile (88%) male and (12%) female meio- and mitogynogenics. 4 meiogynogenic ♀♀, when backcrossed to normal ♂♂, 33–98% F_1 progenies survived; female ratio among these progenics was 6, 12, 22 and 55% for 4 females tested
Paramisgurnus dabryanus You et al. (2008)	7 inductions of gynogensis resulted in 27–94% females; of 135 gynogenics produced, 64% were females. When ♀ gynogenics were crossed with ♂ gynogenics or with normal ♂, either 50% or 100% females were generated
Acipenser transmontanus Van Eenennaam et al. (1999)	4 meiogynogenic inductions produced 50–82% females. They were all sexually differentiated male and female gynogenics
A. brevirostrum Flynn et al. (2006)	56% females and 44% males
Esox masquinongy Rinchard et al. (2002)	Gynogenic induction produced 59% females, 39% males and 3% intersexes
Cynoglossus semilaevis Ji et al. (2010)	10% hatching; 7% females; on amplification with *Cse* 205 F_1 and *Cse* F 205 R, PCR produced only female specific 205 bp product in four out of 24 females

88% with a viability of WW superfemales (Van Eenennaam et al., 1999). In *Poecilia sphenops*, hormonally sex reversed superfemale ($Z^1 Z^2$) is viable (George and Pandian, 1997) but gynogenic supermale is not viable even in combination with $Z^1W^1W^1$ in *Rhodeus ocellatus ocellatus* (Kawamura, 1998). For the heterozygotic females, the reported female ratio ranges from 12% in *R. ocellatus ocellatus* to 27–94% in *Paramisgurnus dabryanuas* (Table 20). With values of > 88 % female ratio in *P. dabryanus*, its superfemale (W^1w^1) is likely to have survived.

A few available values for sex ratio within a species among many female heterogametic fishes detected by gynogenesis show that there are two females for every male (Table 21), suggesting that the sex determining factor is independently assorted of the centromere.

Tables 19 and 22 list the occurrence of unexpected sex in the gynogenics, and consequent departures from the expected sex ratios. Reasons for these departures are traced to (i) paternal genetic contamination and (ii) presumed presence of minor genes. Using DNA fingerprinting technique, Volckaert et al. (1994) recorded such contamination in two batches of pressure-shocked gynogenics of *Clarias gariepinus*. Using RAPD DNA primers known to generate sire specific markers, Van Eenennaam et al. (1996) screened 108 putative gynogenics of known parentage of *Acipenser*

Table 21 Normal sex ratios of female heterogametic fishes.

Species	Sex ratio		Reference
	Normal ♀ : ♂	Gynogenics ♀ : ♂	
Barbus barbus [1]	0.50 : 0.50	0.63 : 0.27	Castelli et al. (1994)
Oreochromis aureus [2]	0.54 : 0.46	0.67 : 0.33	Mair et al. (1991b)
Rhodeus ocellatus ocellatus	0.47 : 0.53	–	Kawamura (1998)
Esox masquinony [3]	?	0.70 : 0.30	Rinchard et al. (2002)
Paramisgurnus dabryanus [4]	0.89 : 0.11	0.68 : 0.32	You et al. (2008)
Female : male ratio*	1 : 1	2 : 1	

*Excluding *P. dabryanus*, 1. see their Table 9.4.3, 2. see their Table 2; 3. only one control value is given, 4. see their Table 2

transmontanus and identified two putative gynogenics, whose genomes were contaminated by paternal genomes. A second reason is the presumed presence of minor genes responsible for the observed departures from the expected sex ratio. For instance, *mas* by Komen and Richter (1990), *Mm* by Castelli (1994), *Ff* by Mair et al. (1991b) and *W"* by Kawamura and Hosoya (2000). Notably, the presence of these presumed minor genes sitting on the sex chromosome or on one or another autosome seems to be ubiquitous among fishes and plays a diverse role in modifying the expected sex ratio from slight to significant levels.

Carassius auratus gibelio exists in its all female triploid gynogenic form in northern Eurasia. However 5–25% male progenies are also known to occur (Gui et al., 1992). In an attempt to trace the reason for such an unusual occurrence, male progenies produced by the gynogenic *C. auratus gibelio*, Jia et al. (2008) found a usual paternal genomic leakage (contamination) among these males. Induction of gynogenics produced 23% unexpected male progenies. The male ratios of these gynogenics were variable and ranged from 0 to 37%. Microsatellite genotyping at 15 loci showed 100% progenies of four mothers shared the same genotype, as their respective mothers, but those of the 5th mother shared only 97% genotype with their mother, conforming that gynogenesis is the normal mode of reproduction in the fish. However, 0.63% of all progenies did show incorporation of paternal genetic material, which *albeit* is not responsible for sex determination in males.

4.4 Fertility and F_1 sex ratio

Information on fertility of gynogenics is available for a number of species belonging to Cichlidae (Table 22), Salmonidae and Cyprinidae (Table 23). The mitogynogenic amago salmon produce fertile females, which, in turn, produce viable F_1 all female progenies, but their counterpart meiogynogenics do not survive to attain sexual maturity (Kobayashi et al., 1994). In other

Table 22 Female ratios of gynogenics and their F_1 progenics arising from different crosses between normal, hormone—treated gynogenic males and females belonging to *Oreochromis*.

Species/Reference	Observations
Oreochromis mossambicus Penman et al. (1987)	160 meiogynogenics arising from 4 mothers were all females
O. mossambicus Varadaraj and Pandian (1989a)	100% ♀♀ in 7 crosses of F_1 (X¹x¹) ♀ x (X¹Y²) ♂; irradiated sperm
O. mossambicus Pandian and Varadaraj (1990a)	100% ♀♀ in 11crosses of F (X¹x¹) ♀ x F (X¹ Y²) ♂; irradiated sperm 100% ♀♀ in 11 crosses of F (X¹x¹) ♀ x F (X¹ x¹) sex reversed ♂
O. mossambicus Marian and Pandian (1992)	Survival of mitogynogenic increased from 22% in F_0 to 64 % F_2
O. niloticus Muller-Belecke and Horstgen-Schwark (1995)	Meiogynogenics arising from 2 ♀♀ were all females ♀♀ Mitogynogenesis in 14 mothers produced 35 % male; 26 out of 27 gynogenic ♀♀ spawned eggs; of them, 10 alone produced fertile eggs; 6 of 42 ♂♂ alone produced quality sperm
O. niloticus Shah (1988)	89% meiogynogenics arising from 2 mothers were all females
O.niloticus Mair et al. (1991a)	86% meiogynogenics arising from 11 mothers were all (but 1 cross) produced 100% meiogynogenic females (see Penman et al., 1987)
O. niloticus Hussain et al. (1991)	Mitogyncgenics arising from one mother produced 20% ♂♂
O. niloticus Sarder et al. (1999)	Mitogynogenics arising from 6 mothers produced 20% ♂♂
O. niloticus Ezaz et al. (2004a)	100% ♀♀ produced by 14 crosses involving 14 (X¹x¹) gynogenic ♀ with 4 (X¹X²) neomales; 100% Y¹Y¹ males from 15 of 16 crosses involving 16 (Y¹y¹) gynogenic ♂ x 4 (Y¹Y¹) ♀
O. niloticus Karayucel et al. (2004)	Backcrosses of 3 out of 6 heterozygous red F_1 ♀♀ (X¹x¹ Rr Ss) with 2 wild coloured males produced 11% ♂♂
O. aureus Mair et al. (1991b)	100% ♀♀ in 3 crosses involving one mitogynogenic (X¹X¹) ♀ x 3 normal (X¹Y²) ♂♂, but 73% ♀♀ in 4th cross between normal (X¹X²) ♀ x mitogynogenic neomale (X¹X¹)

Table 23 Fertility of meio and mitogynogenics of male heterogametic fishes.

Species/References	Observations
Salmonidae	
Oncorhynchus mykiss Krisfalusi et al. (2000)	33% meiogynogenics with aberrant gonadal morphology due to the presence of chromosomal fragments, i.e. incomplete inactivation of maternal genome
O. mykiss Quillet (1994)	Spawning postponed and prolonged; 10% reduction in fecundity and wide variations in egg size
O.rhodurus Kobayashi et al. (1994)	F_1 mitogynogenics are all females. But none of meiogynogenics survive till maturity
O. kisutch Piferrer et al. (1994)	34% gynogenics with various levels of abnormal ovaries
Cyprinidae	
Carassius auratus Oshiro (1987)	9–34% gynogenics have underdeveloped ovaries with no oocyte
Ctenopharyngodon idella Jensen et al. (1978)	20% gynogenics; their ovaries with no oocyte
Cyprinus carpio Cherfas (1981), Komen and Richter (1993)	50% gynogenics with abnormal and reduced ovaries; males and intersexes occur
Others	
Pagrus major Kato et al. (2002)	Only one of 13 gynogenic females successfully reproduces
Paralichthys olivaceus Tabata (1991)	Maturity advanced; fecundity and progeny survival increased

salmonids and cyprinids, different proportions of gynogenics suffer from abnormal, underdeveloped and/or reduced ovaries with small or large areas filled with no oocyte. The proportion of partial infertility ranges from 9 to 34% of females of *Carassius auratus* to 50% females of *Cyprinus carpio* (Table 22). Piferrer et al. (1994) provided descriptions of such partial infertility in *Oncorhynchus kisutch* but found no male germ cells in its ovaries, as has been reported for the cyprinids like *Cyprinis carpio* (Komen and Richter, 1993).

Gynogenesis produces more or less 100% females in male heterogametic fishes (Table 19) but the female heterogametics produce various proportions of females; the proportions vary not only from species to species but also within a species. The founder females produce 50–82% female progenics in *Acipenser transmontanus* and 27–94 % in *Paramisgurnus dabryanus* (Table 20). Among different species, the female ratio ranges from 12% in *R. ocellatus ocellatus* to 54% in *Barbus barbus*. However, all the female heterogametics are fully fertile; partial or complete sterility of these females has so far not been reported; successive generations of gynogenics seem to increase the female ratio; for instance, from 12% in F_0 to 22–55 % in F_1 in *R. ocellatus*

ocellatus and 54% in F_0 to 99.8% in F_1 in *Barbus barbus* (Table 20). But, the same cannot be generalized for others.

4.5 Sex ratio and departures

Generation of all male (monosex) progenies in economically important cichlids has been a real challenge. Expectedly, many European institutions have made considerable efforts to solve the problem using *Oreochromis niloticus* strain from Lake Manzala, Egypt. Table 22 is a comprehensive summary of information compiled from relevant publications. Except for a single report on the unexpected occurrence of 4% $\male\male$ in *O. niloticus* (Mair et al., 1991a), all other attempts to induce meiogynogenics in *O. mossambicus* and *O. niloticus* produced 100% female meiogynogenics. However, induction of mitogynogenics led to unexpected production of 20% males (Hussain et al., 1991; Mair et al., 1997; Sarder et al., 1999), 35% (Muller-Belecke and Horstgen-Schwark, 1995) and 27% in one of the crosses of the female heterogametic mitogynogenetics of *O. aureus* (Mair et al., 1991b). Notably, Hussain et al. (1991) commenced the diploidization shock in one case 27–30 minutes post-fertilization (pf) and in the second 30–40 minutes pf (see Pandian and Koteeswaran, 1998); Muller-Belecke and Horstgen-Schwark (1995) did it 25–35 pf and the shock was prolonged for 4.5 to 7.5 minutes, while the optimal required period of shock is just 2 minutes (see Peruzzi et al., 1993); at least in one case, when it was commenced, within 20 minutes pf and prolonged for 4 minutes only, no male was produced. The timing and duration of diploidization shock should coincide with prophase of first mitotic division. Hence it is likely that the delayed initiation and extended shocking duration to induce diploidization by some of these authors may be a major reason for their reported low survival and unexpected occurrence of males in *O. niloticus*.

It is known that incomplete elimination/inactivation of paternal genome and/or delayed diploidization may lead to genetic contamination of the gynogenic. In infertile gynogenics of *O. mykiss*, the presence of chromosomal fragments was found to be positively correlated with irregular ovarian development (Piferrer et al., 1994). Similarly, Chourrout and Quillet (1982) provided photographic evidence for the presence of paternal chromosomal fragments in the gynogenics of *O. mykiss*, which Quillet (1994) considered as a possible cause for the delayed maturity and infertility of their gynogenics. Ocalewicz et al. (2009) reported the presence of 1–7 chromosomal fragments in healthy androgenic males and females of *O. mykiss* (Fig. 13). However, from a genetic analysis of unexpected males among gynogenic families of *O. mykiss* and transmission of maleness over three generations, Quillet et al. (2002) assumed that the presence of a recessive mutation in the sex determining gene explained the unexpected appearance of males among

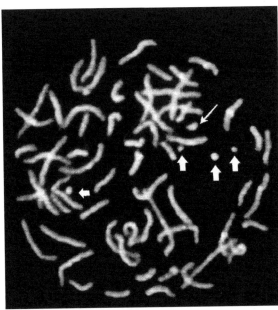

Fig. 13 DAPI stained mitotic chromosomes of androgenic *Oncorhynchus mykiss*. Arrows point to the largest, linear chromosome fragments and arrow heads chromosome fragments showing ring morphology (from Ocalewicz et al., 2009)

the gynogenics and crosses between them. Interestingly, no male appeared, when the gynogenics were crossed with wild strain.

Ji et al. (2009, 2010) successfully induced gynogenesis in female heterogametic tongue sole *Cynoglossus semilaevis*. Microsatellite analysis at locus *C sou* 6 revealed that none of the tested 24 gynogenic has inherited (male specific 192 bp) paternal alleles. However, due to recombination between the *C sou* 6 and the centromere at the rate of 0.625 produced heterozygous gynogenics. Sex ratio was 0.17 ♀ : 0.83 ♂, clearly indicating the presence of lethal gene(s) in the W chromosome.

It is likely that the infertility and possibly the unexpected occurrence of males among founder gynogenics are at least partially due to incomplete elimination of paternal genome and/or delayed initiation of shock for diploidization. This generalization is also supported by Ezaz et al. (2004a), who found 100% females produced by all the tested crosses involving 14 (X^1x^1) gynogenic females and four neomales (X^1X^2). They also suggested that the factors that "cause departures from predicted chromosomal sex determination appears to be autosomal, heritable, polymorphic" and may act in either direction, i.e. feminization of genetic male (X^1Y^2) progeny or masculinization of genetic female (X^1X^2) progeny. Karayucel et al. (2004) provided evidence for at least two unlinked 'sex reversal' loci, one of which is linked to the autosomal red body colour locus.

In female heterogametic fishes, all male (monosex) progenies can be produced by inducing androgenesis, *albeit* a single report alone is available (Liu, 2010). Low survival and reduced fertility observed among gynogenics of male heterogametic species may strike a negative bell on the scope for gynogenesis as a method for generation of all female progenies. Yet, the progressive increase in hatching with successive generations (*C.auratus-C.carpio* mitogynogenic hybrids) and the same observed for elimination of overriding autosomal influence in modifying sex ratios (e.g. *Barbus barbus*, Table 20; *O.mossambicus*, Table 22) provide green signals for achieving increased hatching, survival, fertility and reduced autosomal influence on sex ratio. In fact, Karayucel et al. (2004) have identified two such autosomal factors, which can be eliminated by selective breeding. Sooner or later, British scientists are likely to come up with a strain of *O. niloticus* that may produce all males or all females alone, at least in temperature controlled aquafarms. However, the presence of overriding autosomes appears ubiquitous especially among female heterogametic fishes. Hence sex in fishes may not be determined by sex chromosomes alone especially in female heterogametic fishes.

Though the technique to retain the second polar body is simple and widely practicable, artificial gynogenesis has so far been successfully induced only in about 50 odd species (see also Pandian and Koteeswaran, 1998). Gynogenics are generated during the induction process of gynogenesis in a half dozen male and female heterogametic species (Table 15). Without retaining polar body gynogenics can be produced only in a couple of species, using the diploid eggs produced by the naturally available tetraploids. Occurrence of natural gynogenics is limited to only one male heterogametic crucian carp *Carassius auratus*. These facts seem to indicate that the gynogenesis is not a selected evolutionary phenomenon and may not yield clues to resolve the elusive enigma of sex determination in fishes. Besides, survival and the levels of fertility of surviving gynogenics are low and vary widely, though their survival and fertility significantly and progressively improves-apparently by elimination of deleterious genes with successive generations from F_1 to F_3 in both male and female heterogametic species (Table 20, 22). Incidentally, among gynogenic fishes, the expected sex ratios are 100% females (X^1x^1) for male heterogametic species and 1 ♂ (Z^1z^1) : 1 ♀ (W^1w^1) for female heterogametic species. The 2 ♀♀ : 1 ♂ sex ratio reported for the surviving gynogenics (see Table 21) clearly indicates the significantly low survival of homogametic Z^1z^1 males, in comparison to homogametic, i.e. X^1x^1 females.

Androgenesis and Autosomes

Androgenesis is a developmental process that facilitates exclusive inheritance of the paternal genome. It may prove useful to (i) identify the sex determination system (e.g. Cyprinidae: *Puntius tetrazona*, Kirankumar and Pandian, 2003, *P. conchonius*, Kirankumar and Pandian, 2004a, Characidae: *Hemigrammus caudovittatus*, David and Pandian, 2006a), (ii) produce all-male (monosex) progenies sired by YY androgenics (e.g. Scheerer et al., 1986; the term used instead of androgenotes (Thorgaard et al., 1990), androgens etc, and in line with gynogenics, allogenics, xenogenics, transgenics), (iii) produce inbred isogenic lines (e.g. Bongers et al., 1997b) and (iv) conserve germplasm of desired strain/sperm using its preserved sperm and genome-inactivated eggs of a surrogate species (e.g. Bercsenyi et al., 1998, David and Pandian, 2006b). Unlike in triploids and gynogenics, no natural occurrence of androgenic has been reported. The only claim is by Stanley et al. (1976), who have found from electrophorogram analysis the loss of carp genome in a few hybrids of the cross between the common carp *Cyprinus carpio* ♀ (2n = 50) x grass carp *Ctenopharyngodon idella* ♂ (2n = 48). Polyspermy occurs normally in 5% eggs of the grass carp. The entry of two sperms of grass carp could have accidentally eliminated the carp genome. In the process of producing homozygous transgenic of *Misgurnus mizolepis* by inducing gynogenesis with hetrologous sperm, Nam et al. (2000) too obtained 3% androgenics along with 40% meiogynogenics and 26% mitogynogenics. Almost all earlier publications in this area are concerned with elimination/inactivation of maternal genome from eggs and diploidization of zygotes by thermal or pressure shock. The relevant publications have been reviewed by Pandian and Koteeswaran (1998), Pandian and Kirankumar (2003), Komen and Thorgaard (2007).

5.1 Production methods

In fishes androgenics are produced following different methods resulting in different levels of heterozygosity. Maternal androgenics are produced by inducing gynogenesis in sex reversed (X^1Y^2) neofemale (Fig. 14), and the retention of the second polar body may increase heterozygosity, and it is relatively easier to achieve (see Pandian and Koteeswaran, 1998). Barring this simple method, the other methods involve essentially two steps: (i) elimination of the maternal genome in eggs by fragmentation using ^{60}Co-irradiation (25KR–88KR) or X-rays (Table 24), or inactivation of the maternal genome by UV-irradiation (150–300 mJ/cm^2) and (ii) activation of the irradiated eggs by haploid or diploid gamete; the monospermic activation of the irradiated eggs is to be obligately followed by suppression of the first mitotic cleavage with thermal/pressure shock. In oviparous fishes, mitotic androgenics are produced by diploidization of the irradiated developing eggs by a shock. The same procedure is followed to generate allo-androgenics (i.e. interspecific androgenotes), except for the activation of the irradiated eggs by the heterologous sperm.

To reduce the intense stress involved in diploidization of the process of shocking, Thorgaard et al. (1990) produced paternal androgenics by activation of the irradiated eggs by diploid sperm drawn from tetraploid males (see also Arai et al., 1995). Alternatively, dispermic androgenics are generated by dispermic activation of the irradiated eggs; incubation of milt in polyethylene glycol (PEG) facilitates fusion of the sperm. The same procedure is followed to generate allo-dispermic androgenics, except for the activation by heterologous sperm. Briefly six types of androgenics with different levels of heterozygosity are produced. Table 25 summarizes the advantages and limitations of these methods of producing androgenics.

An unknown but a classical experiment was performed by Anon (1980) under the guidance of Prof TC Tung in China. Nucleus from a just fertilized carp egg was sucked out into a fine glass needle and transferred into an enucleated egg of crucian carp. Essentially, Anon successfully induced androgenesis. The hybrids, as Anon called them, are indeed perfect androgenics. They possessed all morphological characteristics of the carp like the barbel pairs, one pair on each side of the mouth, 36 scales on the lateral line and so on, as against the absence of barbel and presence of 28 scales on the lateral line, all of which are characteristics of the crucian carp. Thus Anon has also confirmed the paternity of the androgenic carp that he achieved purely by skilled surgery. Amazingly, the Chinese in the 1980's already achieved inducing androgenics, whatever was achieved in the field of androgenesis in fishes by the non-Chinese during 1990–2010.

Preserved milt: Using cryopreserved sperm, mitotic androgenics have been produced in *O. mykiss* (Scheerer et al., 1991) and *Oreochromis niloticus* (Myers

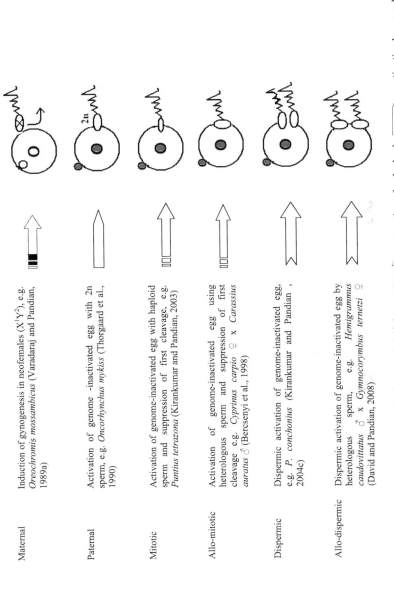

Maternal — Induction of gynogenesis in neofemales (X^1Y^2), e.g. *Oreochromis mossambicus* (Varadaraj and Pandian, 1989a)

Paternal — Activation of genome-inactivated egg with 2n sperm, e.g. *Oncorhynchus mykiss* (Thorgaard et al., 1990)

Mitotic — Activation of genome-inactivated egg with haploid sperm and suppression of first cleavage, e.g. *Puntius tetrazona* (Kirankumar and Pandian, 2003)

Allo-mitotic — Activation of genome-inactivated egg using heterologous sperm and suppression of first cleavage e.g. *Cyprinus carpio* ♀ x *Carassius auratus* ♂ (Bercsenyi et al., 1998)

Dispermic — Dispermic activation of genome-inactivated egg, e.g. *P. conchonius* (Kirankumar and Pandian, 2004c)

Allo-dispermic — Dispermic activation of genome-inactivated egg by heterologous sperm, e.g. *Hemigrammus caudovittatus* ♂ x *Gymnocorymbus ternetzi* ♀ (David and Pandian, 2008)

Fig. 14 Established protocols to induce different types of androgenesis in fishes ➡ = retention of polar body, ⇨ = activation by unreduced diploid sperm, ⇨ = activation followed by thermal shock, ⇨ = dispermic activation. Note the shape and size of sperm heads, which indicate homologous/heterologous sperm and haploid/diploid sperm, also polar body is retained only in maternal androgenics.

Table 24 Methods used to eliminate or inactivate maternal genome and diploidization in androgenesis of fishes (from Pandian and Kirankumar, 2003, modified and updated).

Species	Inactivation of female genome	Genetic marker	Survival (%)	Sperm source/remarks
Oncorhynchus mykiss	^{60}Co; 36kR	Isozyme	9	Outbred
O. mykiss	^{60}Co; 40 kR	–	12	Tetraploid
O. mykiss	^{60}Co; 36 kR	Isozyme	1.3	Cryopreserved
Salvelinus fontinalis	^{60}Co; 88 kR	Allozyme	38 ?	–
Cyprius carpio	X-ray; 25–30 kR	Colour	9	Inadequate genome elimination
C. carpio	UV; 150–300 mJ/cm^2	Colour	15	Irradiation of eggs in ovarian fluid
C. carpio	^{60}Co; 25 kR	Colour, barbel		*Carassius auratus gibelio*
Danio rerio	X-ray; 10 R	RAPD, SSR, MHC	22	*Danio rerio*
Misgurnus anguillicaudatus	UV; 7500 ergs/mm^2	Allozyme	8	100 % elimination of egg genome; 2n sperm
Oreochromis niloticus	UV	Colour	3	Fresh or cryopreserved sperm
Gymnocorymbus ternetzi	UV	Molecular marker, *Gfp*	7	BT tetra cadaveric sperm
Puntius tetrazona	UV	*Gfp*	15	Tiger barb sperm
P. tetrazona	UV	Molecular marker	15	*P. conchonius*; revived from preserved sperm (-20°)
P. tetrazona	UV	Molecular marker	11	Rosy barb cadaveric sperm
P. tetrazona	UV	Molecular marker	1.7	Rosy barb double sperm
WT tetra	UV	*Gfp*	14	WT tetra sperm
WT tetra	UV	Molecular marker	1.8	BT tetra double sperm

BT tetra = *Hemigrammus caudovittatus*, WT tetra = *Gymnocorymbus ternetzi*

Table 25 Advantage and limitations of different methods adopted to generate androgenics in fishes.

Androgenic	Advantages	Limitations
Maternal Y^1Y^1	Zygotes suffer relatively less stress of shock to retain polar body, 50% of treated zygotes become androgenics	Hormone treatment to generate X^1Y^2 ♀; Requires relatively a longer duration; Heterozygosity
Paternal 2Y	High survival; Embryos don't suffer shock for diploidization	Homozygosity; Eggs suffer irradiation; Aneuploid sperm-microplaylar opening too small for entry of 2n sperm
Mitotic Y^2Y^2	50% treated zygotes become androgenic; Instantaneously produced	Fairly high level homozygosity; Zygotes suffer shock treatment to arrest I mitotic cleavage
Dispermic $Y^{2a}Y^{2b}$	Heterozygosity enhanced; No need for zygote to suffer shock to arrest I mitotic cleavage	Milt undergoes PEG incubation; Head to tail fused sperm rare
Allo-mitotic Y^2Y^2	Conservation of strain from preserved sperm	Requires compatible surrogate; Zygotes suffer shock to arrest I mitotic cleavage
Allo-dispermic $Y^{2a}Y^{2b}$	Conservation of species from preserved sperm	Requires compatible surrogate species; Milt undergoes PEG incubation; Hazardous fusion of sperm; Head to tail fused sperm rare

et al., 1995). Bercsenyi et al. (1998) have also used cryopreserved sperm of the goldfish *Carassius auratus* to induce allo-mitotic androgenics. Survival of these androgenics ranges from 1.3 to 22% (Table 24) and does not differ significantly from those produced using fresh sperm. Live, fertile sperm drawn from specimens that have earlier been post-mortem preserved at –20°C for periods of 30 days in the rosy barb *Puntius conchonius* (Kirankumar and Pandian, 2004a), 40 days in the Buenos Aires (BT) tetra *Hemigrammus caudovittatus* (David and Pandian, 2006b) and 240 days in the Indian catfish *Heteropneustes fossilis* (Koteeswaran and Pandian, 2002) may also be used to activate or fertilize eggs of the respective species. Sex ratio of F_1 progenies sired by the cadaveric androgenics has revealed that no damage or mortality is caused to X-specific or Y-specific cadaveric sperm. The sperm count is 8.6×10^3/ml with 19% fertilizability for the cadaveric sperm drawn from the 30-day post-mortem preserved specimens of BT tetra. But it increases to 6.8×10^4/ml with 24% fertilizability, when the specimens are glycerol packed prior to preservation at –20°C (David and Pandian, 2006b). This simple method of using cadaveric sperm may be widely practicable in developing countries. With the use of cadaveric sperm, survival of allo-mitotic androgenics decreases from 14 to 7% in the rosy barb and 11 to 1.8 % in the Bueno aires (BT) tetra (Fig. 15).

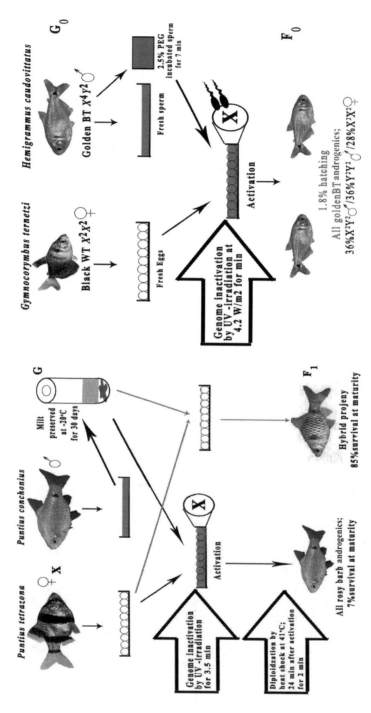

Fig. 15 Left panel = Induction of allo-androgenics of *Puntius conchonius* using its preserved sperm for activation of genome-inactivated (surrogate) eggs of *P. tetrazona* (from Pandian and Kirankumar, 2003). Right panel = Allo-androgenics of *Hemigrammus caudovittatus* using its preserved sperm for activation of genome-inactivated (surrogate) eggs of *Gymnocorymbus ternetzi* (from David 2004).

Colour image of this figure appears in the color plate section at the end of the book.

Bercsenyi et al. (1998) are perhaps the first to generate allo-mitotic androgenic goldfish using eggs of the common carp. Subsequently, Kirankumar and Pandian (2004a), David and Pandian (2006b) have produced allo-mitotic androgenic rosy barb and BT tetra, respectively, *albeit* using cadaveric sperm. These developments have many implications to genome conservation and restoration of desired fish species, using cryopreserved or cadaveric sperm and surrogate eggs.

Markers: Paternity of the androgenics is confirmed by using one or more of the following: Phenotypic, karyotypic, protein (isozyme, allozyme) and molecular markers. Green Fluorescent Protein (*Gfp*) gene, a reporter gene, has been used for the first time as marker to confirm paternity in the tiger barb *P. tetrazona* (Kirankumar and Pandian, 2003) and BT tetra (David and Pandian, 2006a) (Fig. 16). The marker has two advantages: i.e. paternity can be identified from even 16-hour old live or dead embryos.

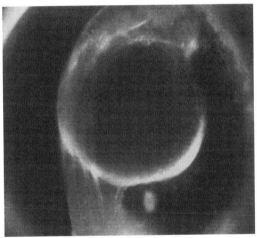

Fig. 16 Green Fluorescent Protein (*Gfp*) gene expression in 16 hr-old haploid androgenic blond *P.conchonius* (from Pandian and Kirankumar, 2003).

Colour image of this figure appears in the color plate section at the end of the book.

However, the use of a reliable marker to identify and confirm the paternity in allo-mitotic androgenics is critically important particularly to show that the androgenics are not contaminated from 'bits and pieces' of maternal genome (see Ocalewicz et al., 2009, 2010). PCR analysis using *Tc*-1 transposan specific primers has revealed that the genomic DNA of the donor rosy barb produced 300 and 800 bp products but the recipient surrogate tiger barb 300 bp only (Fig. 17). Hybrids of these barbs produce both the 800 bp and 300 bp products. However, the allo-mitotic androgenic rosy barbs following monospermic or dispermic activation produce the expected 800 bp product and thereby confirms the paternity in the allo-mitotic androgenic rosy barb (Pandian and Kirankumar, 2003).

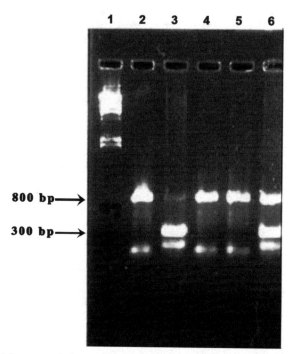

Fig. 17 Agarose gel electrophorogram showing species-specific *Tc*-1 amplification. Lane 1 = λ *Hind* III marker, lane 2 = golden rosy barb *P.conchonius*, lane 3 = tiger barb *P. tetrazona*, lane 6 = hybrid barb, lane 5 = androgenic clones induced by a single heterologous sperm and lane 4 = double heterologous sperm of the rosy barb (from Kirankumar, 2003).

Supermales: With internal fertilization and development of the embryo inside the mother's body, the eggs of 500 viviparous fishes are not amenable for induction of androgenesis. The oft-used experimental group of viviparous poeciliid alone is composed of 200 species (Lucinda, 2003) with a single oviparous *Tomeurus gracilis* (Hrbek et al., 2007). The only method opened for the production of Y^1Y^2 supermales (not androgenics) is to cross the F_1 hormonally sex reversed neofemales (X^1Y^2) with normal males (X^1Y^2) and to select 25% supermales (Y^1Y^2) among the F_2 progenies by progeny test (Fig. 18). In oviparous fishes too, this hormonal method of producing supermales (e.g. *O. rhodurus*, Onozato, 1993; *P. conchonius*, Kirankumar et al., 2003) has been adopted. In this hormone-cum-progeny testing method, the eggs suffer neither irradiation nor the zygotes thermal or pressure shock. But a longer duration is required to produce 25% supermales among the F_2 progenies. However, the duration could be reduced by identifying the hormonally sex reversed neofemales (X^1Y^2) with PCR analysis. The analysis of the estrogen-treated rosy barb *P. conchonius* showed that male specific 588, 333 and 200 bp products were amplified by the genetic males, which were functioning as sex reversed neofemales (X^1Y^2), while the genetic females amplified only

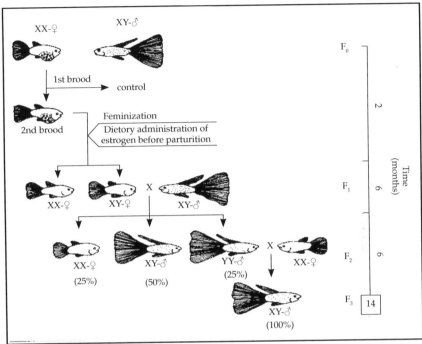

Fig. 18 Protocol for production of Y^1Y^2 supermale in a male heterogametic viviparous guppy *Poecilia reticulata* (from Kavumpurath, 1992).

200 base product (Fig. 19, see also Fig. 20). These hormonal supermales are also considered in the context of monosex production.

Incidentally, it may be noted that at the molecular level, no difference seems to exist between the supermales (Y^1Y^2) and androgenics (Y^2Y^2) of *P.conchonius*. Using the *SRY* -specific primers, PCR analysis of the genomic

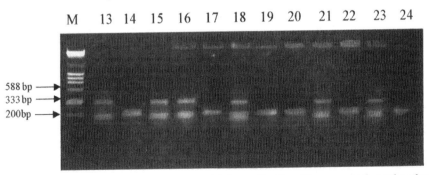

Fig. 19 PCR analysis of the estrogen-treated 3.5 month-old golden rosy barb neofemales. M= marker, lanes 13, 15, 16, 18, 21 and 23 are sex reversed neofemales (X^1Y^2) showing PCR amplified products of 588, 333 and 200 bp. Lanes 14, 17, 19, 20, 22 and 24 are females (X^1X^2) showing PCR amplified product of 200 bp only (from Pandian and Kirankumar, 2003)

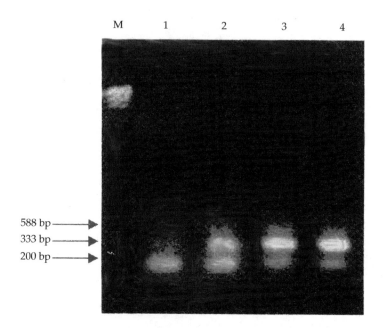

Fig. 20 PCR products amplified by *SRY* primers in the genomic DNA of the golden rosy barb *Puntius conchonius* carrying different sex genotypes. Lane M-λ *Hind* III marker, lane 1 – normal ♀ (X^1X^2), lane 2–normal ♂ (X^1Y^2), lane 3–hormonally induced supermale (Y^1Y^2), lane 4–androgenic supermale (Y^2Y^2) (from Pandian and Kirankumar, 2003).

DNA of the rosy barb male produced three products, 588, 333 and 200 bp. However, the 200 bp product is alone amplified in the female genome (Fig. 20). Hence it is possible to use the first two products as molecular markers to identify the barb possessing a Y-chomosome. The presence of the same 333 and 588 bp fragments in both the supermale (Y^1Y^2) and mitotic androgenic (Y^2Y^2) male has clearly revealed the absence of any detectable difference between these two males (Kirankumar et al., 2003).

5.2 Survival

Reliable data available on survival of androgenics are limited to a dozen and odd species only. Most of these values are abysmally low and fall between 1 and 7% (Table 24), although a few are in the range of 10 to 22%. The major causes for the low survival of androgenics have been traced to (i) homozygous expression of lethal genes, (ii) egg quality, (iii) damage inflicted by thermal/pressure shock treatment to suppress the first mitotic cleavage, (iv) possible damage caused by the irradiation of eggs (however, see Thorgaard et al., 1990) and (v) size incompatibility between micropyle and sperm head of 2n diploid sperm and high frequency of anueploid sperm produced by tetraploid males.

Survival of androgenics decreases with progress of embryonic development; for instance, it decreases to 15% at hatching in the barbs. The chronological course of embryonic mortality has been depicted for the haploid and diploid embryos of the rosy barb in heterologous eggs of tiger barb. The embryonic stages between activation and 18th somite, and just before and after hatching are critical stages (Fig. 21). Interestingly, despite activation by heterologous sperm, the development proceeds as precisely in the time scale, as in the eggs fertilized by homologous sperm.

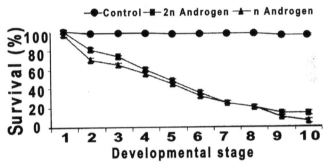

Fig. 21 Chronological sequence of embryogenic development of haploid and diploid androgenic of the golden rosy barb *Puntius conchonius* in the genome-inactivated eggs of the tiger barb *P. tetrazona* (from Pandian and Kirankumar, 2003).

More interestingly, haploid nucleus of the heterologous sperm also directs the chronological sequence of embryonic development, as precisely as diploid homologous sperm nucleus does it (see Pandian and Kirankumar, 2003). This observation confirms the earlier report by Stanley (1983), who has shown that the gene expression operates normally in haploid embryos of *Salmo salmo* but mass mortality occurs at hatching, possibly due to expression of lethal genes.

Homozygosity: The lethal expression level of the two maternal chromosomes ($X^1 x^1$), one drawn from the pronucleus of the egg proper, and the other from the second polar body (x^1) in meiotic gynogenics ($X^1 x^1$) seems to be low and results in 70% egg hatchability (Pandian and Koteeswaran, 1998). But that of the paternal two chromosomes ($Y^2 Y^2$; $Y^1 Y^2$) in androgenics with exclusive paternal genomic contribution appears incompatible and leads to 7% hatchability only. Even the presence of a single X^1 maternal genome in a paternal triploid ($X^1 X^2 Y^2$; $X^1 Y^2 Y^2$) is unable to increase the hatchability to higher than 7% (David and Pandian, 2006c). The lethal expression level of $Y^{2a} Y^{2b}$ is assessed to cause about 55% embryonic mortality (David and Pandian, 2008). Homozygosity in the supermales ($Z^1 Z^2$) of *Betta splendens* results in 100% mortality (George et al., 1994).

Double sperm: Efforts have been made, however, to improve survival of androgenics; for instance, the need for restoration of diploidization by shocking the zygote by the activation of irradiated eggs with 2n sperm (Thorgaard et al., 1990; Arai et al., 1995). Alternatively, the dispermic activation may not only exclude the shock treatment but also may introduce more heterozygosity (Nagoya et al., 2010). But the data available on survival of dispermic androgenics are low as well: 0.1% for *Onchorhymchus mykiss* (Araki et al., 1995), 1.7% for *P. conchonius* (Kirankumar and Pandian, 2004c) and 1.8% for *H. caudovittatus* (David and Pandian, 2008). To obtain fused double sperm for dispermic activation, milt is first incubated for an optimal duration in polyethylene glycol (PEG). Being an extremely hydrophilic molecule, PEG has the property of fusing similarly charged membranes. Since cell membrane of sperm is negatively charged, PEG induces fusion of adjoining biological membranes of cells like spermatozoa (MacDonald, 1985). Unfortunately, the PEG induces also wrong fusions like head opposing the head (Fig. 22) and incomplete fusions like the fusion of heads but not the flagella; secondly, within the 'optimal incubation period', a large number of spermatozoa may remain singly, and when allowed, activate the irradiated eggs, leading to haploid syndrome. These hazardously wrong and incomplete fusions and sperm left to be fused perceptibly reduce the number of properly fused sperm and thereby critically reduce the androgenic survival.

For instance, David and Pandian (2008) traced the causes for the low survival of allo-dispermic androgenics in *H. caudovittatus*. Of the PEG-incubated sperm, 31% displayed not the usual zig-zag but circular movement; hence they were motile but not mobile, and failed to activate eggs. Another 37% unfused haploid sperm activated the irradiated eggs, and all of them succumbed due to the haploid syndrome. Yet another 27%, due to wrong or incomplete fusion (Fig. 22), could not activate the eggs. Less than 1% fused triple sperm activated the eggs, which did not hatch. Of the remaining 4% completely and properly fused double sperm activated the irradiated eggs, of which 2.2% (i.e. 55% of the 4%) failed to hatch, perhaps due genetic incompatibility. Finally, only 1.8% of the fused double sperm activated the eggs, which successfully hatched out. Thus the low success dispermic activation is more due to problems associated with PEG incubation rather than with enhanced heterozygosity.

Diploid sperm: To induce paternal androgenics, a diploid female is crossed with a tetraploid male, which implies that a large diploid sperm has to enter through a small micropyle of the haploid egg and initiate development. With the significantly (1.4 times) larger head, slow (a tenth of the normal) motility and 83% reduced motility duration, a diploid sperm of the loach *Misgurnus anguillicaudatus* (Yoshikawa et al., 2007) holds extremely, low chances of the activation of haploid eggs. Not surprisingly, very low survival values for

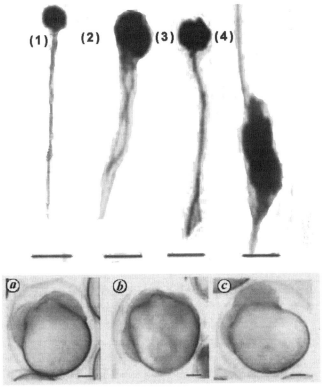

Fig. 22 Effect of PEG incubation of milt on fusion of sperm of BT tetra *Hemigrammus caudovittatus*. Upper panel 1 = normal sperm, 2 = incompletely fused sperm with to independently moving tails, 3 = completely fused double sperm and 4 = head to head wrongly fused sperm. Lower panel : Photograph showing the first division after entry of (a) double, (b) triple BT sperm, (c) abnormal first mitotic division after entry of double sperm (from David and Pandian, 2008).

Colour image of this figure appears in the color plate section at the end of the book.

the paternal androgenics (2Y) have been reported. There are many passing remarks on the production of aneuploid sperm by induced tetraploids but quantitative data are needed.

Egg quality: The poor quality of eggs has been adduced as a major cause for the low survival of androgenics (e.g. Bongers et al., 1995; Komen and Thorgaard, 2007). Hence a brief departure is made to list some factors that are known to affect egg quality. Factors affecting egg quality are intrinsic properties of the egg itself and the environment, in which the egg is fertilized and then incubated (Kamler, 2005). Rainbow trout females that produce better quality eggs continue to do so during subsequent spawnings; this information suggests that the parental genes strongly influence egg quality. Some information is available on gene expression and mRNA translation in fish embryo. Hormones in eggs play a role in embryonic development; thyroid hormones of maternal origin, for instance, play a role in successful embryonic development (Lam, 1994). Cortisol and a host of steroids present in fertilized eggs may promote larval survival (Brooks et al., 1997). Female size is an important contributive factor to egg size and holds a significant positive correlation with egg size. "The bigger is better paradigm applies well to egg size but is not ubiquitous" (Kamler, 2005).

In captivity, confinement and crowding of fish may affect egg quality. In the Atlantic cod *Gadus morhua*, crowding stress leads to low fertilization success and increased frequency of abnormal embryos (Wilson et al., 1995). Delaying ovulation in the Arctic charr *Salvelinus alpinus* by changing photoperiod regime improves egg quality (Gillet, 1994). Conversely, delaying spawning by photoperiod manipulation increases alevin mortality from 5 to 80% in *O. mykiss* (Dabrowski and Blom, 1994). Overripening may also affect egg quality, as indicated by 38% reduction in hatching success of *Silurus glanis* (Lindhart and Billard, 1995). Due to crowding in farmed marine fishes, egg quality, as determined by hatching success, decreases to 50% in salmonids (Bromage et al., 1992), 10–15% in the European sea bass *Dicentrarchus labrax* (Carrillo et al. 1989) and as low as <1% in the Atlantic halibut *Hippoglossus hippoglossus* (Norberg et al., 1991).

5.3 Fertility

Much of the available information on androgenics, especially on long living salmonids is limited to production and survival. Working on cyprinids and characids characterized by less than 120 days of generation time, relevant information on fertility and reproductive performance of androgenic males and females has been made available. Table 26 is a representative example for reproductive performance of androgenic *H. caudovittatus*. Similar information is also available for *P. tetrazona* (Kirankumar and Pandian, 2003), *P. conchonius* (Kirankumar and Pandian, 2004a), *Gymnocorymbus ternetzi* (David 2004) and supermales of cichlids (e.g. Tuan et al., 1999). In general, androgenic males have higher sperm count and more fertile sperm. Conversely, the androgenic females are characterized by delayed maturity; about 100% extended interspawning intervals, reduced Gonado-Somatic Index (GSI), fecundity and hatching success. This observation also confirms the earlier report of Fineman et al. (1975), who found that a hormonlly sex reversed Y^2Y^2 of superfemale *Oryzias latipes* produced 200 eggs, against 250 by a normal female (X^1X^2). Incidentally, Bongers et al. (1997a) found that in the F_1 carp progenies of the cross between homozygous androgenic (Y^2Y^2) male and gynogenic females (X^1x^1), the GSI is largely determined by the female parent. They also reported 13–94% sterility in males and greatly reduced fertility in females; of 48 putative females, only four produced

Table 26 Reproductive performance of F$_1$ and F$_2$ mitotic androgenic *Hemigrammus caudovittatus* generated using fresh or cadaveric sperm (from David, 2004).

Parameter	Control (X^{1}X^2)	Fresh F$_1$ androgenic (Y^{2}Y^2)	Fresh F$_2$ androgenic (Y^{2}Y^2)	F$_1$ cadaveric androgenic (Y^{2}Y^2)	F$_2$ cadaveric androgenic (Y^{2}Y^2)
Androgenic (Y^{2}Y^2) ♂					
Sexual maturity (day)	120	134	135	139	137
Gonado somatic index	0.5	0.5	0.5	0.4	0.5
Sperm count (no/ml)	7.6 x 10^5	7.2 x 10^6	7.8 x 10^6	6.8 x 10^6	7.1 x 10^6
Fertilizability (%)	94	93	90	89	91
Androgenic ♀ (X^{2}X^2)					
Sexual maturity (day)	120	145	148	151	150
Interspawning period (day)	15	24	32	28	32
Gonado somatic index	0.5	0.4	0.4	0.4	0.4
Fecundity (no/spawn)	160	140	120	120	130
Hatchability (%)	96	79	72	73	72

Fresh and cadaveric androgenics mean androgenics produced using sperm from live and cadaver fishes, respectively viable eggs. Incidentally it must also be indicated that Ocalewicz et al. (2009) reported the presence of one of seven chromosomal fragments in healthy one dozen androgenic males (10) and females (two) of the rainbow trout *O. mykiss*. The presence of these chromosomal fragments was related to irregular ovarian development in gynogenic *O. mykiss* (Krisfalusi et al., 2000).

In general, naturally occurring X^1X^2 males (e.g. *Poecilia reticulata*, Winge, 1930) and sex reversed (X^1Y^2) neofemales (e.g. *P.reticulata*, Kavumpurath and Pandian, 1992a) are fertile. Hence it is likely that fertility specific gene(s) are located either on the X-chromosome or on autosome(s) and that the fertility specialization in Y-chromosome in fishes is not as pronounced, as in mammals. Many gene loci controlling colour pattern associated with mate choice are located in or near the unpairing sex determining region of the sex chromosome (Volff and Schartl, 2001; Lindholm and Breden 2002). Some examples of Y-specific loci are *Ma* (*Macropodus* pigmentation), *Ar* (*Aramatus* pigmentation and caudal fin) and *Pa* (*Pauper* pigmentation) in *P. reticulata*; supermale Y^1Y^2 guppies with different combinations of chromosomes namely $Y^{Ma}Y^{Ar}$, $Y^{Ma}Y^{Pa}$ or $Y^{Pa}Y^{Ar}$ are all viable and fertile (Kirpichnikov, 1981).

5.4 Overriding autosomes

Listing seven methods of producing all male progenies Beardmore et al. (2001) noted that androgenesis is not commercially used as a hormonal sex reversal method. For instance, Mair et al. (1997) described a method of producing 100% male progenies in *O. niloticus*; the hormone-treated supermale (Y^2Y^2) sired a supermale, which was crossed to the homogametic Y^2Y^2 female. Production of the homogametic Y^2Y^2 females involved twice feminizations each time followed by progeny selection through F_1 to F_5 generations. Crosses between these supermales (Y^1Y^2) and superfemales produced 100% male progenies. More recently, Muller-Belecke and Horstgen-Schwark (2007) followed a similar five hormonal sex reversal step to produce a large number of supermales (Y^1Y^2) of *O. niloticus* and crossed them with superfemales (Y^1Y^2). They claimed that in none of these crosses, the unexpected female ever appeared. The required technical skill to induce androgenesis and their low survival are major hurdles in adopting androgenesis as a method of large scale production of all male progenies in aquaculture farms.

To know the genotype of these unexpected females sired by the androgenics, Kirankumar et al. (2003) performed PCR analysis using the previously mentioned *SRY* specific primers. The male (X^1Y^2) specific pattern with 333 and 588 bp products in all the unexpected females revealed that these unexpected female rosy barbs sired by the hormone-induced supermales are indeed genotypic males (Fig.19). Incidentally, it also explains the natural occurrence of XY female *P. reticulata* (see Winge, 1930).

In *H. caudovittatus* too, the findings of Kirankumar et al. (2003) hold good. The PCR products of the genomic DNA of normal and androgenetic males amplified by the *DMRT-1* specific primers are of 237 and 300 bp length but that of the females is 100 bp only (Fig. 23). The consistent presence of the two products and the absence of 100 bp fragment in the unexpected females confirm that these unexpected females are indeed genetic males but are functional females, perhaps due to the overriding minor factor (David and Pandian, 2006b).

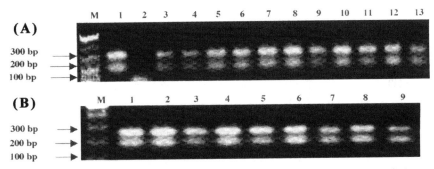

Fig. 23 PCR products of the genomic DNA of golden coloured tetra *Hemigrammus caudovittatus* as amplified by DMRT1 primer. A upper panel: Lane 1 = male, Lane 2 = female, Lane 3–4 = F_0 fresh androgenics, Lane 5–6 = F_1 fresh androgenics, Lane 7–9 = F_0 cadaveric androgenics, Lane 10–13 = F_1 cadaveric androgenics. B Lower panel: Lanes 1–7 = F_1 unexpected female progenies and Lane 8–9 = F_1 male progenies (from David, 2004)

Table 27 lists the frequency of unexpected females sired by androgenics and supermales. The selected examples provide representations from (i) all the methods used to induce androgenesis, which result in the production of androgenics with different levels of heterozygosity, (ii) eight commercially important species belonging to Salmonidae, Cyprinidae, Cichlidae and Characidae, (iii) homozygous (X^2X^2) and heterozygous (X^1X^2) females as mating partners and (v) to F_1 to F_3 generations of androgenics. Rather surprisingly, in almost all the tested species, the unexpected females did appear, mostly at the frequencies of 1 to 12% but rarely at the extremes of 0 and 100%. The results arising from these tested fishes provide arguably strong evidence for the ubiquitous but diverse overriding role played by one or more minor modifying gene(s) in sex determination. However it must be indicated that no androgenic has thus far been generated in female heterogametic fishes. George and Pandian (1995) too produced only ZZ neofemale of *Poecilia sphenops* and there is only circumstantial evidence for the production of W'w' superfemales in *Paramisgurnus dabryanus*. Studies on gynogenesis brought evidence for the ubiquitous but diverse

Table 27 Unexpected female progenies generated by Y^1Y^2 supermales and androgenics Y^2Y^2 crossed with X^1X^2 and X^2X^2 females in fishes.

YY genotype	YY sire used (no)	Dame genotype	Dame used ♀	F of YY	Total cross (no)	Unexpected ♀♀ (%)	Decisive YY ♂♂ (%)
Oncorhynchus tshawytscha Devlin et al. (2001)							
MaternalY^1y^1	–	X^1X^2	–	–	–	0	100
Oreochromis mossambicus (Varadaraj and Pandian, 1989a)							
MaternalY^1y^2	5	X^1X^2	3	F_1	15	0	100*
O. niloticus (Tuan et al., 1999)							
MaternalY^1y^2	11	X^1X^2	2	F_1	19	1–15	0
O. niloticus (Ezaz et al., 2004a)							
MaternalY^1y^2		Y^1y^1	5	F_1	14	2.5	97
Puntius conchonius (Kirankumar et al., 2003)							
MaternalY^1y^2	4	X^1X^2	4	F_0	4	5–7	50
O. mykiss (Scheerer et al., 1986)							
Paternal 2YY	9	X^2X^2	4?	F_1	9	6–100	56
Puntius tetrazona (Kirankumar and Pandian, 2003)							
MitoticY^2Y^2	5	X^1X^2	8	F_1	15	2–3	0
Gymnocorymbus ternetzi (David, 2004)							
MitoticY^2Y^2	7	X^1X^2	8	F_1	42	2–4	29
MitoticY^2Y^2	7	X^2X^2	7	F_1	14	2–5	57
MitoticY^2Y^2	7	X^1X^2	11	F_2	42	1–3	43
MitoticY^2Y^2	5	X^2X^2	5	F_2	14	2–3	40

Hemigrammus caudovittatus (David and Pandian, 2006a)							
Mitotic Y²Y²	7	X^1X^1	F_1	9	42	3–9	29
Mitotic Y²Y²	5	X^2X^2	F_2	7	14	2–3	86
Allo androgenic Y^aY^a	5	X^1X^1	F_1	9	20	2–4	40
Allo androgenic Y^aY^a	5	X^2X^2	F_1	5	10	2–3	80
Puntius conchonius (Kirankumar and Pandian 2004b)							
Allo androgenic Y^aY^a	8	X^1X^2	F_1	8	48	6–25	40
Allo androgenic Y^aY^a	4	X^1X^2	F_2	8	22	12–25	50
Allo androgenic Y^aY^a	4	X^1X^2	F_3	7	22	12	43
H. caudovittatus (David and Pandian, 2008)							
Dispermic allo-androgenic $Y^{2a}Y^{2a}$	4	–	F_0	9	16	2–4	50

*On a field trial 3–12 % female progenies appeared (Pandian, 1993) role of one or another modifying minor gene on sex determination in female heterogametic fishes. The studies on androgenesis have provided complementary evidence for the same in male heterogametic fishes. Yet it is not clear whether these minor genes modify the role of primary sex determining genes or one or another in the downstream of genetic cascade that completes sexualization of fish.

Triploidy and Sterility

The nucleus of an egg in the ovary of fish has 4n chromosomes. Ovulation is commenced with the disappearance of nuclear membrane in the egg and the appearance of chromosomes, leading to the first meiosis and reduction of chromosomes to 2n (Fig. 24). The first polar body is formed and ejected from the egg. The follicles attached to the eggs within the ovary split and partially dissolve, releasing the eggs. Then the post-ovulated eggs are spawned. Following entry of a sperm into an egg through its micropyle (see Fig. 3), the 2n chromosomes in the egg is again reduced to n, and discarded chromosomes, the mirror image of those retained in the egg, are held in the second polar body. The pronuclei of the egg and sperm fuse (syngamy) to form a zygote. In most cases, triploids are produced by preventing the completion of the second meiotic division and retaining the second body (however, see also Fig. 24). Therefore, two sets of chromosomes are contributed by a female and one set by a male. Most of the triploid fishes are sterile due to failure of the three homologous/heterologous chromosome sets to pair correctly prior to crossing over early in meiosis (Benfey, 1999); hence, the triploids are regarded as dead ends in reproductive lineages. Sexually sterile triploids are useful to: 1. Control reproduction of exotic species and contain transgenics in the events of accidental or targeted introduction (Cotter et al., 2000; Pandian, 2001; Wong and Van Eenennaam 2008). 2. Prevent potential backcross of hybrids with either parent species resulting in intermingling of genetic material and introgression and 3. Accelerate growth of culturable species by eliminating the diversion of energy for reproduction. The ease with which the second polar body can be retained has led to numerous publications, which have been reviewed from time to time by Purdom (1983), Pandian and Koteeswaran (1998) and Piferrer et al. (2009).

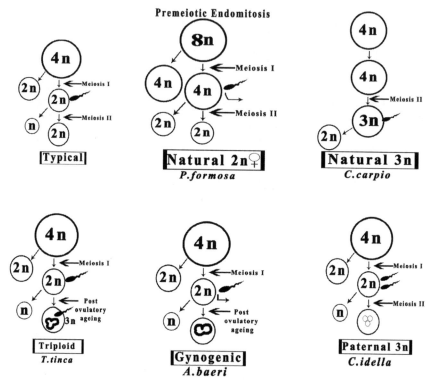

Fig. 24 Oogenesis and its modifications in natural and induced gynogenic diploids-triploid fishes. Note the premeiotic endomeiosis in natural 2n♀, the elimination of Meiosis I in *Cyprinus carpio* (Horvarth and Orban, 1995), the addition of n chromosomes due to post-ovulatory ageing in *Tinca tinca*, gynogenic *Acipenser baeri* and the entry of 2 sperm in paternal triploids of *Ctenopharyngodon idella*.

6.1 Types of triploids

Natural triploids occur in six gynogenic unisexuals: e.g. *Poecilia formosa*, *Cobitis* complex, and *Carassius auratus* (Pandian, 2010; see also Chapter 10). They are developed from unreduced triploid clonal eggs produced through premeiotic endomitosis (Fig. 24, 25). Among gonochores, spontaneous triploids *Curimata modesta* (Venere and Galetti (1985), *C. auratus langsdorfi* (Umino et al., 1997), *Noemacheiius barbatulus* (Collares-Pereira et al., 1995), *Astyanax scabripinnis* (Maistro et al., 1995), *Gymnotus carapo* (Fernandes-Matioli et al., 1998), *Trichomycterus davisi* (Borin et al., 2002); *Misgurnus anguillicaudatus* generate maternal triploids by producing triploid clonal eggs. From a collection of 242 specimens of the loach, Arai et al. (1991a) first recorded 4.1% incidences of natural triploids. In a subsequent survey covering 33 localities of Japan, Zhang and Arai (1999b) recorded a wide

Ploidy type	Induction protocol	Species / Reference	Gametes / Fertilization
Spontaneous	Clonal eggs by premeiotic endomitosis	*C. auratus gibelio* Umino et al. (1997)	
Natural	Clonal eggs by premeiotic endomitosis	*Poecilia formosa* Rasch and Balsano (1989)	
Maternal	Activation of 3n unreduced egg	*M. auguillicaudatus* Arai et al. (1991a)	
Auto-triploid	Fertilization by homologous sperm and retention of polar body	*Betta splendens* Kavumpurath and Pandian (1992a)	
Allo-triploid	Fertilization by heterologous sperm and retention of polar body	*O. mossambicus* ♀ & red tilapia ♂ Varadaraj and Pandian (1989b)	
Trigenomic	Unreduced hybrid 2n egg fertilized by a 3rd species	(*S. salar* ♀ x S. *trutta* ♂) F₁ ♀ x *S.fontinalis* ♂ Johnson and Wright (1986)	
Paternal	Fertilization by 2 homologous sperm	*H. caudovittatus* x *G. ternetizi* David and Pandian (2006c)	
Hybrid paternal	Fertilization by 2 homologous sperm	*C.idella*♀ x *H nobilis*♂ Cassani et al. (1984)	
Interploid	Crossing 4n ♀ with 2n ♂	*M. anguillicaudatus* Arai et al. (1993)	

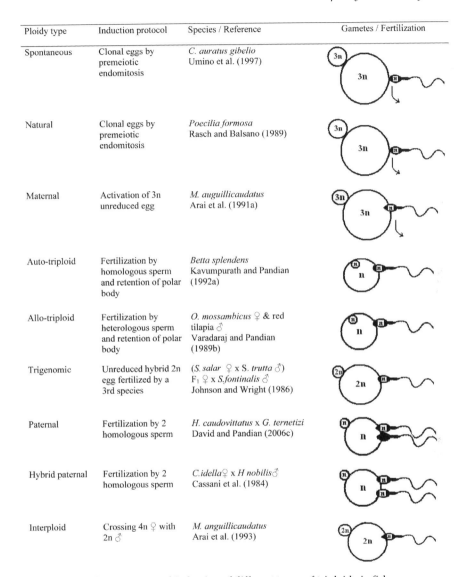

Fig. 25 Natural occurrence and induction of different types of triploidy in fishes.

range of incidence from 7.7% in Ichimiya town to 15.8% in Hirokami village. From India, natural occurrence of triploid *Heteropneustes fossilis* has been reported; from Tamilnadu and Kerala states, a 4.2% incidence of triploid catfish was recorded with the sex ratio of 4 ♂ : 1 ♀ (Pandian and Koteeswaran, 1999). Besides, natural hybrid triploids are also claimed to occur in many centrarchid species (Dawley, 1987).

Paternal triploids are also recorded to occur in nature. For instance, eggs of many fishes, especially the overripened eggs, are amenable to dispermy or even polyspermy. Five percent of just-fertilized eggs of *Ctenopharyngodon idella* contain three or more pronuclei (Mantelman, 1969). In carps too, there are records for dispermy and production of natural paternal triploids (e.g. Wu et al., 1993). When the *Fundulus heteroclitus* female is crossed with the *Menidia notata* male, more than 50% of the hybrid eggs are found to have been fertilized by two sperms and the hybrids are all natural paternal triploids (Moenkhaus, 1904). From a study of the cross between *C. idella* ♀ and *Hypophthalmichthys nobilis* ♂, Cassani et al. (1984) found that the triploid is a product of dispermic fertilization and the hybrids carried two sets of paternal and one set of maternal chromosomes.

Spontaneous diploidization of the maternal genome in 0.1 to 5.5% diploid gynogenic fishes has already been noted (Fig. 24). The phenomenon of spontaneous triploidization is also widespread in fishes, and arises, when an egg with diploid chromosomes is fertilized by a haploid sperm. Post-ovulatory ageing of oocytes is known to result in suppression of the second meiotic division (Aegerter and Jalabert, 2004), besides causing chromosomal aberrations (Uedo, 1996; Varkonyi et al., 1998). The heritable nature of spontaneous triploidization has been experimentally shown in *Cyprinus carpio* (Cherfas et al., 1995). Some fishes appear to be genetically predisposed for spontaneous diploidization of their eggs. Flajshans (1997) has considered that the high triploid incidences in hatchery stocks of *Tinca tinca* are perhaps associated with ageing of oocytes during the process of hypophysation. Post-ovulatory ageing of oocytes within the ovarian cavity of the European catfish *Silurus glanis* induces not only triploidy but also tetraploidy, besides aneuploidy (Varkonyi et al., 1998). Besides ageing duration, temperature accelerates the triploidization frequency in aged oocytes. Subjecting oocytes of *T.tinca* to selected durations (1 to 5 hours) and to temperatures (17, 22 and 24°C) both *in vitro* and *in vivo*, Flajshans et al. (2007) showed that both ageing duration and temperature accelerated spontaneous triplodization, especially under *in vitro* condition; the frequency of triploidization increased from 0.8 to 5.3% in oocytes aged for 1 hour and 5 hours durations at 24°C and it increased from 0.3% at 17°C to 5.3% at 24°C. Under *in vivo* condition, the frequency however, increased from 0.3 to 0.9% in oocytes aged for 3 and 5 hours, respectively.

Incidentally, it is known that the eggs of fish, for instance, *Acanthurus nigrofasciatus*, remain fertilizable for longer than 5 hours after they are spawned (Kiflawi et al., 1998; see also Pandian, 2010). Notably, the process of ageing of post-ovulated oocytes within the female or overripening of spawned/stripped eggs seems to facilitate either (i) the entry of two or more sperm inducing paternal triploidization or (ii) the spontaneous diploidization of maternal genome inducing spontaneous gynogenesis or (iii) the induction of spontaneous triploidization, when the egg is fertilized by a haploid sperm (Fig. 24). Sex ratio of these spontaneous gynogenics, paternal and maternal triploids may provide some insights into the sex determining mechanism in fishes; research in this area may prove rewarding.

Most induced triploids are autotriploids. Retention of the second polar body by a shock results in the production of autotriploids (Fig. 25). The credit of inducing triploidy goes to an Indian, who produced the first triploid stickleback *Gasterosteus aculeatus* (Swarup, 1959). Triploidy is confirmed by (i) phenotypic markers (e.g. David and Pandain, 2006b) (ii) karyotype, sometimes with a marker chromosome (Fig. 26) (iii) cytometric assay, say, the size of a red blood corpuscle (iv) protein marker and/or (v)

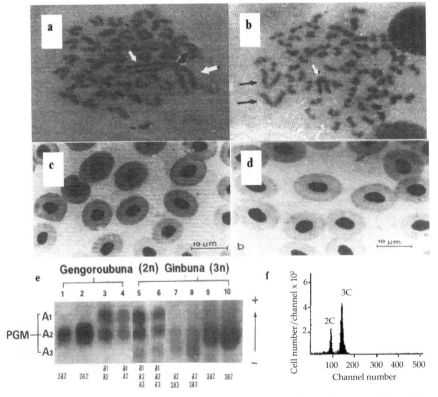

Fig. 26 Upper panel: Karyotype with the marker chromosomes (shown by arrows) in diploid (a) and (b) triploid *Oreochromis mossambicus* (Varadaraj and Pandian, 1989b). Middle panel: Erythrocytes of diploid (c) and triploid (d) *Betta splendens* (Kavumpurath, 1992). Lower left panel: (e) Typical electrophoretic patterns of phosphoglucomutase (PGM-A) from the liver of diploid gengoroubuna (lane 1–4) and triploid ginbuna (lane 5–10) (from Zhang et al., 1992a). Lower right panel: (f) Flow cytometry histogram for diploid and triploid loach *Misgurnus anguillicaudatus* on the basis of DNA content (from Zhang and Arai, 1999b).

flow cytometric analysis for estimation of genome size (Fig. 26). In view of the economic importance of monosex in aquaculture, considerable efforts have been made to induce all male or all female triploids. In male heterogametics, the protocol for production of all male triploids involves three steps: (i) hormonal masculinization followed by progeny testing to select neofemales (X^1Y^2) (ii) crossing the sex reversed neofemale with normal $\male$ to produce 50% X^1X^1 females and 50% Y^2Y^2 males and (iii) crossing the Y^2Y^2 males with normal X^1X^2 females followed by a shock to retain the second polar body to produce 100% $X^1x^1Y^2$ males (Fig. 27). Production of all-female triploids also requires three steps : (i) production of all-female F_1 progenies by induction of gynogenesis (ii) hormonal masculinization followed by progeny testing to select neomales (X^1X^2) and (iii) crossing them with the normal females (X^1X^2) followed by retention of polar body with a shock. These manipulations result in the production of all female ($X^1x^1X^2$) triploid F_2 progenies (Fig. 27).

Allo-triploids are produced following heterospecific insemination and subsequent shock to retain the polar body. Hence they carry two chromosome sets of female of one species and one chromosome set of another species (Fig. 25). In fact, heterospecific insemination leads to the success of hybridization in many fishes, in which hybridization between two diploid species failed (Table 28). The triploid hybridization between salmonids, which can adopt to salinity at an early feeding stage like *Oncorhynchus keta, O. gorbuscha* and those, which cannot tolerate transfer to sea water until age 1+ (e.g. *Salmo salar, O. tshawytscha, O. kisutch*), is commercially important. Hence considerable attempts have been made to hybridize these two groups of salmonids by many authors (e.g. Seeb et al., 1993, Habicht et al., 1994, see Pandian and Koteeswaran, 1998). Their studies indicate that with the exception of the *S. fontinalis* $\female$ x *O. kisutch* or *S. trutta* or *O. keta* $\male$ cross, all combinations produced viable triploid hybrids and survival at feeding stage ranged from 76 to 96% . Incidentally, the induction of triploid hybridization between land–locked (e.g. *S. fontinalis*) and anadromous *S. salar* resulted in the production of a high percentage abnormal embryos and larvae (Sutterlin et al., 1987).

Another restoration route is the production of trigenomic triploid e.g. *O. apache-O. mykiss-O.clarki* hybrid (Wilkins et al., 1993). Crosses between (2n Atlantic salmon $\female$ x 2n brown trout $\male$) 2n $F_1\female$ x 2n brook trout $\male$ resulted in the production of trigenomic triploids; isozyme studies indicated the expression of all three species. Apparently, the hybrids between Atlantic salmon and brown trout produced unreduced diploid eggs, with a haploid genome of Atlantic salmon and the other of the brown trout (Johnson and Wright, 1986). Among the adrianichthyids, meronids, cyprinodontids and centrarchids, the occurrence of trigenomic triploids has been reported (Table 29), although some of these observations may require confirmation. Many

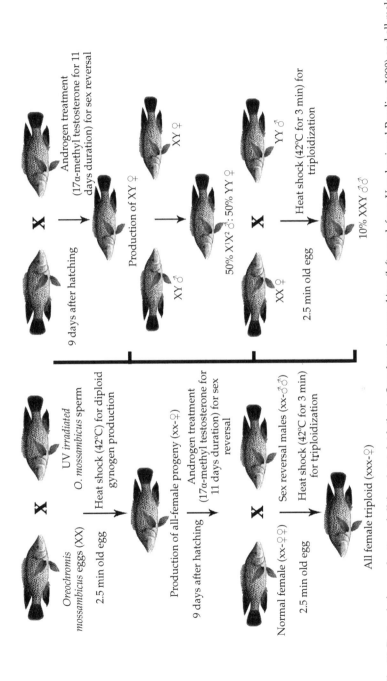

Fig. 27 Protocols for production of all-female triploid tilapia *Oreochromis mossambicus* (left panel, from Varadaraj and Pandian, 1990) and all-male triploid tilapia *O. mossambicus* (right panel, partly from Varadaraj and Pandian, 1989a).

Table 28 Survival of hybrid triploids salmonids, in which diploid hybrids do not survive. Numbers in bracket against each species represent 2n chromosome number.

Female		Male	Survival (%)
Chum salmon *Oncorhynchus keta* (74)	x	Brook trout *Salvelinus fontinalis* (84)	76**
Chum salmon *O. keta* (74)	x	Pink salmon *O. gorbuscha* (52)	87*
Pink salmon *S. gorbuscha* (52)	x	Chinook salmon *O. tshawytsha* (68)	83*
Pink salmon *O. gorbuscha* (52)	x	Chum salmon *O. keta* (74)	96*
Chimook salmon *O. tshawytscha* (68)	x	Pink salmon *O. gorbuscha* (52)	81*
Masu salmon *O. masou* (66)	x	Brook trout *S. fontinalis* (84)	23+
Rainbow trout *O. mykiss* (60)	x	Brook trout *S. fontinalis* (84)	43ᵒᵒ
Brook trout *S. fontinalis* (84)	x	Coho salmon *O. kisutch* (60)	8++
Brook trout *S. fontinalis* (84)	x	Brown trout *Salmo trutta* (80)	22++
Brown trout *S. trutta* (80)	x	Chum salmon *O. keta* (74)	14++

*Joyce et al. (see Pandian and Koteeswaran, 1998), **. Arai. (1986), +Oshiro et al. (1991), + Quillet et al. (1988), oo Deng et al. (1992), ++ Gray et al. (1993)

Table 29 Natural occurrence of trigenomic triploids in fishes.

Trigenomic cross and composition	Reference
(2n *Oryzias latipes* ♀ x 2n *O. curvinotus* ♂) 2n F₁ ♀ x 2n *C. luzonensis* ♂	Sakaizumi et al. (1992)
(2n *O. latipes* ♀ x 2n *O. curvinotus* ♂) 2n F₁ ♀ x 2n *C. celebensis* ♂	Kurita et al. (1992)
(2n *Merone punctatus* ♀ x 2n *M. dolomeiu* ♂) 2n F₁ ♀ x 2n *M. salmoides* ♂	Trautman (1981)
(2n *Lepomis megalotis* ♀ x *L. auritus* ♂) 2n F₁ ♀ x 2n *L. microlophus* ♂	Guest (1984)
(2n *L. cyanellus* ♀ x 2n *L. macrochirus* ♂) 2n F₁ ♀ x 2n *L. auritus* ♂	Smitherman and Hester (1962)
(2n *L. gibbosus* ♀ x *L. cyanellus* ♂) 2n F₁ ♀ x 2n *L. macrochirus* ♂	Dawley (1987)

of these authors have not reported information on sex ratio and fertility status. The trigenomic triploids may prove interesting from the point of sex determination, especially when the hybridizations in all possible combinations are made between male heterogametic species like *Oreochromis niloticus*, *O. mossambicus* and female heterogametic *O. aureus*.

As indicated, when haploid eggs are fertilized simultaneously by two haploid homologous and heterologous sperm, paternal and hybrid paternal triploids are generated, respectively. Although natural occurrence of paternal triploids has been reported as early as in 1904 by Moenkhaus, publications reporting successful induction of paternal triploids are limited to Uedo et al. (1986) and David and Pandian (2006c). Within a species, morphs bearing diploid, triploid and tetraploid karyotypes have been reported in many fishes (e.g. *Misgurnus anguillicaudatus*, Arai, 2001). Triploids have been generated crossing 4n ♀ x 2n ♂, 2n ♀ x 4n ♂, and 4n ♀ x 3n ♂ (Zhang and Arai, 1996). A large number of publications are available on triploids from interploid crossings between 4n ♀ x 2n ♂ but with no data on sex ratio of these triploids (e.g. Li et al., 2006).

6.2 Survival and sex ratio

Survival: The presence of the three sets of chromosomes in autotriploids implies that homologous chromosomes have to compete for synapsis during meiosis. Triploids segregate the chromosomes of trivalents randomly, which leads to the production of aneuploid gametes. Following fertilization, the gametes produce zygotes with multiple trisomy, which are frequently not viable (see Cunado et al., 2002). Table 30 summarizes representative values for survival of triploids produced by following different protocols. Within each group, the values range so widely that it is difficult to identify the cause(s) for such wide variations. One reason is that the window for the optimal shocking temperature and duration is very narrow that any deviation drastically reduces the survival and yield of triploids. At 32°C, the optimal temperature for shocking to retain a second polar body, triploid survival of *Acipenser transmontanus* was the highest (63%); a deviation of either 1°C to 33°C or 2°C to 30°C drastically reduced it (Van Eenennaam et al., 1996). Likewise, the yield for viable triploids of *Oncorhynchus kisutch* was 38, 54 and 96%, when the thermal shock lasted for 20 minutes in 20-minutes old zygotes (Parsons et al., 1986), for 20 minutes in 10- minutes old zygotes (Seeb et al., 1993) and for 10 minutes in 10- minutes old zygotes (Habicht et al., 1994), respectively. With a change in inductor, the window for the shocking duration is also shifted from one to another; for instance, the window is opened at 5-minutes old zygotes of *Oreochromis niloticus* for thermal shock but is shifted to 2-minutes old ones for pressure shock (Hussain et al., 1991). In general, the durations for cold shock last about 10 times longer for cyprinids cichlids and silurids but ensure higher yield. The yield is higher (67–92%) for the all-female triploids than for the all-male triploids (45%). Induction of paternal triploid seems to result in the lowest survival of 5%.

Sex ratio: Data on sex ratio, especially those on departures from the expected ratio may provide a clue to understand the sex determining mechanism. However, very few authors have reported the relevant data on sex ratio of the induced triploids. In their recent review, Piferrer et al. (2009) have chosen not to speak on it. Strikingly, barring the ratio reported for *O. aureus* based on only a couple of observations, the sex ratio in triploids is skewed in favour of males, irrespective of whether the concerned species

Table 30 Survival of different triploids in fishes.

Types of triploids/Species/Cross	Survival (%)	References
Triploidy in ♂ heterogametic Autotriploidy		
Cyprinus carpio	11	Cherfas et al. (1993)
Onchorhynchus mykiss	63	Chorrout and Quillet (1982)
Oreochromis mossambicus	65	Pandian and Varadaraj (1988b)
All–male triploidy		
Betta splendens	45	Kavumpurath and Pandian (1992b)
All–female triploidy		
O. mossambicus	67	Varadaraj and Pandian (1990)
Silurus asotus	92	Nam et al. (2001a)
Allo–triploidy		
O. mossambicus ♀ x red tilabia ♂	85	Varadaraj and Pandian (1989b)
Gymnocorymbus ternetzi ♀ x Hemigrammus caudovittatus ♂	17, 25	David and Pandian (2006c)
Paternal triploidy		
G.ternetzi ♀ x H. caudovittatus ♂	5	David and Pandian (2006c)
Triploidy in ♀ heterogametic		
Acipenser transmontanus		Van Eenennaam et al. (1996)
Thermal shock at 3° C	17	
34° C	43	
33° C	53	
32° C	63	
Interploid–triploidy		
Misgurnus anguillicaudatus ♀ x ♂		Zhang and Arai (1999b), Arai et al.(1993)
4n 2n	18, 54, 64, 64, 67+	
4n 3n	3, 4, 20	
2n 4n	41, 64, 67 +	
Trigenomic triploidy		
(Carassius auratus cuvieri ♀ x Cyprinus carpio ♂) F₁ ♀ x C. auratus cuvieri ♂	32	Zhang et al. (1992b)
(C. auratus cuvieri ♀ x C. carpio ♂) F₁ ♀ x C. carpio ♂)	11	Zhang et al. (1992b)

is a male or female heterogametic one i.e. irrespective of carrying $X^1x^1Y^2$ or $W^1w^1Z^2/W^1Z^1z^2$ genotype, the triploid females do not survive to attain sexual maturity (Table 31). Lincoln's (1981) data provide an explanation; the male ratio shifts from 75% in the one year old plaice *Pleuronectes platessa* to 95% in the two year old plaice, suggesting sex-dependent mortality. Clearly,

Table 31 Sex ratio in induced triploids in male and female heterogametic fishes.

Gamety/Species	Sex ratio $\female : \male$	Reference
Male heterogametic triploids ($X^1x^1X^2 : X^1x^1Y^2$)		
Oreochromis mossambicus	35 : 65	Penman et al. (1987)
O. mossambicus	49 : 51	Pandian and Varadaraj (1988b)
O. niloticus	50 : 50	Penman et al. (1987)
Pleuronectes platessa		Lincoln (1981)
I year	25 : 75	
II year	5 : 95	
Cyprinus carpio	40 : 60	Cherfas et al. (1994b)
(*Carassius auratus cuvieri* $\female$ x *C. carpio* $\male$) F_1 $\female$ x *C. carpio* $\male$	33 : 67	Zhang et al. (1992b)
Hybrid triploids		
Gymnocorymbus ternetzi (WT) $\female$ x *Hemigrammus caudovittatus* (BT) $\male$		
WT (gray) $\female$ x BT $\male$ (maternal)	55* : 45	David and Pandian (2006c)
WT (albino) $\female$ x BT $\male$	15 : 85	
WT (gray) $\female$ x BT $\male$ (paternal)	0 : 100	
Female heterogametic triploids ($W^1w^1z^2/ : Z^2z^1Z^1$)		
Rhodeus ocellatus ocellatus	0 : 100	Uneo and Arimoto (1982)
O. aureus	6 : 94	Mair et al. (1991b)
Others		
Danio rerio	0 : 100	Kavumpurath and Pandian (1990)
Pagrus major (protandrous)	3 : 97	Kitamura et al. (1991b)
Acanthopagrus schlegeli (protandrous)	0 : 100	Kitamura et al. (1992)
Barbus conchonius	0 : 100	George and Pandian (unpublished)

* Sterile $\female$

some combinations of $X^1x^1X^2$ and $W^1w^1Z^2$ seem lethal and cause progressive mortality of females during embryonic and post-embryonic stages to reduce the female ratio to less than 5%, especially among paternal triploids. While 100% male ratio is expected in the protandrous *Acanthopagrus schlegeli* triploids, it is difficult to reconcile with 97% male ratio in the protogynous *Pagrus pagrus* triploids. Yet, the expected sex ratio of about 0.5 $\female$: 0.5 $\male$ in other fishes suggests that the sex-dependent mortality may be the sole reason for the observed departures in sex ratio among the triploids (see also Cherfas et al., 1994b).

6.3 Male's fertilizability

Sterility: Triploidy interferes with gametogenesis in almost all the fishes thus far tested. Gametogenic dysfunction is imposed by the presence of three sets of homologous/heterologous chromosomes creating incompatibilities during gametogenesis; in triploids of the turbot *Scophthalmus maximus*, synapsis occurs at a higher frequency in female than in male zygotes (Cunado et al., 2002); not only synapsis but also recombination have been shown to occur at least between some homologous chromosomes in triploid crucian carp (Fig. 28); incidentally, recombination is also suggested to occur between paternal and maternal mtDNA in triploid hybrid crucian carp (Guo et al., 2006). Expectedly, full gonadal sterility in both sexes of fishes like the turbot *Scophthalmus maximus* has been reported (Cal et al., 2006). However, at least the gonadal development process in triploid males may be similar

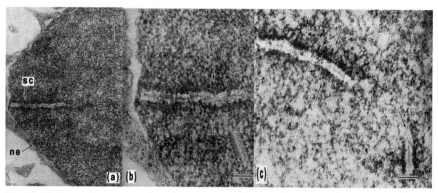

Fig. 28 Electron microscopy of a primary oocyte of triploid *Carassius auratus langsdorfii* (a) showing the attachment of synaptonemal complexes (sc) to the nuclear envelope (ne). Scale indicates 0.4 μm. (b) and (c) are higher magnifications of (a). Arrows show the sc. Bar indicates 0.2 μm (from Zhang et al., 1992a).

to that in diploid males (Flajshans et al., 1993), as spermatogonial mitosis, cyst formation and division of steroidogenic cells are premeiotic events in males; since male germ cells do not enter meiosis until sexual maturity, testes development is generally not impaired in triploid males (Krisfalusi et al., 2000). There is evidence from several species that triploid females are hormonally incompetent to develop gonads, or when developed, produce only a few oogonia and oocytes. In contrast, triploid males show normal hormonal profile (see Pandian and Koteeswaran, 1998) and are capable of developing testis. Triploid males of masu salmon *Oncorhynchus masou rhodurus* develop external secondary sexual characteristics and display normal courtship (however, see also Kavumpurath and Pandian, 1992b) and induce egg release in diploid females (Kitamura et al., 1991a). Peruzzi et al. (2009) found no difference in the ability between diploid and triploid males to induce spawning in diploid females of *Gadus morhua* held in a tank. In many fishes, the triploid males do produce fertilizable spermatozoa (Table 32).

In *Carassius auratus gibelio*, triploid males and females are fertile. But in the hybrid triploids of goldfish x common carp, males are sterile, although

Table 32 Spermatogenesis in triploid fishes.

Species	Spermatocyte	Spermatid	Spermatozoa	Remarks
Cyprinus carpio	*	*	+	Large, non-motile spermatozoa
Ctenopharyngodon idella	*	*	+	Aneuploid spermatozoa, 27 % eggs fertilized
Carassius auratus	+++	++	-	No spermatogenesis
C. gibelio[1]	*	*	+	Fertile haploid spermatozoa; aneuploid sperm present
Danio rerio	+++	++	+	Very dilute milt with non-fertilizable sperm
Tinca tinca[2]	*	*	+	Fertile spermatozoa with 1n and 1.47n DNA
Oncorhynchus masou	+++	++	+	Abnormal sperm with 2 tails, 2 heads of various sizes
O. mykiss	++	++	+	Fertile sperm; 58% fertilization; aneuploid sperm
Ictalurus punctatus	*	*	-	No Spermatozoa
Betta splendens	+++	++	+	Sperm fertilize eggs
Oreochromis mossambicus	++	++	+	Different sized sperm
Perca flavescens	+++	-	-	No spermatozoa
Misgurnus mizolepis	+++	++	+	Different sized sperm
M. anguillicaudatus[3]	+	+	+	Aneuploid sperm; in natural 3n ♂, aneuploid spermatozoa with different DNA content
Pagrus major	*	*	+	Spermatozoa with 1–7n; also with 2 heads, 2 tails; mitochondrial abnormalities
Gymnocorymbus ternetzi ♀* x *Hemigrammus caudovittatus* ♂[4]	+	+	-	No motile spermatozoa; spermatids with n, 2n or 3n

1. Flajshans et al. (2008), 2. Linhart et al. (2006), 3. Gratefully from Prof. K. Arai's group, 4. David and Pandian (2006c); for others, references can be obtained from Pandian and Koteeswaran (1998); +++, ++, + relative presence; - absence; * data not available.

females are fertile. An overview of male sterility in triploids indicates that the level of sterility ranges from production of no sperm to aneuploid sperm and to euploid sperm capable of fertilizing eggs of diploids. For instance, in seabass (*Dicentrarchus labrax*), turbot, (*Psetta maxima*), sea bream (*Calamus bajonado*) and Arctic charr (*Salvelinus alpinus*), triploid males are completely sterile and do not produce sperm (Otto and Whiton, 2000). In most triploids, spermatocytes cease cell division at the synaptic stage of the first meiosis (Kobayashi et al., 1993). In others such as goldfish *Carassius auratus* (Yamaha and Onozato, 1985) and the hybrids of WT x BT tetra (David and Pandian, 2006c) spermatids are produced but no spermatozoa (Table 32). Histological sections of testis of triploid hybrids of WT x BT tetra show the presence of haploid, diploid and triploid spermatids alone (Fig. 29). The second group of triploids such as carp (*Cyprinus carpio*, Cherfas et al., 1994b), Atlantic salmon (*Salmo salar*, Benfey, 1999), loach (*Misgurnus angullicandatus*, Zhang and Arai, 1999a), barfin flounder (*Verasper moseri*, Mori et al., 2006) and Atlantic cod (*Gadus morhua*, Perruzi et al., 2009) produce aneuploid

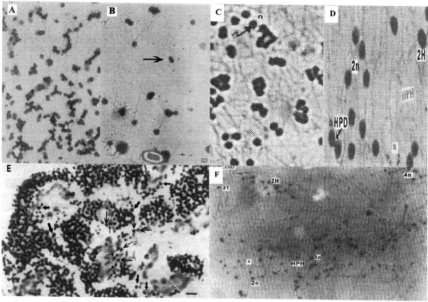

Fig. 29 Upper right panel: Spermatozoa of (A) diploid and (B) triploid *Oncorhynchus masou*. Note the abnormal sperm with 2 heads, 2 tails or smaller head (from Nakamura et al., 1993) Lower right panel: Histological section of testis of 6-month old hybrid paternal triploid of *Hemigrammus caudovittatus*. Scale = 20 μm. → 3n spermatid, → 2n spermatid and → n spermatid (from David and Pandian, 2006c). Upper right panel: sperm of natural diploid and tetraploid and Lower right panel: natural triploid of *Heteropneustes fossilis* (x 200 X). Note the presence of hyperdiploid (HPD) and hypodiploid (HPH) sperm in tetraploids. Also note the presence of n, 2n, 3n and 4n sperm in 2n and 4n males. Sperm with 2 heads (2H) and 2 tails (2T) are indicated by arrows (from Pandian and Koteeswaran, 1999).

sperm. These triploids produce reduced number of aneuploid sperm with wide variations in the DNA content, larger head and in many species with abnormal two heads and two tails (Fig. 29). Considering the Atlantic cod as a representative for the triploid males producing aneuploid sperm, some features of spermatogenesis and progeny production capacity are compiled in Table 33. Notably, the aneuploid sperm of the cod is also capable of fertilizing the eggs of diploids. The hazardous cytological course, through which these anueploid sperm are produced, is reflected in wide variations in the DNA content of the sperm. Correspondingly, the alevins sired by the triploid males too carried widely varying DNA contents. However, fertilization of 40, 000 and 1,90,000 eggs yielded no surviving alevin that ever attained the feeding stage. Hence these triploids producing aneuploid sperm may prove a valuable resource in containing transgenic cod (see Peruzzi et al., 2009).

Table 33 Aneuploid sperm characteristics and larval productivity of diploid and triploid cod *Gadus morhua* (compiled from Perruzi et al., 2009).

Parameter	Diploid ♂	Triploid ♂
Mature males (%)	55	12.5
Milting frequency (time/week)	2	2
Milt volume (ml/milting)	750	500
Sperm count (n x 10^9)	12.34	9.71
Sperm motility (µm/sec)	124	113
DNA content of sperm (n)	1	1.4 (1.2–1.6)
DNA content of alevin(n)	1	2.42 (2.1–2.75)
Fertilization success (%)	16–82	16–69
Hatching success (%)*	63	12
Survival at feeding (%)**	90	0

*as percentage of fertilization success, **as percentage of hatching success

Fertile sperm: Natural triploid males do produce euploid and aneuploid sperm (Fig. 29). The estimated sperm produced by natural triploid males of the Indian catfish *Heteropneustes fossilis* is in the ratio of 57% euploids and 43% aneuploids (Pandian and Koteeswaren, 1999). From their histological studies on testis of triploid male *O. kisutcu*, Piferrer et al. (1994) found that the most important difference between the diploid and triploid is the size of spermatogonium. In triploids, it measures 14.0 µm in diameter and 1, 501 µm^3 in volume, in comparison to that of 11.9 µm and 964 µm^3 diploid spermatogonium. Briefly, on this account alone, there is a 21–36% decrease in the spermatogonial number in triploids. This may account for the 25% reduction observed in the gonado somatic index of the tench *Tinca tinca* (Table 34). Hence the reduction in sperm count in triploid males is due to (i) reduction in the number of spermatogonial cells and

Table 34 Testis size, sperm characteristics and larval productivity of diploid and triploid tinca *Tinca tinca* (compiled from Lindhart et al., 2006).

Parameter	Diploid	Triploid
Age at sexual maturity (years)	4	4
Testis size (g)	2.2 ± 0.8	1.7 ± 1.3
Gonadosomatic index (%)	1.2 ± 0.4	0.9 ± 0.6
Sperm content : volume (ml/♂)	0.58 ± 0.28	0.05 ± 0.03
Sperm content : count (n x 10^9/♂)	2.1 ± 1.3	0.1 ± 0.7
Sperm head size (µm)	1.60 ± 0.18	1.86 ± 0.02
DNA content of a sperm (n)*	1.0 ± 5.7% cv	1.5 ± 8.3% cv
DNA content of a Somatic cell (n)*	1.9 ± 2.3% cv	2.8 ± 3.0% cv
Viable sperm on activation (%)	92.01	94.28
Motile sperm (%)**	93–900	37–77
Sperm motility (µm/s)**	82–110	57–90
Sperm motility at 120 s after activation	45–120	15–20
Fertilization success (%)	60.3 ± 18.8	15.9 ± 24.4
Hatching success (%)	54.0 ± 22.2	15.5 ± 24.5
Malformed larvae (%)	3.9 ± 6.5	11.6 ± 18.0

*Relative to a haploid somatic cell, ** at 15 s after action, cv coefficient of variation

(ii) cytological incompatibility due to random segregation of trivalents. Triploids of plaice (*Pleuronectes platessa*, Lincoln, 1981), Fengzheng crucian carp (*Carassius auratus gibelio*, Shen et al., 1983), stone loach (*Noemacheilus barbatulus*, Collares-Pereira et al., 1995), the Indian catfish (*Heteropneustes fossilis*, Pandian and Koteeswaran, 1999), loach (*Misgurnus anguillicaudatus*, Morishma et al., 2002, Oshimo et al., 2005) and tench (*Tinca tinca*, Linhart et al., 2006) are capable of producing fertile, mostly euploid sperm. Fertilization success of triploid sperm is <7 % in tench (Linhart et al., 2006), < 6% in grass carp (Van Eenennaam et al., 1990), 58% in rainbow trout (Lincoln and Scott, 1984) but 0.06% only in rose bitterling (Kawamura et al., 1999); incidentally, the 3n rose bitterling produces only 2% sperm, in comparison to that (100%) of 2n bitterling. Reduction in fertilization success of the tench may rather be due to reduced sperm count (see Table 34) but not due to the inability of entry of larger 3n sperm. The interior and exterior diameter of micropyle canal openings in 2n eggs of the tench are 4.2 µm and 2.5 µm and are sufficient to accommodate the head of 3n sperm. In a representative example for triploids producing mostly euploid sperm, the tench has been chosen and relevant information from Linhart et al. (2006) is compiled in Table 34. The testis size, sperm characteristics and progeny production capacity of diploid and triploid males are compared. Incidentally, these values may also be compared with those presented for the triploid producing aneuploid sperm. Notably, the sperm produced by triploids have been reduced to less than 5% of that produced by diploid male. Yet, the triploid is able to successfully sire more than 2.5% of alevin, in comparison to that (100%) produced by diploid males.

In the Prussian carp *Carassius gibelio*, the triploid is known to arise from an oocyte, which has not undergone the first meiotic division but extrudes the polar body following the second meiotic division (Fig. 24, see Horvath and Orban, 1995, see also Sun et al., 2007). Kobayashi (1976) reported that the triploid female *Cyprinus carpio* did not undergo the first meiotic division (Fig. 24). On the basis of electrophoretic analysis of the monomeric enzyme PGM-A, Zhang et al. (1992a) showed that synapsis and recombination did occur at least between some homologous chromosomes during oogenesis of these triploids (Fig. 28). Hence in these fertile triploid carps, oogenesis is successfully completed at least in some oocytes. From a collection of 16 diploids, five triploids and one tetraploid *C. gibelio* from the Prussian waters, Flajshans et al. (2008) reported the relative DNA content of the sperm of diploid, triploid and tetraploid as 1n, 1.5n and 1.6n, respectively besides both the triploid and tetraploid males were found to produce sperm with highly variable DNA content. This gave evidence for the presence of aneuploid sperm in the triploid and tetraploid males. Linhart et al. (2006), who investigated the ploidy level in sperm of induced triploid males of tench *T. tinca*, found that the diploid, triplod and tetraploid tench males produced sperm with relative DNA content of 1.1n, 1.5n and 1.6n, respectively.

Relative DNA content of F_1 progenies sired by triploid *O. mykiss* was 3.5n (tetraploidy, Ueda et al., 1987, 1991) and by triploid *R. ocellatus ocellatus* 2.5n (triploid, Kawamura et al., 1999). Ploidy level and relative DNA content F_1 progenies sired by triploid *T. tinca* crossed to 2n female were 64% diploids each with 1.9n, 34% triploids each with 2.8n and 2% tetraploids each with 3.7n. Not expectedly, a control mating of diploid male with diploid female too produced 86% diploids each with 1.9n, 13% triplods each with 2.8n and 1% tetraploid with 3.7n (Linhart et al., 2006). As indicated elsewhere, some of these fishes like the Prussian carp and tench have more than one means to diploidize or triploidize their genome by (i) inducing spontaneous gynogenesis/triploidy in the aged post-ovulatory oocytes, (ii) eliminating the first meiotic division followed by fertilization and extrusion of the second polar body and (iii) facilitating the entry of two or more sperms into the overripened eggs. Hence it is not surprising that even the mating between diploid male and female tench produces triploid and tetraploid F_1 progenies. Hulak et al. (2009) provided experimental evidence for the triploid progenies sired by the diploid male tench mated to diploid female and traced it to post-ovulatory ageing of oocytes. Molecular pedigree analysis of control (2n ♀ x 2n ♂) crosses showed that the diploid female crossed to diploid male yielded 93% juveniles with diploid genotypes and 7% triploid genotypes, due to the aged oocytes. In contrast, individual crosses between 3n ♂ x

2n ♀ produced 73% juveniles with triploid genotypes and 27% diploid genotypes. Whereas both the diploids and triploids belonging to the first genotype contained only maternal alleles, the second one had maternal alleles and an additional allele derived from the male.

6.4 Female sterility

Most triploid females are hormonally and cytogenetically sterile, whereas triploid males are hormonally fertile, but cytogenetically they too are partially or completely sterile. Many authors have recorded considerable delay in sexual maturity in induced triploids. For instance, Kavumpurath and Pandian (1990) observed that at the age of 3 months, diploid females of *Danio rerio* were fully mature but the triploids did not show any morphological, anatomical and behavioural sign of sexual maturity. In the protagynous hermaphrodite *Pagrus major* too, maturity was delayed for longer than one year (Kitamura et al., 1991b). In the protandrous hermaphroditic gilthead seabream *Sparus aurata*, all the triploids were sexed as males at the first reproductive season. All of them matured as males during the second reproductive season. But none of them changed sex, as 70% of the diploid bream did. In the testis of triploid bream, spermatocytes proceeded upto meiosis II and did not develop further (Haffray et al., 2005). In the European catfish *Silurus glanis*, Kirasznai and Marian (1986) observed that in 2 year old triploid oogonial cells attained stage III, but those of diploids were already in stage V. In the Indian catfish *Heteropneustes fossilis*, oocytes of triploids grew from 98 μm diameter size to 1.36 and 202 μm size, while those of diploids developed from 111 μm size to 348 and 225 μm size during the corresponding period; however, with vitellogenesis, the diploid oocyte grew to 586 μm size, while the triploid oocyte (189 μm size), with the failure of vitellogenesis, began to undergo atresia (Tiwari et al., 2000).

Sterility in female triploids has received much attention, especially the salmonids: Atlantic salmon (Johnson and Wright, 1986), masu salmon (Nakamura et al., 1993), coho salmon (Piferrer et al., 1994), rainbow trout (Krisfalusi et al., 2000) and brook trout (Schafhauser-Smith and Benfey, 2001). An important finding is that the ovaries of triploids remain at the prophase of the first meiotic stage even during the peak breeding season (Manning and Burton, 2003; Han et al., 2010). Consequent to this failure of meiosis, many abnormal developments occur: atresia, oogonia with vacuolated and/or ruptured nuclei, binucleated oocyte, collapse of cysts, formation of cytocyst and hermaphroditism as well as appearance of non-follicular elements like tubules resembling renal tubules, fat deposits, vascular lacunae and so on. With failures of oocyte growth and vitellogenesis, the triploid ovaries become much smaller than that of diploid ovary, as in rainbow trout (Fig. 30c). Consequently, the gonado somatic indices of triploid gonads

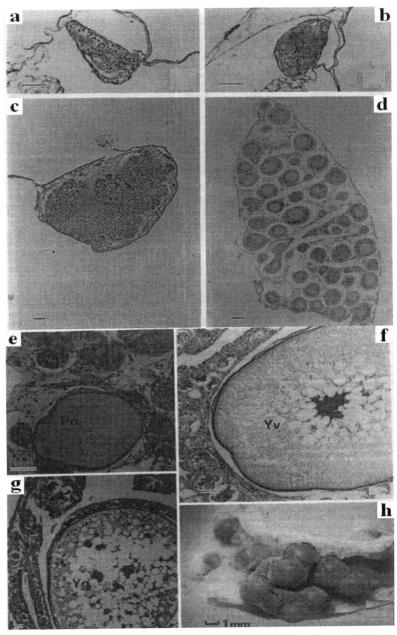

Fig. 30 Early oocyte development in (b and d) diploid and (a and e) triploid *Oncorhynchus mykiss* and ovarian eggs of 3-year old female: (e) peri-nuclear stage, (f) yolk vesicular stage (g) yolk-globular stage. and (h) ovarian eggs grown to 3 mm in diameter scale: a-g : bar = 10 μm (from Kobayashi et al., 1998).

are 10 to 100 times less than those of diploids (see Table 25 of Pandian and Koteeswaran, 1998).There are sporadic reports on the occurrence of mature vitellogenetic oocytes but usually in older triploids. Kobayashi et al. (1998) were perhaps the first to provide unequivocal evidence for the presence of vitellogenic oocytes (see also Kim et al., 1994) that were passing from perinucleolar, through yolk vesicular and to yolk globular stages (Fig. 30) in the 3-year old triploid rainbow trout. In the 1, 2 and 3-year old triploid brook trout *Salvelinus fontinalis*, the estimated production was 2 ± 2, 15 ± 8 and 72 oocytes, in comparison to 453 ± 23, $1,348 \pm 65$ and $1,941 \pm 66$ oocytes of the diploids, respectively. Notable is the wide variations in the number of oocytes produced by the triploids. Of 19 triploid females, six had no oocytes, and the remaining 13 had oocytes only in the left ovary. Though not directly comparable, the wide variations in the number of oocytes and an abnormal ovary of the triploids are similar to those observed for the variations in DNA content of aneuploid and euploid sperm of triploid males (Schafhauser-Smith and Benfey, 2001). Incidentally, in rainbow trout *Oncorhynchus mykiss*, hormonally sex reversed XXY females with ovaries and XXX males with testes showed similar characteristics to their non-sex inversed counterparts (Krisfalusi and Cloud, 1999).

The Japanese loach: Arai et al. (1993) generated triploid loach *Misgurnus anguillicaudatus* by reciprocal hybridizations between normal diploids and natural tetraploids. The triploid males are partially (see Zhang and Arai, 1999a) sterile, but the females are completely fertile, a rarity among the induced triploids, and simultaneously produce large (3n) and small (n) eggs (see Fig.1 of Matsubara et al., 1995, Zhang and Arai, 1996). On fertilization by haploid sperm of a normal diploid, further development of these eggs results in the production of tetraploid and diploid progenies, respectively. Hence the reproductive mechanism of this Japanese loach remains an enigmatic system. Incidentally, the triploid female hybrid between (*C.a. cuvieri* ♀ x *C. carpio* ♂) F_1 x *C. carpio* ♂ too simultaneously produce eggs of different sizes (Fig. 31).

Using multilocus DNA fingerprinting, Arai and Mukaino (1997) showed the clonal nature of large eggs produced by the gynogenic triploid, as indicated by high BSI values, which ranged from 0.962 to 0.991. The presence of 150 chromosomes specifically in an ovary of the triploid loach (but only 75 chromosomes in the cells of gills and spleen) suggested that the large eggs were developed from these hexaploid oogonia, as the percentage of hexaploid cells (10, i.e. 34%) were approximately equal to the proportion of large eggs (11, i.e. 37%) laid by the triploid females (Matsubara et al., 1995). From their cytogenetic studies, Zhang et al. (1998) too found both triploid (3n = 75) and hexaploid (6n = 150) metaphase in immature triploid ovary. The small oocytes showed about 50 chromosomes, comprising 25 thick densely

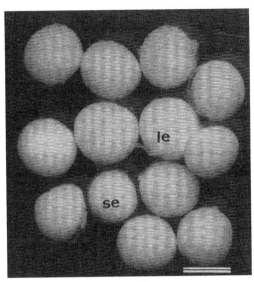

Fig. 31 Simultaneously produced smaller (se) and large (le) eggs of the triploid female, a hybrid between (*Carassius auratus cuvieri* ♀ x *Cyprinus carpio* ♂) F₁ ♀ x *C. carpio* ♂ (from Zhang et al., 1992b).

stained bivalent and 25 thin faintly stained univalents, i.e. bivalent-univalent complexes (see Fig. 4 of Zhang et al., 1998). The large ones showed more than 60 thick elements, taken as 75 bivalents. *In vitro* culture of these oocytes revealed that both large and small oocytes underwent two conventional successive meiotic cycles, i.e. formation of a bipolar spindle in meiosis I and equal segregation of homologues, extrusion of the first polar body and the appearance of a bipolar spindle in meiosis II. In the small, full grown oocytes, some chromosomes got detached from the spindle and remained in the surrounding cytoplasm and ultimately these unpaired univalents were eliminated. Similar elimination of uniparental chromosomes is known from unisexual (e.g. *Poeciliopsis,* Cimino, 1972) and bisexual (e.g. *Oncorhynchus mykiss,* Fujiwara et al., 1997) fishes. Thus the triploid eggs of the loach are formed from oocytes of a large size, which are generated by premeiotic endomitosis, while the small haploid eggs were produced from bivalents after elimination of univalents in meiosis. In subsequent publications, Prof. K. Arai's group reported that the loach collected from Hokkaido Island, Japan produces at least four types of eggs, i.e. haploid, diploid, triploid, and aneuploid eggs (see Momotani et al., 2002, Oshima et al., 2005).

In the induced sterile triploid male loach, spermatogonial cells mitotically divided after endomitosis but never entered into meiotic process to form spermatozoa. Cells with 6C DNA content were observed in the testis of triploid loach (Fig. 32) but haploid spermatozoan nuclei were not detected (Zhang and Arai, 1996). Haploid spermatogonia were arrested in

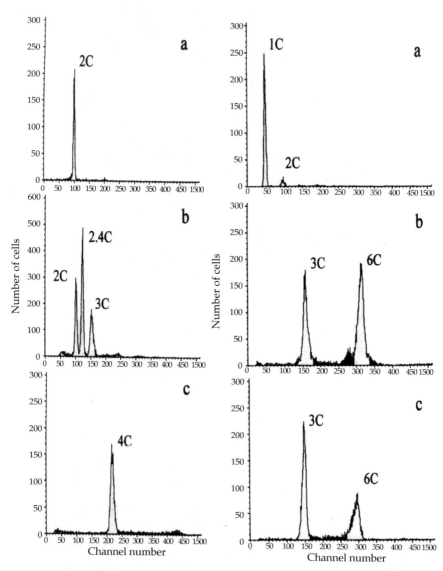

Fig. 32 *Misgurnus anguillicaudatus:* Flow cytometry histograms for ploidy determination testicular cells (right panel) and alevin (left panel) on the basis of DNA content. Right panel: (a) n sperm of 2n male, (b, c) 3n and 6n cells with 3C and 6C DNA content of 3n male. Left panel: (a) 2n progeny with 2C DNA of 2n female, (b) 2.4n C DNA aneuploide progeny of the cross 3n ♀ x 2n ♂, and (c) 4n ♀ progeny with 4C DNA of the cross 3n ♀ x 2n ♂ (from Oshima et al., 2005).

the course of spermatogenesis. Thus the cytogenetic events in both sexes of the triploid loach led to the simultaneous production of large (3n) and small (n) eggs in the female but production of no sperm in the male (Zhang et al., 1998). Investigating the reproductive capacity of natural triploids of Hokkaido Islands, Oshima et al. (2005) reported the presence of sperm, rather sperm-like cells at the counts ranging from 2.62×10^3/ml to 4.2×10^3/ml. These sperm-like cells have a larger head and smaller tail and are inactive, when activated with water. The poor fertility of these sperm-like cells is due to their (i) shortage of adequately active sperm to fertilize eggs, (ii) lack of active motility and too large head to enter through micropyle. Hamaguchi and Sakaizumi (1992) reported medaka hybrids too produced such sperm-like cells. From a culture of a single spermatocyte isolated from the hybrid medaka, Shimizu et al. (1997) found that the spermatocyte differentiated into one sperm-like cell without meiotic division on an equal time schedule to the parent medaka. Their other studies led them to suggest that spermeiogenesis is intrinsically independent of the proceeding meiosis. In semen of the triploid males from Hokkaido population, sperm with 3C and 6C DNA were detected; at least two males showed the presence of haploid cells (Fig. 32). Thus, the Hokkaido triploid loaches too have a small potential to produce very few haploid sperm (Oshima et al., 2005).

Natural occurrence of triploids is limited to six species each among unisexuals and gonochores. Irrespective of carrying $X^1x^1X^2$ or $W^1w^1Z^2$/$W^1Z^2z^2$ genotype, females do not survive until sexual maturity, resulting in male biased sex ratio among both male and female heterogametics. With higher genomic heterozygosity than diploids, triploids are expected to be more viable. The causes for such sex dependent mortality are not known. However studies on triploidy do not seem to provide any clue on sex determination.

Tetraploidy and Polyploidy

Polyploidization is the increase in genome size caused by the inheritance of additional set (or sets) of chromosomes. It represents one of the most dramatic mutations known to occur (Otto, 2007). In plants, the rate at which polyploids have arisen and are sustained, is in the order of 0.01 per lineage per million years, i.e. approximately a tenth of speciation (Meyers and Levin, 2006). Although polyploidization is less prevalent in animals, many independent examples of polyploidization in fishes have been reported (Gregory and Mable, 2005). Both unreduced gametes and polyspermy contribute to the production of polyploidy in animals. In fishes, "unreduced gametes are produced by (i) derangements of gametogenesis caused by alterations of meiosis such as (a) the premeiotic endoduplication of the chromosome set (Fig. 24), (b) the suppression of the first and/or second meiosis or (c) the non-disjunction of mitotic chromosomes following cleavages and (ii) by the suppression of the second meiotic division in the post-ovulatory aged oocytes" (Piferrer et al., 2009). Polyspermy is also known to occur in fishes .

7.1 Survey and history

If polyploid fishes were an evolutionary dead end, such taxa may be expected to occur near the tips of tree of life and may have only a few species. On the contrary, polyploidy is more common among the lower, ancient teleosts than higher teleosts (Leggatt and Iwama, 2003). Of the extant of fish species in 57 orders, the majority (63%) of 9 orders, especially Cypriniformes (2,662 species with > 90 polyploid species), Siluriformes (2,400 species with > 27 polyploid species), Characiformes (1,343 species with > 5 polyploid species) and Perciformes (9,293 species with > 1 polyploid species) are known to include polyploids (Le Comber and Smith, 2004). Indeed, the most speciose genus *Barbus* (with > 800 species) alone includes more than 81 polyploid species and the ploidy level differs from species to species with chromosome numbers of 50, 100, 150 and > 200 (Berrebi, 1995). Thus, all but one family of lower teleost order Cypriniformes are represented by at least one and in some species multiple occurrence of polyploidy (Table 35). Comparisons of chromosome numbers suggest the occurrence of 7–20 polyploidization events in extant of ray finned fish lineages (Mank and Avise, 2006a). Polyploidization has been common among sturgeons with estimates of one of the highest ploidy levels at either 8x or 16x (Gregory and Mable, 2005). It is generally accepted that Salmonidae and Catostomidae are polyploids. Other examples for polyploidy are scattered around other families including the Callichthyidae, Cobitidae and Poeciliidae (Smith and Gregory, 2009). Hence polyploidy is a more wide spread phenomenon in the evolution of fishes than is usually appreciated.

Table 35 Chromosome number and ploidies in selected fishes (from 1. Rock et al. (1996), 2. Birstein et al. (1997), 3. Suzuki and Taki (1981), 4. Berrebi (1995), 5. Guegan and Morand (1996), 6. Golubtsov and Krysanov (1993), 7. Berrebi and Rab (1998), 8. Guegan et al. (1995), 9. Tsigenopoulos et al. (2002), 10. Ruiguang et al. (1986), 11. Onozato et al. (1983), 12. Shimizu et al. (1993), 13. Jianxun et al. (1991), 14. Raicu et al. (1981), 15. Yu et al. (1987), 16. Vasil'yev (1981), 17. Benfey (1989), 18. Borin et al. (2002), 19. Galetti (1998), 20. Rab (1981), 21. Pandian and Koteeswaran (1999), 22. Ohno (1970), 23. Schultz (1989 see also, Leggatt and Iwama, 2003).

Family/Species	2n	Ploidy
Protopteridae		
Protopterus dollie[1]	68	4N
Polydontidae		
Polyodon spathula[2]	120	4N
Acipenseridae		
Acipenser spp (9 species)	230–234	8N
Acipenser brevirostrum, A. mikadoi[2]	360–500	16N
Cyprinidae		
Acrossocheilus sumatransis[3]	98	4N
Barbus (3 species) *Bynni* (17 species)[4-9]	148–150	6N
Barbus (27 species)	100	4N
Percocypris (3 species)	98	4N
Pseudolabrus spp	100	4N
Sinocyclocheilus (2 species)[10]	96	4N
Tor (3 species)	100	4N
Varicorhinus (2 species)	150	6N
Aulopyge huegelii	100	4N

Table 35 contd....

Table 35 contd....

Family/Species	2n	Ploidy
Carassius cantonensis	100	4N
C. auratus auratus, C. auratus langsdorfi[11, 12]	100, 156–162, 206	4N, 6N, 8N
C. auratus buergeri, C. auratus gibelio[13]	94–104, 150–162	4N, 6N
C. auratus cuvieri, C. a. grandoculis	100	4N
C. carassius[14]	50–100	2N, 4N
C. carassius gibelio	156	6N
Ctenopharyngodon idella	–	3N
Cyprinus carpio chilia[15]	100–150	4N, 6N
Leuciscinae (7 species)[8]	78	2N, 3N, 4N
Diptychus dipogon[13]	446	14N
Schizothorax spp (5 species)[15, 16]	148	6N
Cobitidae		
Misgurnus fossilis	100	4N
Cobitis complexes (6 species)[16]	40–48, 98–100	2N, 3N, 4N
M. anguillicaudatus	50–100	2N, 3N, 4N
Gymnotiformes (2 species)[17]	81	3N
Characiformes (5 species)[17, 18]	76	3N
Curimatidae (2 species)[17, 19]	50–100	2N/4N
Callichthyidae		
Corydoras aeneus[20]	120–132	4N, 6N
Corydoras (3 species)	92–98	4N
Heteropneustidae		
Heteropneustes fossilis[21]	30, 56, 87, 116	N, 2N, 3N, 4N
Siluridae		
Wallago attu[20]	86	4N
Salmoniformes[17, 22]		
Coregonus lavaretus		3N, 6N
Oncorhynchus (2 species)		3N, 6N
Salvelinus fontinalis		3N, 6N
Poeciliidae[23]		
Poeciliopsis, Poecilia		3N
Channidae[3]		
Channa stewarti	104	4N

Two genome-doubling events are considered to have occurred before the tetraploids split from the fish 360 MYa (Zhou et al., 2002). The ancestors of vertebrates underwent one genome doubling prior to the Cambrian explosion and second in the early Devonian, after the divergence of the lobe-finned Scarcopterygii. The fish genome was doubled for the third time in the ray-finned Actinopterygii (Wittbrodt et al., 1998). This is based on the recent finding that many teleosts have an additional set of *Hox* gene clusters compared to Scarcopterygii. *Hox* genes encode DNA binding proteins that specify the cell fate along the anterior-posterior axis of bilaterean animal embryos and occur in one or more clusters of upto 13 genes-clusters (Gehring, 1998). Thus the cephalochordate *Amphioxus* possess a single *Hox*-cluster and the Scarcopterygiians such as coelacanth and lungfishes four clusters. Recently, additional *Hox* gene clusters have been discovered in many Actinopterygiians such as *Danio rerio, Oryzias latipes, Oreochromis niloticus* and *Takifugu rubripes* (van der Peer et al., 2003). There is evidence that the third genome-doubling event occurred after the eels split from euteleosts but before the Polyodontiformes such as *Polypterus palmas* split from the telosts (Leggatt and Iwama, 2003).

Ohno (1970) hypothesized that big leaps in evolution required the generation of new gene loci with previously non-existent function and emphasized genome duplication via tetraploidy as the mechanism for production of such new genes. Comparisons of gene number do provide support for large scale gene multiplication. In their genome survey, Van der Peer et al. (2003) have often found more multiple gene copies in fishes for single genes than in other vertebrates but almost never the opposite. In fact, the success and diversity of fishes is a direct consequence of gene copy number (Meyer and Schartl, 1999). Ohno (1974) and Kirpichnikov (1981) have traced the origin and evolution of diploid-tetraploid systems among several teleostean families. Tetraploidy in the loach *Cobitis biwae*, for instance, increases the DNA content of an erythrocyte by 2.1 times (from 3.15 pg/nucleus in diploid to 6.62 pg/nucleus in tetraploid) but by 1.7 times only in hepatocytes, characterized by relatively low mitotic activity (Juchno et al., 2009). "Two potential advantages of tetraploidy are the overall increased heterozygosity leading to heterosis and gene redundancy, which masks recessive alleles, and provide evolutionary potential for diversification of gene function" (Piferrer et al., 2009). The disadvantages of tetraploidy are: (i) decrease in cell numbers to maintain similar body size in diploids; for instance, 54% reduction in cumulative cell number in the 5-hours old 4n alevin of *Ctenopharyngodon idella*, in comparison to that of the diploid (Cassani, 1990), (ii) reduced fertility of tetraploid, as their diploid sperm have difficulty in entering through micropyle of the haploid egg (Blanc et al., 1993) and (iii) mechanics of pairing and separation of chromosomal homologues during mitosis and meiosis that lead to aneuploidy (McCombie et al., 2005), especially due to the increased frequency of chromosome non-disjunction (Storchova et al., 2006) and to contain it by spindle (DeCordier et al., 2008).

7.2 Scope for polyploidization

Delineation of the history of polyploidy evolution was necessary for a better understanding that (i) the scope for successful induction of tetraploidy is limited to the series Ostariophysi and within the order Cypriniformes (e.g. *Carassius, Megalobrama, Misgurnus*), and (ii) despite more than 121 polyploid species are known to have originated and occur possessing 16n (2 species), octoploid (> 10 species), hexaploid (>31 species), and tetraploid (> 78 species), it has been possible to successfully induce a maximum of hexaploid in *Misgurnus anguillicaudatus* and *Oncorhynchus mykiss*. The causes for this limited scope for induction of tetraploid are not yet known.

Repeated attempts to induce tetraploids in cichlids have not met with any success (see Pandian and Koteeswaran, 1998 for all references). For instance, no tetraploid survived beyond post-hatching in *Oreochromis mossambicus* and *O. niloticus*. In *O. aureus*, the yield was less than 1–2%; only two females survived until the age of 18 months but they were smaller and did not show any behavioural sign of breeding, whereas their diploid siblings commenced to reproduce at the age of 6 months. Many surviving polyploids were ventrally or laterally conjoined twin embryos and succumbed sooner or later (Owusu-Frimpong and Hargreaves, 2000). Likewise attempts to produce polyploids even in other salmonids like *Salmo salar* and *S. gairdneri* resulted in the low yield of > 2% mosaic embryos, which did not survive beyond post-hatch. Pressure induction of tetraploidy in *S. trutta* led to the 4n yield of 17% but all of them died by the hatching stage. From *Clarias gariepinus* too, no surviving tetraploid fry could be obtained (Varadi et al., 1999). Although triploid yield was high in the channel catfish *Ictalurus punctatus*, the tetraploid mosaics did not survive too long. Muniayandi et al. (2006) also successfully induced triploids in *Channa punctatus* but reported the non-viability of mitotic tetraploid; surprisingly, from this same locality, Pandian and Koteeswaran (1999) reported the natural occurrence of triploids and tetraploids of the catfish *Heteropneustes fossilis*. More surprisingly, attempts to produce fertile tetraploids in the bighead carp *Hypophthalmichthys nobilis* (Yunhan, 1990) and grass carp *Ctenopharyngodon idella*, both belonging to the Cyprinidae, a family known for the presence of the highest number of polyploidy species, have not succeeded. The known and assumed causes for failure of these induced tetraploids may be briefly listed: (i) mosaicism, (ii) aneuploidy, (iii) reduced cell surface, (iv) 'wrong' cytological events and (v) homozygosity.

Two publications need to be considered. Nam et al. (2001b) successfully induced mitotic tetraploid mud loach *Misgurnus mizolepis* and 86% of the progenies were reared until the age of 9 months. Of 48 tetraploid males, 12 were sterile; 26 produced only haploid sperm but three of them produced diploid sperm, which when crossed with diploid females, produced triploid mud loach (Nam and Kim, 2004). Mitotic allotetroploids were produced by crossing *Megalobrama ambylocephala* ♀ with *M. terminalis* ♂. About 30% abnormal fry were produced and only about 8% proven tetraploid progenies survived and attained delayed sexual maturity. Sex ratio was 0.44 ♀ : 0.56 ♂. At the age of 2+, ovaries of females were shrunken and string-like but at 4+, the males were successfully induced to spawn. Experiments are in progress (Zou et al., 2008).

7.3 Tetraploid types

Natural tetraploids are generated by mating natural tetraploid males and females, as in *Misgurnus anguillicaudatus* (Fig. 33). Gynogenic tetraploids can be induced by activating diploid eggs (of natural tetraploid female) followed by retention of the second polar body (e.g. *M. anguillicaudatus*). Mitotic tetraploids are produced by successfully arresting the first mitotic cleavage, as in *Megalobrama amblycephala* (Zou et al., 2004) Meiotic or the so called second generation of tetraploids are sired by tetraploid males crossed

Ploidy type	Induction procedure	Species	Gametes	Reference
Natural	Crossing 4n with 4n	*M. anguillicaudatus*	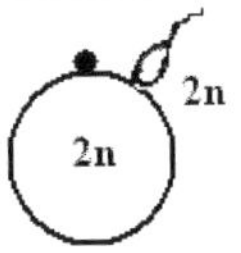	Arai et al. (1993)
Meiotic	Fertilization of 2n egg by 4n sperm and retention of polar body	*O. mykiss*	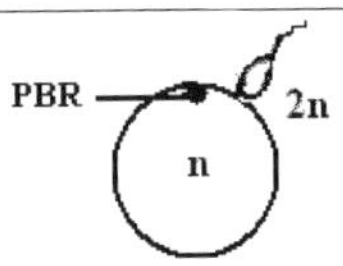	Blanc et al. (1993)
Mitotic	Suppression of cleavage of fertilized egg	*O. mykiss*	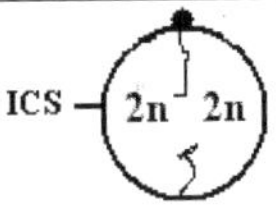	Thorgaard et al. (1981)
Gynogenetic	Activation by sperm (IR) leads to spontaneous duplication of 2n eggs	*M. anguillicaudatus*	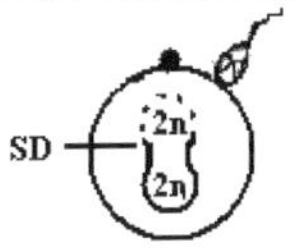	Arai et al. (1991b)
	Activation of 4n eggs by sperm (IR) followed by shock	*M. anguillicaudatus*	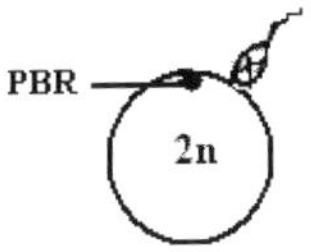	Arai et al. (1993)
Hybrid	a) Fertilization of unreduced hybrid eggs (2n) by unreduced sperm (2n)	[*C. auratus* (♀) x *C. carpio* (♂)] F$_1$ (♀) x F$_2$ (♂)	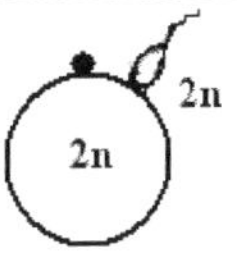	Cherfas et al. (1994a)

Fig. 33 Types of natural and induced tetraploids in fishes. PBR = polar body retention, ICS = I cleavage suppression, SD = Spontaneous duplication.

to diploid females and followed by retention of the second polar body (e.g. *Oncorhynchus mykiss*). Allotetraploid are generated by crossing the diploid *Carassius auratus* ♀ with *Megalobrama amblycephala* ♂, or by fertilizing unreduced diploid eggs and sperm produced by the F_2 hybrid *C. auratus gibelio* ♀ and *Cyprinus carpio* ♂. Claims have been made for production of viable tetraploids in about 10 species (Arai, 2001). However, fertile bisexual tetraploids have been produced only in five species, of which three are created in China, one in Japan and the other in the USA. In view of the inherent academic and commercial importance, these five are described a little more elaborately.

7.4 True induced tetraploids

Oncorhynchus mykiss: Thorgaard et al. (1981) were the first to successfully induce mitotic tetraploids in the rainbow trout. Since then autotetraploids have been induced by Chourrout et al. (1986), Chourrout and Nakayama (1987) and Horstgen-Schwark (1993). By comparing single pair and group matings, representing pure and cross-bred progenies, respectively, Horstgen-Schwark showed that the single pair matings produced tetraploids alone, while the group matings yielded some diploids also. Survival and growth were severely depressed in the first generation of mitotic tetraploids; only 40% progenies survived and at the age of 260 days, body weight attained by the tetraploid was half of that of diploids. In an attempt to enhance genetic improvement through maximization heterozygosity, a second generation of meiotic tetraploids was generated by Blanc et al. (1993) through a cross between 2n female and 4n male followed by retention of the second polar body. However, a major limiting factor was the low ($\approx$ 40%) fertility of tetraploid sires, which Chourrout et al. (1986) ascribed to diploid spermatozoa being too large to enter through the micropylar canal of the haploid egg. With increasing diameter of the inner canal of the micropyle from 1.4 μm to 2.2 μm, the fertilization success increased from 20 to 75%.

The process of increasing ploidy by a shock or no shock considerably affects the control and regulation of disjunction and migration of daughter chromosomes. For instance, crosses between 4n ♀ and 4n ♂ rainbow trout produced 82% 4n progenies and others in one cross, but in another cross only 5% 4n, 24% 2n and 51% 5n progenies. Similarly a cross between 4n ♀ and 2n ♂ followed by a thermal shock to retain the second polar body, produced 53% pentaploids progenies and others, while 100% pentaploids were expected (Chourrout and Nakayama, 1987). However, there were as high as 40% mortality but it is not known whether cytological problems led to such high mortality and whether the mortality was sex-dependent. Yet successful induction of tetraploidy resulted in the production of viable fertile progenies with sex ratios of 0.04 ♀ : 0.96 ♂ for the mitotic tetraploids and 0.07 ♀ : 0.93 ♂ for the meiotic tetraploids and the cause for a highly skewed sex ratio remains unknown.

The Chinese creations: Between 2001 and 2007, Chinese scientists created at least three 'new species' by 'synthesizing' bisexual tetraploids. Amazingly, the time scale for the creation of these new bisexual tetraploids has been comprised to a very short period, which otherwise nature would have required many years to pass through the process of hybridization → production of unreduced diploid eggs and sperm → establish a true tetraploid. The first creation of the bisexual tetraploid is by Dr. S.J. Liu and his colleagues, who have established a hybrid line using 4.7% fertile males and 44.3% fertile females produced by crossing *Carassius auratus gibelio* ♀ with *Cyprinus carpio* ♂ (see Fig. 4 of Liu and Zhou, 1986). Unusually some F_3 hybrid females simultaneously produced haploid (0.13 cm) and diploid (0.17 cm) euploid eggs and 4n (2.0 cm) aneuploid eggs as well (see Fig. 4 of Liu et al., 2001). Likewise some F_3 hybrid males also produced two times larger volume of milt and the milt consisted of 40% haploid, 49% diploid and 11% tetraploid octoploid and aneuploid sperm. Fertilization of diploid eggs by diploid sperm produced bisexual (0.5♀ : 0.5♂) tetraploid hybrids in F_3. Until now up to F_{16} progenies are proved to be tetraploid hybrids, showing that their tetraploidy is stably inherited from one generation to another (Liu et al., 2001, 2003, 2004c; Sun et al., 2003).

Incidentally, induction of gynogenesis in the fertile hybrids (red crucian carp ♀ x common carp ♂) produced all female progenies. Instead, the induction of androgenesis resulted in the production of 50% fertile females (X^1x^1) and 50% fertile males (Y^2y^2). Crosses between these androgenic males and females (A_0) produced 85% 4n, 10% 3n and 2n (A_1) progenies. Apparently, A_0 progenies, on attainment of sexual maturity at the age 2+, produced diploid gametes. G_1- 4n progenies possessed normal ovaries and testes. They grew faster, possessed the desired smaller heads and shorter tails, and resisted diseases more vigorously (Duan et al., 2007).

As a new species with 200 chromosomes, the allotetraploid has become important not only economically but also as a new source of research material. The allotetraploid males (F_3–F_{16}) have been mated with the Japanese diploid female crucian carp *C. auratus cuvieri* to produce triploids characterized by high survival and faster growth. In fact, these trigenomic triploids possess one genome, each of *C. auratus gibelio*, *C. carpio* and *C. auratus cuvieri*. During recent years, > 100,000 tetraploids and > 300 million triploids are annually produced in China. The complete mtDNA nucleotide identity between the triploid crucian carp and its male parent allotetraploid is higher (98%) than that (93%) between the triploid crucian carp and its female parent Japanese crucian carp. Further analysis of mtDNA has suggested that the newly developed triploid crucian carp possess unusual recombination mtDNA fragment of 12, 759 bp derived from the tetraploid male (Guo et al., 2006).

The second species, a bisexual mitotic autotetraploid blunt snout bream *Megalobrama amblycephala* has been generated by Dr. S. Zou and his colleagues at the Shanghai Fisheries University. Using thermal shock to suppress the first mitotic cleavage, they have produced tetraploids. About 28% of the F_1 progenies were abnormal. However, in others, tetraploidy has been confirmed by the presence of 96 chromosomes and erythrocyte size of 18.5 µm³, which are exactly double the number and size of those of diploids. Sexual maturity is delayed; about 11% and 71% females mature at the age of 2 and 3 years, respectively. Sex ratio is skewed 0.3 ♀: 0.7 ♂, representing a significant departure from the expected 0.5 : 0.5 ratio. Table 36 provides comprehensive information on the reproductive performance of the tetraploid *M.amblycephala* (Zou et al., 2004).

Table 36 Reproductive performance of diploid and tetraploid blunt snout bream *Megalobrama amblycephala* (data as given to the author by Dr. S. Zou to review and improve the MS and with his permission).

Ploidy (n) and sex	Age (year)	Relative fecundity (no/g fish)	Egg size (mm)	Hatchability (%)	Net yield of fry (no/g fish)	Net yield of 4n fry as % of 2n cross
2 ♀ x 2 ♂	2	65.2	0.99	85	55.4	–
2 ♀ x 2 ♂	3	146.2	1.11	90	131.6	–
4 ♀ x 4 ♂	2	47.8	1.23	61	29.2	53
4 ♀ x 4 ♂	3	128.6	1.38	66	84.9	65

The third in this series of success stories is the 'creation' of a bisexual allotetraploid generated by a simple but ingeniously designed distant cross between the omnivorous cyprininaean red crucian carp (RCC) 2n *Carassius auratus* ♀ and the herbivorous cultinaean blunt snout bream (BSP) 2n *Megalobrama amblycephala* ♂ by Dr. S. Liu and his colleagues at the Hunan National University, Changsha. Fertilization and hatching success of RB hybrids were 60 and 50%, respectively. Among surviving F_1 progenies, 23 and 77% accounted for 3nRB and 4nRB; there were no 2nRB. The reciprocal cross between RCC ♂ and BSB ♀ did not yield any progeny. The 4n RB hybrids were easily recognized by morphological features, namely the presence of a pair of oral barbels and 31–33 lateral scales (Table 37) and histological features of two large sub-metacentric marker chromosomes and erythrocytes with double nuclei; with increase in ploidy level, the percentage of multinuclear erythrocytes increased (Fig. 34).

Using primers of *HMG* of *Sox* genes and sequencing, the PCR analysis showed that there were three DNA fragments (215, 617 and 1,958 bp) in RCC, two (215 and 712 bp) in BSB, three (215, 616 and 1,955 bp) in 3n RB and four (215, 616, 918 and 1,956 bp) in 4nRB (Fig. 34, Table 37). Thus the hybrid 4n RB hybrids not only changed chromosomal ploidy level but also

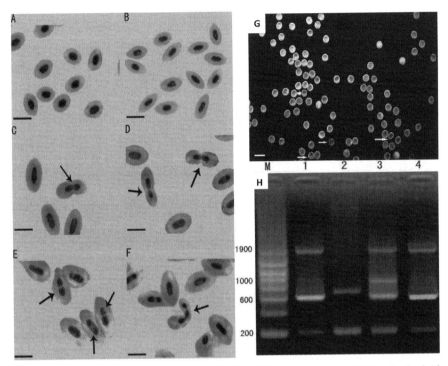

Fig. 34 Erythrocytes of the red crucian carp (RCC) and blunt snout bream (BSB) and polyploid hybrids. (A) normal erythrocytes with one nucleus in RCC (B) normal erythrocytes with one nucleus in BSB (C) normal erythrocytes with one nucleus and unusual erythrocyte with two nyclei (arrows) in a 3n RB hybrid (D) normal erythrocytes with one nucleus and unusual erythrocyte with two nuclei (arrows) in a 4n RB hybrid (E) normal erythrocytes with one nucleus and unusual erythrocyte with two nuclei (arrows) in a 5n RB hybrid (F) normal erythrocytes with one nucleus and unusual erythrocyte with two or three nuclei (arrows) in a 5n RB hybrid. Bar in A–F = 0.01 mm. (G) Simultaneous production of two sized eggs by 4n RB hybrids. Smaller eggs are shown by arrows. Bar = 0.4 cm. (H) Amplified DNA fragments resulting from PCR based on the primers of HMG of *Sox* genes in BSB, RCC, 3n RB and 4n RB hybrids. M = DNA ladder marker. Lane 1 = 3DNA fragments in RCC, Lane 2 = 2 DNA fragments in BSB, Lane 3 = 4 DNA fragments in 4n RB hybrids. Lane 4 = 3 DNA fragments in 3n RB hybrids (from Liu et al., 2007a).

Colour image of this figure appears in the color plate section at the end of the book.

changed DNA sequences. Further, the 213 bp fragment of 4n RB belonged to *Sox* 1 rather than *Sox* 11, as in the case of RCC, BSB and 3n RB. Clearly, more DNA variations occurred in 4n RB hybrids than in 3n RB hybrids. Triploid RB hybrids were sterile. But 4n RB hybrid females were fertile. They simultaneously produced 95% large (2.0 mm size) unreduced tetraploid eggs and 5% smaller (1.7 mm size) diploid eggs (Fig. 34). When mated with 2n RB ♂, the 4n RB female pentaploids, the 4n RB males produced water-like semen, which when used to fertilize eggs of 4n RB females, produced only eight live F_2 progenies (Liu et al., 2007a; Yan et al., 2009).

Table 37 Characteristics of the red crucian carp (RCC) and blunt snout bream (BSB) and their triploid (3nRB), tetraploid (4nRB) and pentaploid (5nRB) hybrids (compiled from Liu et al., 2007a).

Parameter	RCC	BSB	3nRB	4nRB	5nRB
Chromosome(no)	100	48	124	148	172
Ploidy level (n)	2	2	3	4	5
Barbels	0	0	0	+	?
Lateral scale (no)	–	–	28–30	31–33	–
Marker chromosome	0	1	1	2	3
Composition chromosome	2n RCC	2n BSP	2n RCC + 1BSP	2RCC + 2n BSB	2nRCC + 3BSP
Erythrocyte (% nuclei no)			3.4, 2	9.4, 2	21.8, 2–3
DNA level	52.19	42.92	72.63	97.05	117.46
Diagnostic DNA fragments (bp)	215, 617, 1,958	215, 712	215, 616, 1,955	215, 616, 918, 1,556	

– = not reported, + = present

Misgurnus anguillicaudatus: Spontaneous natural occurrence of tetraploids of this loach has been reported from Japan (Arai et al., 1991a). The loach is amenable for tetraploidization using at least four different protocols: (i) autotetraploid (a) crossing 4n males with 4n females (Arai et al., 1993) and (b) fertilizing large unreduced triploid eggs of triploid female with haploid sperm of diploid male (Matsubara et al., 1995), (ii) gynogenic tetraploid: by activation of diploid eggs of tetraploid female or unreduced diploid eggs of diploid female followed by a shock to retain the second polar body (Arai et al., 1991b, 1993) and (iii) meiotic tetraploid: fertilizing haploid egg of a diploid female by a diploid sperm of a tetraploid male followed by a shock to retain the second polar body. All these tetraploids, except the gynogenic tetraploid with 100% female progenies, produced about 50% males and 50% female progenies. On sexual maturity, both male and female tetraploid produced diploid sperm and diploid eggs, respectively. However, the gynogenic tetraploids produced unreduced tetraploid eggs.

7.5 Polyploidy

Natural occurrence of hexaploids in *Barbus* and *Cyprinus carpio* and octoploid in *Ceratophyrys dorsata* has been reported (see Pandian and Koteeswaran, 1998). Recently the natural occurrence of hexaploid *Misgurnus anguillicaudatus* has been reported from China (Abbas et al., 2009) and tetraploid catfishes *Clarias batrachus* (Pandey and Lakra, 1997) and *Heteropneustes fossilis* (Pandian and Koteeswaran, 1999) from India. *M. anguillicaudatus* may be considered to represent some of these polyploids, as its gynogenetic diploids, triploids and tetraploids serve as stepping stones to induce pentaploid and hexaploid. Triploids can be produced by four different crossings of 2n ♀ x 4n ♂, 4n ♀ x 2n ♂, 3n ♀ x 2n ♂ and 2n ♀ x 3n ♂ (Table 38), tetraploids by five types of crossings of 4n♀ x 4n♂, 4n♂ x 4n♀, 2n♀ x 6n ♂, 6n ♀ x 2n ♂, 3n ♀ x 2n ♂, pentaploids by the crosses of (i) 3n ♀ x 4n ♂, (ii) 4n ♀ x 6n ♂, (iii) 6n ♀ x 4n ♂, (iv) 4n ♀ x 2n ♂ followed by a shock to retain the polar body and hexaploids by the crosses of (i) 6n ♀ x 6n ♂, (ii) 6n ♂ x 6n ♀ and (iii) 3n♀ with activation by irradiated sperm and followed by a shock to retain the polar body (Table 38). Being male heterogametic, *M. anguillicaudatus* maintain sex ratio around 0.5 ♀ : 0.5 ♂ in all these polyploids (Arai et al., 1999).

Table 38 Fecundity and survival of fry produced by interploid crosses of *Misgurnus anguillicaudatus*.

Cross	Large eggs			Small eggs		
	Ploid	no	Surviving fry (%)	Ploid	no	Surviving fry (%)
Matsubara et al. (1995)						
2n ♀ x 4n ♂	3n	1,557	–			
4n ♀ x 2n ♂	3n	105	–			
3n ♀ x 2n ♂	3n	109*	42	n	687	14.0
3n ♀ x 2n ♂PS	3n	59*	25	n	459	10.0
3n ♀ x 2n ♂UV		99*	16	n	410	0.2
3n ♀ x 2n ♂ UV/PS		137*	16	n	409	0.3
Zhang and Arai (1999a)						
3n ♀ x 2n ♂	3n	18*	–	n	802	13.0
Arai and Imamori (1999)						
2n ♀ x 3n ♂	3n	1,083	14	–	–	–
Arai et al. (1993)						
4n ♀ x 4n ♂	4n	739	67			
4n ♀ x 2n ♂	3n	549	43			
2n ♀ x 4n ♂	3n	863	78			
2n ♀ x 4n ♂	3n	184	31			
Arai et al. (1999)						
2n ♀ x 2n ♂	2n	4,107	16			
2n ♀ x 4n ♂	3n	–	19			
2n ♀ x 6n ♂	4n	–	12			
4n ♀ x 2n ♂	3n	–	16			
4n ♀ x 4n ♂	4n	1,834	20			
4n ♀ x 6n ♂	5n	–	9			
6n ♀ x 2n ♂	4n	–	11			
6n ♀ x 4n ♂	5n	–	47			
6n ♀ x 6n ♂	6n	1,544	45			

2n ♀ x 4n ♂ produced 70% 3n eggs against 3% by 4n♀ x 2n ♂; 3+ hypertriploid eggs; *eggs/ clutch

M. anguillicaudatus is one of the few fishes that are known to simultaneously spawn eggs of different sizes (Matsubara et al., 1995). Triploid females simultaneously spawn different proportions of small and large eggs; 70% large eggs are sired by 4n ♂ but less than 3% large eggs alone are sired by 2n ♂. Pentaploid females produce intermediate eggs only (Matsubara et al., 1995). Smaller eggs, when activated or fertilized by 2n ♂, hardly survive (Table 38). Large eggs are able to accommodate 3–4 sets of chromosomes and are still hatch successfully. Hence large eggs are indeed the stepping stones for induction of polyploidy. Clearly, the number of chromosome sets present in an egg appears to play a decisive role in accommodating greater or less amount of yolk. Following successful fertilization, small, intermediate and larger eggs and their respective polar bodies carried each a haploid, diploid and triploid sets of chromosomes. It is amazing but not clear how during oogenesis, the passage of n, 2n or 3n genome regulates vitellogenesis to deposit small, intermediate and larger amounts of yolk in the respective eggs.

With increasing ploidy level, both body weight and fecundity of *Misgurnus anguillicaudatus* decrease (Table 39) due to increase in size of somatic cells including oogonium. On account of the increase in oogonium size alone in triploid rainbow trout, the decrease in fecundity can be as much as 25% (Piferrer et al., 1994). Hence higher percentage reductions in fecundity of tetraploids and hexaploids are expected. However, the reductions in body weight are 20% in tetraploids and 31% in hexaploids, but those in fecundity are 53 and 62% in tetraploids and hexaploids, respectively. Due to doubling and tripling of genome size, somatic cells of the body seem to suffer less than that of gonadal cells. On the other hand, the progeny production per hexaploid female is in fact more than that of a diploid female (Table 39), *albeit* diploid females are more fecund but suffer heavy embryonic mortality. It is not clear why diploid eggs with one pair of balanced chromosomes suffer 75% mortality, and also why the tetraploids

Table 39 Ploidy, body weight, fecundity and survival of *Misgurnus anguillicaudatus* (compiled from Arai et al., 1999).

Parameter	Diploid	Tetraploid	Hexaploid
Body weight (g)	17.6	14.02	12.01
Reduction in weight (%)	0.0	20.00	31.02
Fecundity (no)	4,106.0	1,834.00	1,544.00
Reduction in fecundity (%)	0.0	53.03	62.04
Relative fecundity (no/g)	238.0	140.00	127.00
Egg size (mm)	1.0	1.25	1.44
Hatching (%)	25.4	30.08	54.08
Survival as % of hatching	16.3	20.04	45.02
Net progeny production (no/♀)	669.0	374.00	698.00

with two pairs of balanced chromosomes are unable to produce viable progenies in large numbers, while the diploids and hexaploids with one and three pairs of homologous chromosomes could do it.

7.6 Interploid hybridizations

To contain exotics and transgenics, sterile hybrids are required in large numbers; the demand for 'broiler' triploids is also very high. For instance, 300 million crucian carp-common carp hybrid triploids are annually produced in China, as already stated. Interploid hybridization is the easiest and widely practicable method to mass produce the desired sterile triploids. Interploid triploids have been produced in many fishes. Table 40 lists the wide range of interploids generation in *M. anguillicadatus* with problems encountered by sperm of 4n male entering into the micropyle of 2n female. Li et al. (2006) designated it as 3n- and that of the reciprocal cross as 3n⁺. However, 3n- *Megalobrama amblycephala* survive better and grow faster.

Table 40 Interploid hybridization in fishes. Se = small eggs; Le = large eggs.

Interploid crosses	Ploidy (n) of the progeny
Misgurnus anguillicaudatus (Zhang and Arai, 1996)	
4n ♀ x 4n ♂	4
4n ♀ x 2n ♂	3
2n ♀ x 4n ♂	3
3n ♀ Se x 2n ♂	2
3n ♀ Le x 2n ♂	4
5n ♀ x 2n ♂	3
4n ♀ x 2n ♂ + HS	5
2n ♀ x 4n ♂ + HS	4
4n ♀ UV	2
4n ♀ Le + UV	3
4n ♀ x 2n ♂ + PS	5
4n ♀ x 4n ♂ + PS	6
2n ♀ x 2n ♂ + PS	3
3n ♀ Se x 4n ♂ + PS	4
2n ♀ Se x 2n ♂	2
Megalobrama amblycephala (Li et al., 2006)	
4n ♀ x 2n ♂	3⁺
2n ♀ x 4n ♂	3⁻

PS = Pressure shock and HS = Heat shock to retain polar body, UV = irradiated sperm

7.7 Unreduced gamete bank

To classify and catalogue the unreduced gametes of fishes, available information on unreduced diploid, triploid and tetraploid eggs (Table 41) and sperm (Table 42) is compiled. As a kind of data bank, this compendium will facilitate further research and exchange of gametes in promotion of commercial venture and operations. The following are notable : 1.

Table 41 Recorded multiploid eggs of fishes.

Classification of ploidy in eggs	References
I 2n unreduced eggs	
1a Bisexuals	
2n *Misgurnus anguillicaudatus*	Oshima et al. (2005), Itono et al. (2007)
3n *M. anguillicaudatus*	Momotani et al. (2002)
4n *Cobitis biwae*	Kusunoki et al. (1994a)
2n *Carassius gibelio*	Horvath and Orban (1995)
1b Unisexuals	
Poecilia formosa	Hubbs et al. (1959)
Menidia clarkhubbsi	Echelle et al. (1988)
Carassius auratus	Kojima et al. (1984)
Phoxinus eos neogaeus	Goddard and Dawley (1990)
Squalius alburnoides	Alves et al. (1999)
1c Hybrids	
3n (*Oryzias luzonensis* ♀ x 2n *O. latipes* ♂)	Sakaizumi et al. (1992)
3n (*O. latipes* ♀ x 2n *O. curvinotus* ♂)	Kurita et al. (1992, 1995)
2n (*Fundus heteroclitus* ♀ x 2n *F. diaphanus* ♂)	Davies et al. (1990)
3n (*Salmo salar* ♀ x 2n *S. trutta* ♂)	Galbreath and Thorgaard (1995)
3n (*S. trutta* ♀ x 2n *S. salar* ♂)	Johnson and Wright (1986)
3n (*C. auratus gibelio* ♀ x *Cyprinus carpio* ♂)	Cherfas et al. (1994b)
3n (*C. carpio* ♀ x 2n *C. auratus gibelio* ♂)	Cherfas et al. (1994b)
3n (*Lepomis gibbosus* ♀ x *L. cyanellus* ♂)	Dawley (1987)
3n (*L. gibbosus* ♀ x *L. macrochirus* ♂)	Dawley (1987)
II 3n unreduced eggs	
2a Bisexuals	
3n *M. anguillicaudatus*	Zhang and Arai (1999b), Morishma et al. (2008)
Carassius lansgdorfi	Umino et al. (1997)
Tinca tinca	Flajshans et al. (1993)
2b Unisexuals	
Poecilia formosa	Rasch and Balsano (1989)
Poeciliopsis monacha lucida	Schultz (1989)

Table 41 contd....

Table 41 contd....

Classification of ploidy in eggs	References
Carassius auratus	Kojima et al. (1984)
Cobitis complex	Vasilyev et al. (1990)
Squalius alburnoides	Alves et al. (1999)
III 4n Unreduced eggs *3a bisexuals*	
4n *M. anguillicaudatus*	Momotani et al. (2002)
4n *C. biwae*	Kusunoki et al. (1994a)
4n *C. auratus lansgdorfi*	Kobayashi et al. (1977)
3b unisexuals	
4n *Poecilia formosa*	Lampert et al. (2008)
4n *Cobitis* complex	Vasilyev' et al. (1990)
IV Simultaneous production of multiploid eggs *4a Haploid-diploid eggs*	
2n *M. anguillicaudatus*	Zhang and Arai (1999b), Arai et al. (2000)
3n (*C. auratus cuvieri* ♀ x *C. carpio* ♂) ♀ F_1x 2n *C. auratus cuvieri* ♂	Zhang et al. (1992b)
4b Haploid-triploids eggs	
3n ♀ *M. anguillicaudatus* ♂	Matsubara et al. (1995)
2n ♀ x 4n ♂ *M. anguillicaudatus*	Arai and Mukaino (1998)
3n ♀ *Squalius alburnoides*	Alves et al. (2004)
3n xenogenic crucian carp	Yamaha et al. (2001)
4c Diploid-triploid eggs	
2n ♀ x 4n ♂ *M. anguillicaudatus*	Arai and Mukaino (1998)
2n ♀ x 3n ♂ *M. anguillicaudatus*	Zhang and Arai (1999b)
4d Diploid-tetraploid eggs	
4n *Megalobrama amblycephala*	Liu et al. (2007a)
4e Multiple ploid eggs	
4n *M. anguillicaudatus* n, 2n, 3n, 4n eggs	Momotani et al. (2002)
Multiple 3n carp 1.5n, 2n, 3n eggs	Lu and Chen (1993)

Amazingly, many fishes are capable of simultaneously producing eggs and sperm characterized by different ploidies, and the combinations can be as many as 1X to 4X namely haploid-diploid, haploid-triploid, diploid-triploid, diploid-tetraploid and multiploid eggs as well as haploid, diploid, tetraploid, hexaploid, octoploid and aneuploid sperm as in 4n *Carassius auratus-Cyprinus carpio* hybrid. This is an area for profitable research. 2. Diploid eggs can be obtained by post-ovulatory ageing, as in some fishes like *Tinca tinca*. 3. The PGC sandwich of triploid crucian carp, transplanted to diploid hybrid sterile carp, can still simultaneously generate unreduced triploid eggs and reduced haploid eggs. 4. All these unreduced diploid, triploid and tetraploid eggs are produced through promeiotic endomitosis,

Table 42 Recorded multiploid sperm of fishes.

Species and its ploidy	Ploidy in sperm	Euploid	Relative DNA (n)	Reference
2n *Tinca tinca*	n	86 %	0.95	Linhart et al. (2006)
	2n	13 %	1.40	
	4n	1 %	1.85	
3n *Tinca tinca*	3n	69 %	1.10	Linhart et al. (2006)
	2n	34 %	1.50	
	4n	2 %	1.60	
3n *Tinca tinca*			1.47	Hulak et al. (2009)
Heteropneustes fossilis		86 %	1.00	Pandian and Koteeswaran (1999)
2n *H. fossilis*	n	94 %	1.00	
3n *H. fossilis*	n	19 %	1.00	
	2n	21 %	1.50	
	3n	12 %	2.00	
4n *H. fossilis*	n	2 %	1.00	
	2n	90 %	2.00	
3n *Rhodeus ocellatus ocellatus*			1.50, 1.90	Kawamura et al. (1999)
3n *Oncorhynchus mykiss*			1.50	Ueda et al. (1991)
3n *Ctenopharyngodon idella*			1.40	Van Enennaam et al. (1990)
3n *Gadus morhua*			1.60	Perruzi et al. (2009)
4n *Misgurnus mizolepis*			2.00	Nam and Kim (2004)
4n *M. anguillicaudatus*			2.00	Yoshikawa et al. (2008)
4n *Carassius auratus-Cyprinus carpio*	n / 2n	40 % / 49 %	1.00 / 2.00	Liu et al. (2001)
2n *Squalius alburnoides*			2.00	Alves et al. (1999)
2n *Oncorhynchus mykiss*			2.00	Ueda et al. (1986)
2n *Puntius conchonius*			2.00	Kirankumar and Pandian (2004c)
2n *Hemigrammus caudovittatus*			2.00	David and Pandian (2008)

except in *Carassius gibelio* (Horvath and Orban, 1995) and possibly in some others (e.g. Sun et al., 2007). In these cases, ejection of the first polar body is eliminated at the meiosis (Fig. 24). 5. Euploid sperm can be with 1.0, 1.5 and 2.0 units of genetic materials. However, the aneuploid nature in some them has been shown by DNA analysis (e.g. *Rhodeus ocellatus ocellatus*, Kawamura et al., 1999). 6. The triploid tench, and the triploid-tetraploid Indian catfish simultaneously produce different proportions of haploid, diploid and tetraploid sperm. Surprisingly, even the normal diploid tench male simultaneously produce haploid, diploid and tetraploid sperm. Hence there is ample scope to study some of these departures that occur during the process of gametogeniesis in fishes. It must also be stated that our knowledge on triploidy and polyploidy has not provided any new clue for a better understanding of sex determination in fishes, as the observed departures from expected sex ratio are mostly due to sex (female) dependent mortality.

In the loach *Misgurnus anguillicaudatus*, crosses between 4n ♀ x 4n ♂ results in survival of 20–67% progenies bearing either $X^1X^1X^2X^2$ or $X^1X^1X^2Y^2$/$X^1X^1Y^2Y^2$ genotypes. Conversely, genome addition by meiotic (followed by polar retention), as in *Oncorhynchus mykiss* or mitotic, as in *Megalobrama amblycephala* tetraploidizations inflicts very high mortality of females, resulting in female ratios of 0.07% in the former and 0.3% in the latter. Progressive mortality of females throughout the life spawn of triploids bearing $X^1x^1X^2$ or $W^1w^1Z^2$/$W^1Z^2z^2$ genotypes indicates that females are not able to tolerate genome addition, as much as males can do it. An enquiry into this may be rewarding and prove useful in monosex aquaculture.

Conversely, hybridization followed by triploidization or tetraploidization seems to enhance female survival; for instance, there were 44.3% fertile 4n females against 4.7% fertile 4n males among the F_3 hybrid progenies of *Carassius auratus* ♀ x *Cyprinus carpio* ♂ (Liu et al., 2001). Dr. S. Zou has indicated that similar sex ratio has been observed among F_1 hybrid between red *Carassius auratus* ♀ x *Megalobrama amblycephala* ♂. Likewise, the high survival reported for many salmonid hybrids may indicate the equal survival of triploid female and male (Table 31). Apparently, hybrid females are able to tolerate genome addition, as much as males can do it. Interestingly, both Liu et al. (2001) and Liu et al. (2007a) have reported that the fertile hybrid tetraploid females simultaneously produce eggs of different sizes (see Fig. 34, see also Fig. 4 of Liu et al., 2001).

8

Allogenesis and Xenogenesis

Animals generated by transplantation of homospecific Primordial Germ Cells (PGCs) or Spermatogonial Germ Cells (SGCs) are called allogenics and those by the transplantation to surrogate heterologous species are called xenogenics. During the 1990s allo-androgenic technique was employed to conserve fish species; progenies of the desired species were generated using a relatively simple technique involving preserved sperm and genome–inactivated (by UV-irradiation) eggs of a compatible surrogate species (Pandian and Kirankumar, 2003). From the beginning of this millennium, Primordial Germ Cells (PGCs) and Spermatogonial Stem Cells (SSCs) have been successfully used to generate allogenics and xenogenics. Some publications in this area have shown that sexual bipotency is retained by the PGCs (e.g. Takeuchi et al., 2003; Yamaha et al., 2003) and SSCs (Okutsu et al., 2006b, 2007), especially the SSCs retain the bisexual potency even after males have attained sexual maturity. The presence of PGCs has been observed from differentiated ovaries, testes and ovotestes of *Liza auratus*, *Serranus hepatus*, *Coris julis* (Brusle, 1988) and *Xiphophrus* sp (Flores and Burns, 1993); however, it is not known whether these fishes have retained the bisexual potency even after sexual maturity. In view of its revolutionary importance to sex determination, aquaculture and conservation of fishes, information on this frontier area of research on allogenics and xenogenics is rather elaborately described.

Shinomiya et al. (2002) generated chimeras of the medaka *Oryzias latipes* by transplanting male (X^1Y^2)- and female (X^1X^2)-specific PGCs into blastulae committed to differentiate into X^1X^2 and X^1Y^2 genotypes, respectively. From these experiments they concluded that it is the germ cell supporting cells (from the blastula) that determine the sex of the medaka progenies.

However, the experimental knockdown of *Cxcr4*, the chemo-PGCs-attractant receptor gene in medaka by Kurokawa et al. (2007) has shown that the somatic supporting cells are predisposed toward male development and the presence of PGCs is essential for sustenance of sexual dimorphism. Thus the PGCs and SSCs supporting somatic cells seem to have a complementary role in sex determination.

8.1 Primordial Germ Cells (PGCs)

The PGCs are the progenitors of germ cell lineage and possess the ability to differentiate into either oogonia or spermatogonia (Wylie, 2000). Hence they carry heritable information to the next generation. The origin of the PGCs can histologically be traced to the somitogenic stage or late blastula at best (Nagai et al., 2001). The PGCs are roundish, relatively larger (10–20 µm in diameter), cells with large nuclei, clear nuclear envelope and granular nuclear chromatin (Timmermans and Taverne, 1989, Fig. 35).

The *vasa* gene, which was the first molecular marker to identify the PGCs, was isolated and cloned, and proved that the gene is specifically expressed in germ cells of the zebrafish *Danio rerio* (Olsen et al., 1997; Yoon et al., 1997). This has made it possible to identify the PGCs including their presumptive ones, which otherwise could not be distinguished. The PGCs are inherited from a maternally supplied mRNA, located at the terminal regions of the early cleavage furrow (Yoon et al., 1997, Fig. 36); this maternal germplasm-like structure called the 'nuage' contains *vasa* and *nanos* genes, the latter is required for proper development of the PGCs, their migration to gonadal anlage and maintenance of germ line stem cells (Kobayashi et al., 1996). On surgical removal of the mRNA, no PGCs are formed in the resultant embryos of zebrafish and thereby Hashimoto et al. (2004) showed that the PGCs originate from the maternally inherited mRNA.

Tracing the germ cell lineage, as identified by *vasa*-mRNA during the embryogenesis in goldfish *Carassius auratus*, Otani et al. (2002) found the maternally supplied *vasa* mRNA localized to the first and second cleavage planes (Fig. 36). Subsequently, it is segregated asymmetrically to one daughter cell and the number of *vasa* mRNA harbouring cells remains at 4 until the mid-blastula stage with 1,000 cells. Then, the germplasm containing *vasa* mRNA is symmetrically segregated into daughter cells. At this stage, the zygotic *vasa* gene commences to be expressed.

Histological studies have shown that trout PGCs are recognized in the lateral mesoderm at neural stage, i.e. on the 7th day post-fertilization (dpf). At the 2-somite stage, most PGCs move into two clusters at the level of the first somite and migrate laterally (see Yoshizaki et al., 2002a) on the dorsal side of the embryo and remain adjacent to mesonephric ducts on the 20–30 dpf at the eyed stage of the embryo. They migrate further to reach the genital

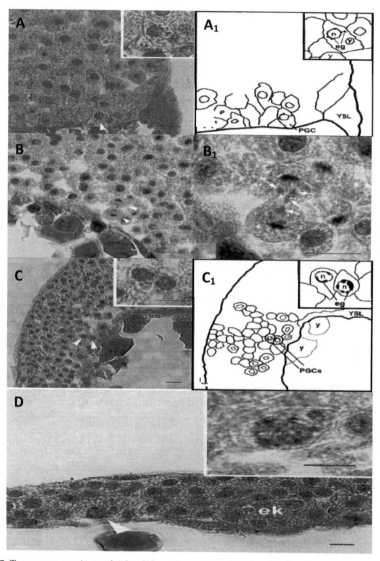

Fig. 35 Transverse sections of zebrafish embryos showing the development of PGC. A and A_1 at sphere stage (4 hpf), B and B_1 epiboly stage (6 hpf) C and C_1 shield stage (8 hpf), and D at bud stage (10 hpf). All inserts are higher magnifications. Arrow heads indicate PGCs. n = nucleus, y = yolk spherule, eg = eosinophilic granules, ek = embryonic keel. Scale bars indicate 10 μm (reproduced from Fig. 1, 2 of Nagai et al., 2001, Zoological Science, 18: 215–223).

Colour image of this figure appears in the color plate section at the end of the book.

ridges along the peritoneal wall and coalesce with the gonad supporting cells. On the hatching day (36th dpf), most PGCs have settled within the genital ridges (Takeuchi et al., 2003). Analysis of several mutant strains of

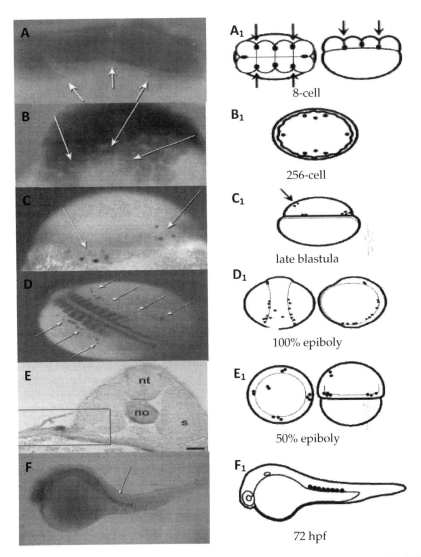

Fig. 36 Whole-mount *in situ* hybridization of cleavage (A,B,C) and segmentation (D, E, F) stages of goldfish embryos A_1, B_1, C_1, D_1, E_1 and F_1 are schematic figures of the corresponding developmental stages. Arrows indicate *vasa* signals. nt = neural tube, no = notochord, s = somite (reproduced from Fig. 1, 3 and 4 of Otani et al., 2002, Zoological Science, 19: 519–526).

Colour image of this figure appears in the color plate section at the end of the book.

zebrafish has shown that the migrations of the PGCs are regulated by the attraction of PGCs towards a somitic target, which later contributes to the formation of pronephros (Weidinger et al., 2002). The chemokine—SDF-1 secreted by presumptive gonadal regions is also known to act as chemo attractant to the migrating PGCs (Doitsidou et al., 2002). The trout PGCs are capable of undertaking the migration in the embryos between the 17th and 29th dpf (Okutsu et al., 2006a).

Extending the findings in zebrafish and goldfish to the rainbow trout, Yoshizaki et al. (2000a) have confirmed that the PGCs are indeed the progenitors of the germ cell lineage and possess the ability to differentiate into oogonia or spermatogonia. A reverse transcription PCR analysis of unfertilized eggs has proved that trout *vasa* is maternally inherited. For a better understanding of the molecular mechanism of determination and development of PGCs, they have isolated and characterized the trout *vasa* cDNA. Further they have also established cell-line-derived PGCs and developed a technique to visualize the PGCs *in vivo* using Green Fluorescent Protein (*Gfp*) gene. The *Gfp*-positive PGC is round, 20 µm in diameter and is with eccentrically placed large nucleus and a granular–rich cytoplasm confirming the previous histological description (see Fig. 35). The RT-PGR-analysis with *vasa* gene specific primers has shown that the *vasa* gene is expressed only in *Gfp*-positive cells, confirming that they are the PGCs (see Yoshizaki et al., 2000b).

Yoshizaki and his team have also generated a transgenic transcript with the *Gfp* gene driven by the *vasa* gene regulatory regions. The transcript carries a 4.7 kb fragment containing 5'-regulatory region, the 3'- untranslated region (0.6 kb) and 1.5 kb fragment of the 3'- flanking region (Fig. 37). Interestingly, a construct without the 3'-untranslated region of the *vasa* gene fails to label the PGC with *Gfp* suggesting that this region is essential for visualizing PGCs in allogenic trout (Yoshizaki et al., 2000b). This visualization system of employing *Gfp* is an ideal strategy to identify and trace the chosen PGCs in the surrogate and their progenies (Yoshizaki et al., 2002b).

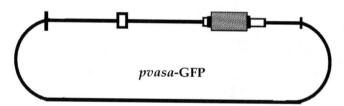

pvasa-GFP

Fig. 37 Structure of the *vasa-Gfp* construct. White and black boxes represent the untranslated region and the amino acid coding region, respectively. The shaded box indicate the *Gfp* gene (from Yoshizaki et al., 2000b).

8.2 PGC transplantation system

With regard to the developmental age of embryonic graft, Yamaha et al. (2007) have identified four types. Table 43 briefly summarizes the skill/ techniques required for these transplantations. Frequency of successful colonization of chimeras/allogenics ranges from 10 to 50% (see also Table 44). Heterochronic transplantation involving aged embryos or hatchlings as recipient may not yield allogenics, as the fully developed immune system of the recipient may inhibit the migration of PGCs. It is in this context that the choice of rainbow trout by Yoshizaki has many advantages: 1. Due to it large size (15 mm body length), hatched embryo or preferably called

Table 43 Age of transplantation graft used to generate germ line chimeras/allogenics.

Transplantation graft	Remarks
Blastomere	Sucked/surgically isolated pluripotent blastomeres can readily be transplanted into recipient blastula. Isochronic transplantation results in > 28% chimeras, of which 50% are donor-derived (e.g. Takeuchi et al., 2001).
Blastoderm	With asymmetric distribution of PGCs (Fig. 36), transplantation of lower part of blastoderm alone successfully generates upto 50% germ line chimeras (e.g. Yamaha et al., 2003). Microsurgical transplantation requires skill.
Genital ridge	Isolation and transplantation at more advanced stage of development leads to low alevin survival to ≈ 8%, of which 32% are germ line chimeras (e.g. Takeuchi et al., 2001). Salmonids have large embryos, and render the removal of genital ridge and its accommodation in relatively large peritoneal cavity. In smaller and more fragile species, the transplantation is a difficult task (e.g. Saito et al., 2008)
Heterochronic	Transplantation into older alevin, e.g. 45 dpf, PGCs fails to migrate, as the fry has already more fully developed immune system (e.g. Takeuchi et al., 2003)

alevin readily renders the removal of the genital ridge containing the PGCs with forceps using a dissection microscope. 2. The large space available in the peritoneal cavity of the recipient alevin can also accommodate the transplanted PGCs. 3. High fecundity of the trout allows thousands of alevins containing *Gfp*-labelled PGCs to be produced from a single spawning and 4. The availability of grey, albino and orange coloured phenotypes serve as readily recognizable markers.

To identify the most appropriate age combination of donor and recipient, Takeuchi et al. (2001) selected blastulae of different ages belonging to dominant orange coloured mutants as donors and recessive grey coloured wilds as recipients. The combination of donor cells (presumably containing PGCs) drawn from mid-blastula (2.5 dpf) and the recipient early blastula (1.5 dpf) generated the highest 32% donor-derived orange coloured chimeras (Fig. 38). PCR analysis of spermatozoa of mature males confirmed that the donor-derived blastula cells were indeed incorporated into the recipient and differentiated into spermatozoa.

Table 44 Colonizing efficiency, sex ratio and transmission frequency of allogenics/ xenogenics derived from transplanted PGCs.

Description	Data
I. Rainbow trout	
1. Transplantation of blastula cells	
a) PGC colonizing efficiency (%)	
i. at fry stage by colour marker	50
ii. germ line chimera	36
b) Sex ratio	0.15 ♀ : 0.85 ♂
c) Transmission frequency (%)	15
2. Transplantation of 5–10 PGCs from alevin to blastula	
a) PGC colonizing efficiency (%)	
i. at mature stage for ♀	14
ii. at mature stage for ♂	17
b) Sex ratio	0.55 ♀: 0.45 ♂
c) Transmission frequency (%)	
i. for ♀	2.6
ii. for ♂	4.1
3. Transplantation of 15–20 cryopreserved PGCs	
a) PGC colonizing efficiency (%)	
i. for 1-day preserved PGCs	10
ii. for 10 month preserved PGCs	21
b) Transmission frequency (%)	
i. from 1-day preserved PGCs-♂	8
from 5-day preserved PGCs-♂	6
ii. from 1-day preserved PGCs-♀	9
from 5-day preserved PGCs-♀	7
4. Transplantation of 20–30 PGCs into the masu alevin	
a) Colonizing efficiency (%)	17
b) Transmission frequency for ♂ C %	0.4
II. Pearl danio	
5. Transplantation of 1 PGC into zebrafish blastula	
a) Colonizing efficiency (%)	46
b) Transmission frequency (%)	
i. for males ♂	94
ii. for females ♀	66

Source: 1. Takeuchi et al. (2001); 2. Takeuchi et al. (2003); 3. Kobayashi et al. (2007); 4. Takeuchi et al. (2004); 5. Saito et al. (2008)

Takeuchi et al. (2003) visually traced the *Gfp*-positive 5–10 PGCs drawn from the genital ridges of the 35- dpf-old dominant orange coloured mutant trout and injected into the lower part of the blastoderm of the blastula of recessive grey coloured wild trout recipient. The transplanted PGCs were

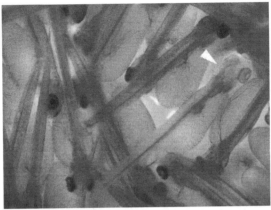

Fig. 38 Germ-line transmission of donor-derived orange coloured progenies of rainbow trout (from Takeuchi et al., 2003).
Colour image of this figure appears in the color plate section at the end of the book.

competent to migrate through the normal route and on time, and colonized the genital ridges of the recipient by the 30th dpf (Fig. 39); they appeared as a single cluster, indicating that the cluster originated perhaps from a single donor PGC. Yet, the PGC proliferated and differentiated into mature eggs and spermatozoa in the gonads of allogenics and produced normal progenies. Interestingly, the donor derived PGCs survived in a foreign gonad, owing to the relative immature status of the immune system until the time of hatching in many fish species (Manning and Nakanishi, 1996). Hence, the PGC transplantation technique developed by Takeuchi et al. (2003) may prove suitable for generation of allogenics in fishes.

PCR analysis of the spermatozoa drawn from PGC recipients showed that two (17%) of 12 mature males were allogenics. One of these males sired 115 (2%) orange coloured allogenic fry among the 5,825 F_1 progenies, clearly indicating that 98% F_1 progenies were drawn from the endogenous PGC of the recipient, and the co-existence endogenous and exogenous PGCs in the same gonad. Of the 115 fry, 63 (55%) displayed *Gfp* expression specifically in their PGCs. As the transferred PGCs were hemizygous, the transmission frequency of the donor- derived spermatozoa to the germline is estimated as 4% (Table 44). Of 14 mature female allogenics, two (14%) were donor-derived progenies. The transmission frequencies of eggs to the germ line of these two females were 3 and 2%.

To reduce contamination by gonadal somatic cells of the donor, Takeuchi et al. (2002) devised a flow cytometric separation technique to sort out *Gfp*-positive (see Fig. 5 of Yoshizaki et al., 2002a) from *Gfp*-negative somatic cells. With this purification, the colonizing efficiency still remained around 10% only (see also Kobayashi et al., 2004a, p. 139) .

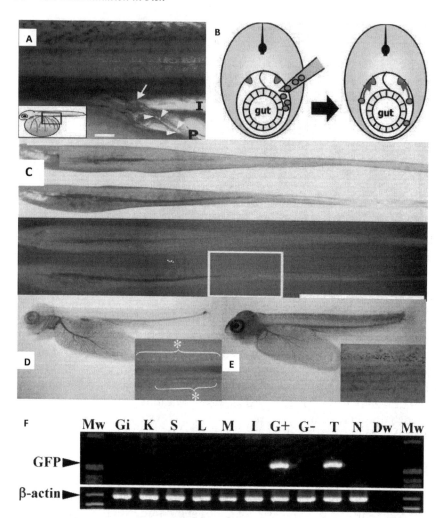

Fig. 39 Transplantation of *Gfp*-labelled PGCs into rainbow trout embryo. (A) Donor PGCs injected into the peritoneal cavity of an alevin. Arrow indicates the position at which the PGCs were injected. Arrow heads represent the PGCs. I = intestine; p = micropipette. (B) Schematic representation of the migration of donor PGCs on the peritoneal wall. (C) Bright view (top) and fluorescent view (bottom) of a pair of testes excised from a male parent. (D) Orange coloured progenies with *Gfp* expression, which were sired by donor-derived spermatozoa and (E) the wild control without *Gfp* expression (right). (F) PCR analysis of the allogenic DNA extracted from 180 days old fry : Gi = gill, K = kidney, S = spleen, L = liver, M = muscle, I = intestine, G+ = donor-derived gonad; G- = gonad control, T = adipose tissue of an allogenic, N = control adipose (from Takeuchi et al., 2003).

Colour image of this figure appears in the color plate section at the end of the book.

As fish eggs and embryos are not amenable for cryopreservation (Chao et al.,1987), a protocol to cryopreserve the limited number of available PGCs has become a necessity. Kobayashi et al. (2003a, 2007) developed a protocol to cryopreserve the trout PGCs carrying *pvasa-Gfp* transcript and restoration of allogenics. Genital Ridges (GR), the embryonic tissue containing PGCs of the trout (30 dpf), can successfully be cryopreserved in a medium containing 1.8 M ethylene glycol (EG). About 15–20 PGCs, cryopreserved for different durations (1 and 5 days) and then thawed, were transplanted into the peritoneal cavity of 32–34 dfp hatchling. The colonization efficiency was 10 and 21% for the 1 and 5 days cryopreserved donor PGCs, respectively. Progeny testing showed that the transmission frequencies of male derived from 1-day and 5-days cryopreserved PGCs were 7.8 and 5.6% , respectively. The corresponding values for the females were 9.1 and 6.5% (Table 44).

8.3 PGCs in hybrid embryos

To facilitate colonization efficiency of the transplanted PGCs in the gonadal ridges of the recipient, the endogenous PGCs may have to be minimized or totally eliminated (see p. 141). To achieve it, Yamaha et al. (2003) used a sterile hybrid of the cross between goldfish (X^1X^2) ♀ and supermale (Y^1Y^2) common carp *Cyprinus carpio* as recipient. They transplanted the lower part of the blastoderm of goldfish *Carassius auratus* (X^1X^2 donor), presumably containing PGCs, as a 'sandwich' between the upper and bottom parts of the blastoderm of the hybrids blastula (Fig. 40). In the sperm of the 'xenogenic goldfish', goldfish specific repetitive sequences alone were detected. Apparently, the PGCs presumed to be present in the 'sandwich' were competent to migrate normally in the hybrid embryo and it is likely that they may do so in alien recipients too. Secondly, the transplanted 'sandwich' presumably containing PGCs of the goldfish carrying X^1X^2 genotype differentiated into both males (60%) and females (40%). This observation confirms the earlier report of Kurokawa et al. (2007) that it is the presence of the germ cells (PGCs) that sustain sexual dimorphism in the F_1 progenies.

Triploid crucian carp *Carassius auratus langsdorfi* reproduce gynogenetically, producing mostly unreduced clonal triploid eggs (Yamashita et al., 1991). Transplanting biotin-labelled PGCs containing the lower part of 2-celled embryo of triploid crucian carp into the blastoderm of bisexual goldfish *C. auratus*, Yamaha et al. (2001) successfully induced germline allogenics. Both labelled and unlabelled PGCs were detected around the genital ridge of the recipient on the 10th dpf, indicating the competence of the donor derived PGCs to migrate in the recipient and to competitively colonize the gonadal anlage of the recipient.

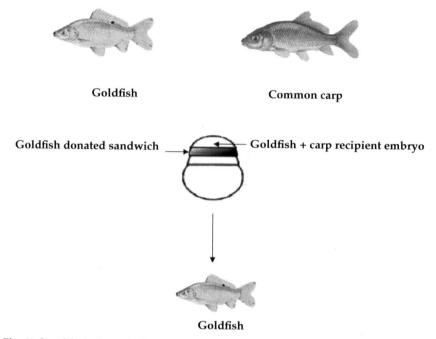

Goldfish   Common carp

Goldfish donated sandwich _____ → ← _____ Goldfish + carp recipient embryo

Goldfish

Fig. 40 Simplified schematic illustration of the transplanted 'sandwich' from blastoderm of fertile goldfish blastula into the blastoderm of sterile hybrid between goldfish ♀ and common carp ♂. Note the similarity in external appearance between the progeny and the fertile donor goldfish (compiled from Yamaha et al., 2003).

Spermatogenesis is reported to proceed almost normally in induced triploid crucian carp (Yamaha and Onozato, 1985). However, the PGCs of the donor triploid did not produce mature sperm. Hence no sperm of the donor could be detected in allogenic males. Five of the founder allogenic females produced simultaneously large and smaller eggs. On being activated by irradiated sperm, 30% larger eggs alone developed into normal triploid, and they were derived from the donor PGCs, although this mean value of 30% represents individual values ranging from 3 to 28% in four females and 93% in the fifth female. It is not clear whether this range of values indicates the wide differences in colonizing efficiency of the donor PGCs on the recipient gonadal anlagen. The larger eggs on being fertilized developed into tetraploids. When fertilized, 70% smaller eggs developed into normal diploid embryos, indicating that these 70% eggs were haploids. Besides, these diploids and triploids were melanin pigmented, as the donors were (see Fig. 3 of Yamaha et al., 2001).

8.4 PGCs in heterospecific embryos

To explore the scope for migration and colonization of the heterospecific genital ridges by the transplanted PGCs, Takeuchi et al. (2004) selected rainbow trout *Oncorhynchus mykiss* as the donor and masu salmon *O. masou* as the recipient. The former is a repeat spawner, i.e. iteroparous and a native of North America, whereas the masu salmon is semelparous, i.e. die after the first spawning at the age of 1+ and is found in East Asia. These two salmonids have been phylogenetically separated for 8 MYa (McKay et al., 1996). About 20–30 *Gfp*-labelled trout PGCs, derived from an alevin, have been transplanted into the peritoneal cavity of masu alevin and the presence of the *Gfp*-labelled trout PGCs has been detected in 17% of the recipient masu salmon, i.e. the colonizing efficiency is 17% (Table 44) and also with an average count of 7 donor-derived PGCs. Clearly, the transplanted trout PGCs have the competence to migrate and colonize the genital ridges of heterospecific recipient.

In the ovary of a 17-month old recipient female, the donor-derived PGCs were differentiated into oocytes (see Fig. 2 of Takeuchi et al., 2004). Progeny test and PCR analysis with *Gfp*-specific primers were also made. A cross between the recipient trout-masu xenogenic male and trout female generated 2,322 eggs; the eggs were hatched following two different incubation periods: the first batch 0.4% embryos hatched on the 34 dpf (early hatchlings) and all the others between 36 and 41 dpf (late hatchlings). RAPD analyses showed that all the early hatchlings were sired by the donor derived trout sperm and all other hybrids were sired by the masu sperm (see Fig. 2 of Takeuchi et al., 2004). Apparently, the recipient xenogenic possessed a 'hybrid gonad', with a smaller part allocated to the exogenous trout PGCs and a larger part to the endogenous PGCs. Notably, both the endogenous and exogenous PGCs co-existed and without interfering with each other, the PGCs developed and differentiated into trout and salmon sperm at ratio of 0.4 : 99.6. Encountering hormonal and micro environmental paradox, the exogenous trout PGCs generated spermatozoa *albeit* at a very low ratio. Yet this xenogenic technology may prove useful and economical both in terms of time and other inputs to generate trout seedlings at 1 year using surrogate masu instead of 2-year old trout.

In fishes, the large number of germ cells that are found in the gonad, originates from a few dozen PGCs only (see Yoshizaki et al., 2000b; Saito et al., 2008). Yoshizaki and his team have transplanted a minimum of 5–10 PGCs (Takeuchi et al., 2003) and a maximum of 15–20 PGCs (Kobayashi et al., 2007) to generate allogenic trout. Hence the 90–100 PGCs available in a donor trout can at best be transplanted into a dozen recipients only. Incidentally, Takeuchi et al. (2003) indicated the possible origin of donor-derived PGC cluster from a single transplanted PGC. It is in this context, the generation of xenogenics using a single PGC transplantation (SPT) method of Saito et al. (2008) becomes interesting. Saito et al. have heterochronically transplanted only one PGC, isolated from 10–15 somitic embryos, labelled with DsRed or *Gfp* of the pearl danio *Danio albolineatus* into the marginal region of early blastula of the recipient zebrafish *D. rerio*. From an unduly condensed publication of Saito et al. (2008), the following may be summarized: The donor-derived PGC migrated undergoing no or one mitosis from its place of transplantation to the gonadal anlagen in 46% embryos on the 6th dpf (Table 44). The migratory route and behaviour of the PGC in the recipient were similar to those of endogenous PGC. However, the PGC was unable to migrate, if the transplantation was made into the later, i.e. 25th somite stage of the recipient embryos, perhaps due to the maturation of immune system of the recipient.

PCR analysis with the *vasa* 3′ UTR species-specific primers confirmed that the gonad of xenogenic was donor derived, indicating that the single PGC alone contributed to the gonadal development (Fig. 41) of the xenogenic. Most interestingly, all the hatched xenogenics were phenotypic males with one normal sized functional testis. The development of endogenous PGCs was blocked by injecting dead end antisense morpholino oligonucleotide (*dead*–Mo). Incidentally, Slanchev et al. (2005) too did it to sterilize the zebrafish embryos; however, all endogenous PGCs-ablated zebrafish embryos developed into males, implying the dispensability of PGCs for the development of male somatic tissue but is required for the differentiation and maintenance of the gonad. However, inducing the morpholino-mediated knockdown of the chemo attractant (for migration of PGCs) receptor gene in the medaka, Kurokawa et al. (2007) found that the gonadal somatic cells are predisposed towards male development in the absence of germ cells; thus the presence of PGCs is essential for the sustenance of sexual dimorphism. Saito et al. too considered that the number of PGCs may possibly be more important for gonadal differentiation and perhaps sustain sexual dimorphism.

The organs namely brain, liver, muscle and fins of xenogenic pearl danio were of the zebrafish genotype (Fig. 41) and the egg size of the Estrogen (E_2)-treated xenogenic was also equal to that of zebrafish. But the xenogenics looked morphologically like the pearl danio. Thus the xenogenic was more of a chimera. Incidentally, the allogenic trout also resembled morphologically the donor rainbow trout (Fig. 39D, E) but almost all of its the internal organs , except the gonad, were of the recipient's genotype (see Fig. 39F). Hence the allogenic trout too was more of a chimera. Incidentally, the F_1 progenies of the fertility-restored hybrid, whose sterile blastoderm with transplanted 'sandwich' from fertile goldfish also looked more like the donor goldfish and carried the 'sandwich'-derived goldfish testis (Yamaha et al., 2003, see Fig. 40). The 3n and 2n allogenic crucian carp were also

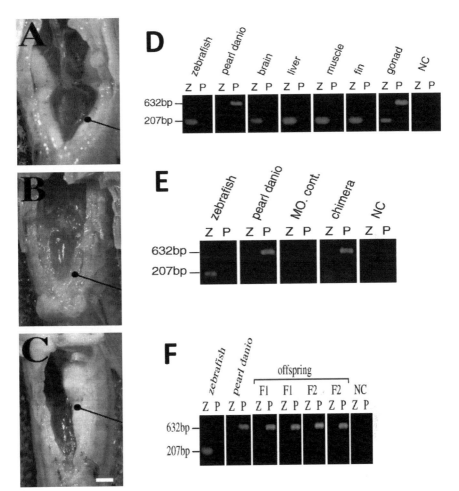

Fig. 41 Xenogenic pearl danio. A, B and C external appearance of gonads in males. A. Wild control, B. Sterile control injected with antigens *dnd* morpholino and C. Xenogenic pearl danio. (D) PCR genotyping of recipient, donor and chimeric fishes. The results of PCR with zebrafish specific primers are in the colum Z and those of pearl danio in the coloumn P. The zebrafish PCR product was expected to be 207 bp and the pearl danio 632 bp. (E) RT-PCR genotyping of gonads of the recipient, donor and their progenies. The gonad of the chimera had only pearl danio specific PCR product. (F) Genotyping of F_1 and F_2 progenies produced by the chimera. Z and P represent amplification of zebrafish and pearl danio specific primers, respectively (from Saito et al., 2008).

Colour image of this figure appears in the color plate section at the end of the book.

pigmented like the donor (Yamaha et al., 2001). Thus the xenogenic trout, goldfish-carp hybrid and pearl danio looked morphologically like their respective donors and possessed donor derived gonad *albeit* all the other internal organs were of the respective recipient's genotype.

It may be recalled that by the time the PGCs break the suspended mitotic division of their 4-cells, each of the 1,000 and odd blastomeres are committed to differentiate into one or other primary germ layer namely ecto-, endo- and mesoderms. When the exogenous PGCs are transplanted, the endogenous PGCs have already restarted mitotic division and commenced migration. Consequently, the expression of exogenous donor-derived PGCs is limited only to gonads of the allogenics, xenogenics. At that stage, the already committed recipients blastomeres have formed the primary germ layers including their respective derivatives namely intestine, liver, spleen, gills from endoderm, brain from ectoderm and muscles from mesoderm of the recipient. Incidentally, Nilsson and Cloud (1993) micro-injected the age-matched isolated blastomeres of diploid trout into triploid blastulae. Using ploidy as a marker, 27% of the trout progenies were found to be chimeric, and donor-derived cells were present in the brain, liver and blood cells, the derivatives of all the three germ layers. In the xenogenic pearl danio too, Saito et al. (2008) have shown that except for the donor derived gonad, all other organs were derived from the ecto-, endo and mesoderms of the recipient. Hence these allogenics and xenogenics remind one of a shadow of the 'Frankenstein monster'; with the recipient's brain and heart, they may physiologically and behaviourally function as the recipient and not as the donor; hence the grand idea of "producing tuna out of mackerel by transplantation of PGCs", however it needs to be cautioned that these xenogenic tunas may smell and taste like mackerel but not of tuna. The use of the allo-androgenic technique to produce "real tuna out of mackerel egg" may prove better and technically simpler (Pandian, 2003). Yet Yoshizaki's team may still realize their ambition of producing truly tasty tunas from surrogate mackerel using the sperm and eggs of chimeric founders (Fo). Indeed, they did produce F_1 and F_2 true trout using the surrogate masu salmon (Okutsu et al., 2007).

About 74% males and 66% Estrogen (E_2)-treated females of xenogenic pearl danio produced donor-derived sperm and eggs, respectively The E_2-treated females too possessed only one normal sized functional ovary (thanks to Dr. Taiju Saito for pers. comm.). Incidentally, the hybrids between pearl danio and zebrafish are naturally sterile. Fertile xenogenics pearl danios produced a large number of progenies over at least 10 reproductive cycles. Using these F_1 progenies, F_2 progenies were also obtained and their gonads were also donor-derived (Fig. 41).

By extending the same SPT method of transplanting a single PGC of goldfish *Carassius auratus* (Cyprinae, Cyprinidae) and loach *Misgurnus anguillicaudatus* (Cobitidae) into the blastoderm of the zebrafish *D. rerio* (Rasborinae, Cyprinidae) as a recipient, Saito et al. (2008) also generated xenogenic goldfish and loach. Morphological and histological observations confirmed that the transplanted PGC of goldfish and loach formed the germline but could be differentiated in to males only, as a consequence of the Mo-treatment by the respective recipients. However the male embryos were amenable to feminization by E_2 treatment. The xenogenics of the goldfish and loach also possessed only a single normal-sized functional testis in males and ovary in females (thanks to Dr Taiju Saito for pers. comm.). Conventional crosses between zebrafish and goldfish as well as zebrafish and loach do not produce hatchable hybrids (Table 45). However, the cross between xenogenic goldfish male and goldfish female did produce 94% hatchlings, indicating that the testis of GZ xenogenic goldfish male was donor-derived. These results were also confirmed by crossing xenogenic loach male (LZ) and loach female, again confirming that the testis of LZ loach was donor-derived. Clearly, the transplanted single PGC of goldfish and loach has the competence to migrate and colonize the genital ridges of their respective recipients and produce fertile sperm derived from the PGC of the respective donors.

Table 45 Hatchlings of the experimental crosses between goldfish and xenogenic goldfish (GZ) and loach and xenogenic loach (LZ), as compared to conspecific and heterospecific crosses (from Saito et al., 2008).

Crosses	Hatchlings (%)
GZ xenogenic ♂ x goldfish ♀	94
Goldfish ♂ x goldfish ♀	91
Zebrafish ♂ x goldfish ♀	0
LZ xenogenic ♂ x loach ♀	89
Loach ♂ x loach ♀	95
Zebrafish ♂ x loach ♀	0

Molecular clock analyses estimate that the phylogenetic separation between Cyprinidae (*D. rerio*) and Cobitidae (*M. angullicaudatus*) occurred at least 133.4 MYa, that between Cyprinae (*C. auratus*) and Rasborinae (*D. rerio*) 50 MYa (Steinke et al., 2006) and that between *O. mykiss* and *O. masou* before 8 MYa (McKay et al., 1996). Surprisingly, successful PGC tranplantations and generation of xenogenics are proved to be feasible crossing the taxonomic borders of family (loach to zebrafish), sub-family (goldfish to zebrafish) and species (pearl danio to zebrafish, rainbow trout to masu salmon) levels, despite respective antiquities of their phylogenetic separation from 8 MYa to 133 MYa.

On transplantation, the transplanted PGCs have to undertake indeed 'a long journey' by lophopodial movement (Braat et al., 1999) to the genital ridges and compete against the endogenous PGCs to colonize the gonadal anlagen. Hence the colonizing efficiency of the PGCs may depend on the age of the donor and recipient, and number and nature of the PGCs. The efficiency remains high at 36%, when midblastula cells are transplanted into an early blastula (Table 44). It perceptibly decreases to less than half of that value, when the donor is an alevin and the recipient a blastula (Takeuchi et al., 2003) or an alevin. With the use of cryopreserved 15–20 PGCs or heterospecific recipient, the efficiency remains within 10–17%, although the number of transplanted PGCs is increased to 20–30 to generate xenogenics. The efficiency remains as high as 46%, when a single PGC was transplanted into an early blastula of the zebrafish. Hence, the age of the donor and recipient appears to be a more important factor in determining the colonizing efficiency. Apparently, the potency of both the donor and recipient seems to dramatically decrease with advancing age, especially during the early embryonic period.

A higher transmission frequency is a reflection of a greater contribution of gametes arising from a much larger part of the 'hybrid gonad' of an allogenic. In the PGC transplantation system developed by Yoshizaki's group, the endogenous PGCs and exogenous transplanted PGCs have to compete against each other to colonize a small or larger part of the recipient's gonad. Consequently, the gonad becomes a 'hybrid' producing gametes of both the recipient and donor. Understandably, the exogenous PGCs are able to colonize only a smaller part of the recipient's gonad, as indicated by 3–15 % transmission frequency (Table 44). The recipient's gonadal part of the masu salmon colonized by exogenous and heterospecific trout PGCs is nearly 250 times smaller, as the transmission frequency of the xenogenic trout was just 0.4% only.

8.5 Spermatogonial Stem Cells (SSCs)

Although the PGC-transplantation and production of allogenic and xenogenic salmonids, danids, goldfish and loach may represent a breakthrough in reproductive biotechnology of fishes, the technique has the following inherent limitations: 1a) The PGCs are accessible only during the short period of embryogenesis, that too prior to the commencement of gonadal differentiation, b) in many fishes, especially in the temperate zone, breeding is limited to a short period only, c) some fishes like the orange roughy *Hoplostethus atlanticus* require a long period of 25 years to mature and d) others like the sturgeons breed once every 3–4 years (see Pandian, 2010), 2. The number of PGCs is limited, for instance, 90–100 in rainbow trout (Yoshizaki et al., 2002b; Okutsu et al., 2006a), 30 in carp (see Braat et al., 1999), 43 in goldfish, 25 in pearl danio, and 16 in loach (Saito et al., 2008), 3. The PGC transplantation has to be restricted to the early embryonic stage alone, i.e. prior to the development of the immune system (Manning and Nakanishi, 1996) and 4. For want of adequate space within the peritoneal cavity, the PGC transplantation may pose a serious technical problem, especially in small fragile fishes (Saito et al., 2008).

On the other hand, Spermatogonial Stem Cell (SSCs) transplantation, as conceived and developed for the mammalian system by RL Brinster, is regarded as a better one for following reasons: 1. A relatively a large number of SSCs are present and are available almost throughout the year and life time of a male fish, i.e. from hatching to death. Spermatogonia are indeed an eternal source of SSCs, and 2. SSC transplantation is not necessarily to be limited to embryonic stages. In fact SSCs have been transplanted into sterile and sterilized adult fish (see Lacerda et al., 2006). However, as many as 10^4 to 10^7 SSCs are required to generate an allogenic/xenogenic, whereas transplantation of 1 to 25 PGCs are more than adequate to do the same (see Table 44, 46).

Table 46 Comparison of methods and techniques employed for the SSC transplantation in the Nile tilapia by Lacerda et al. (2006) and rainbow trout by Okutsu et al. (2006b).

	Franca	Yoshizaki
Donor SSCs	(i) Testis + trypsin + collagenase + DNAse (ii) SSCs labelled by fluorescent cell linker PKH26 (iii) SSCs separated by percoll gradient centrifugation	(i) Immature testis digested by a variety of proteinases (ii) SSCs especially Type A isolated from dominant albino carrying pvasa-G*fp* iii) Labelled SSCs separated and concentrated flow cytometrically
Recipient	Adult ♂ sterilized by injecting busulfan	Alevins of recessive coloured surrogate
Transplantation	1 ml cell suspension with 10^7 SSCs injected by micro-pipette through urinogenital opening into spermatic duct	18,000 cells containing 10,000 SSCs microinjected into peritoneal cavity
Detection of donor deriven SSCs	Appearance of testis in previously sterilized adults	Albino progenies carrying G*fp* identified by PCR
Defect/Disadvantage	Could not rule out total exclusion of endogenous testis	Transplanted SSCs competed with recipient endogenous SSCs
Results	100% allogenics but sacrificed before progeny testing	Proper progeny testing; 5.5% ♂ and 2.1% ♀ allogenics identified

Ever since Brinster (2002) established the powerful and fascinating SSC transplantation system, the strides made by the mammalian biologists are indeed amazing and sometimes unbelievable, for instance, the successful colonization of mouse testis by primate SSCs. Notably, of 19 historical steps made in the development of SSC transplantation to generate xenogenics, 10 have been made by the Japanese (Sofikitis et al., 2003). Not surprisingly, the credit of establishing the hormone responsible for sustenance of SSCs in fishes too goes to the Japanese scientist Dr. Takeshu Miura. Before the topic

SSC transplantation is taken up, a brief description on the development in spermatogenesis in fishes is required.

8.6 Renewal of SSCs

Unlike somatic cells, the germ cells transmit genetic information to the subsequent generation; they are highly specialized to undergo meiosis and to produce eggs and sperm. Their transplantation to a suitable surrogate is presently known to generate progenies of desired strain (synogenesis, allogenesis) and species (xenogenesis). Hence development and banking of their cell lines hold a unique promise for the future. Not surprisingly, molecular studies on their source and development of techniques for their isolation and utilization have become a 'hot area' for research in basic and applied reproductive biology of animals including fishes. Germ cells can be obtained from early post-embryonic stages to the spermatogonia committed to become 'stem cells' in an adult male animal.

Gonadotrophins (FSH and LH) are the primary pituitary hormones involved in regulation of spermatogenesis. They act by stimulating growth factor release or the gonadal biosynthesis of sex steroids. These hormones, in turn, regulate renewal of Spermatogonial Stem Cell (SSC) and sperm production from spermatogonial cells (Schulz and Miura, 2002). In the absence of gonadotrophins, Sertoli cells produce a growth factor, an orthologue of anti-Mullerian hormone, which directs stem cell divisions to the self renewal pathway. On the other hand, gonadotrophin or 11-Ketotesterone (11-KT) stimulation induces spermatogonial proliferation.

It is now established that E_2 regulates the renewal of SSCs (Miura et al., 1996, 1999). 1. E_2 receptor expression is prominent in Sertoli cells. 2. In *Anguilla japonica*, mitosis of SSCs is promoted by the implantation of E_2 but is suppressed by tamoxifan, an E_2 antagonist (Miura and Miura, 2003). 3. In the Japanese huch *Hucho perryi*, *in vitro* treatment of testicular tissues with E_2 promotes renewal of SSCs (Amer et al., 2001). 4. A series of genes (eSRRs) encoding spermatogenesis related substances is known to show unique expression patterns during spermatogenesis. Electrophoretic transfer of some of these genes of *A. japonica* has evidence to confirm that eSSRs 34 gene induces SSC renewal and eSSR 21 induces spermatogenesis inhibiting substances (Miura et al., 2007). These studies of Miura and his team have established that E_2 is an indispensable 'female hormone' in the male and plays an important role in renewal of SSCs.

The complex process of spermatogenesis in fishes can be divided into three major phases (Norbrega et al., 2009): 1 Proliferative mitotic phase. The number of mitotic divisions that spermatogonial cell undergoes prior to forming spermatocyte is: 10 in Anguilliformes, 8 in Characiformes, 8 or 9 in Cypriniformes, 6 or 8 in Salmoniformes, 11 in Gadiformes, 7–10 in Perciformes, 10 in Beloniformes, and 14 in Cyprinidontiformes (Miura, 1999). 2. Meiotic phase, in which the genetic material is duplicated, recombined and segregated and 3. Spermiogenetic phase, during which spermatids are transformed into spermatozoa. The duration of spermatogenesis is species specific and lasts from 12 days in medaka *Oryzias latipes* to 14 days in *Oreochromis niloticus* 14.5 days in guppy *Poecilia reticulata* and to 21 days in *P. sphenops* (Vilela et al., 2003). These durations are relatively shorter than those reported for mammals. One Sertoli cell of fishes is able to support around 100 spermatids, which are about 10 times higher than those of mammals. Briefly, these characteristics of the piscine spermatogonial system provide a large number of sperm to sustain external fertilization in fishes. In fishes, spermatogenesis takes place within spermatocytes, which are completely surrounded by Sertoli cells. Due to incomplete cytokinesis, the cells resulting from the mitotic divisions of a spermatogonium remain interconnected by cytoplasmic bridges; these bridges are responsible for the synchronous development germ cells in a clonal fashion (LeGac and Loir, 1999). But in amniotic vertebrates, non-cystic spermatogenesis takes place. Apoptosis is an integral part of spermatogenesis, occurs in response to temperature, cell density, and eliminates abnormal germ cells. Hence, it plays an important functional role and may reduce the expected theoretical number of spermatic cells.

8.7 SSC transplantation system

The credit of establishing SSC transplantation system in fishes goes to Dr. G. Yoshizaki's group (Okutsu et al., 2006b) but is to be shared with the Brazilian group led by Dr. LR Franca; in fact both the groups have generated allogenics in 2006. They have employed different procedures to produce allogenic salmonid and cichlid, respectively. Yoshizaki's group has employed more sophisticated techniques and developed more reliable procedures to generate SSC derived allogenics (Table 46).

Lacerda et al. (2006) have considered that an ideal recipient for SSC transplantation is a sterile fish, in which endogenous spermatogenesis is excluded but the somatic elements of testis Leydig cells, Sertoli cells and others are functioning normally (see Fig. 42). Sterility of the Nile tilapia *Oreochromis niloticus* was achieved in males with injections of 18 and 15 mg busulfan/kg with a fortnightly interval between the two injections and by maintaining the fish at 35°C (Fig. 42).

They adopted the following procedure to label the germ cells. After testing at selected concentrations of PKH 26, 8 µl PKH/ml of C-diluent for 10 minutes was chosen, taking into consideration the cell viability as tested by tryphan blue and label intensity. The germ cell suspension was

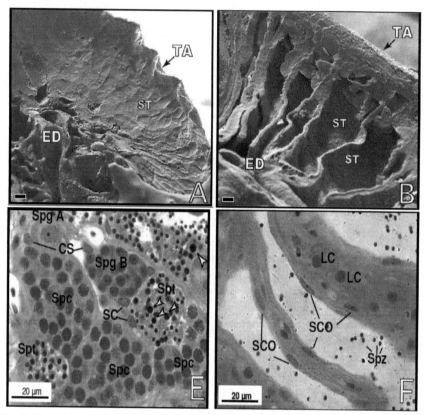

Fig. 42 Transverse sections of the matured testis of *Oreochromis niloticus*. A and B are scanning electron microscopic views of (A) fertile and (B) sterile testes. The latter was treated with busulfan. Light microscopic views showing the spermatogonial cysts in different phases of development in (E) fertile and (F) sterile testes. TA = tunica albuginea, ED = efferent tuctules, ST = seminiferous tubules, spg A and B = types of spermatogonia, spc = spermatocytes, spt = spermatids, SC = Sertoli cells, LC = Leydig cells. SCO = Sertoli cells only; Spz = sperm (from Lacerda et al., 2006).

subjected to percoll gradient by centrifugation (800 g) of the suspension for 30 minutes at 25°C (Fig. 43). Two ml cell suspension containing 5×10^6 SSCs/ml was injected (Lacerda et al., 2006) using a micropipette through urinogenital opening into the short (≈ 2 mm) spermatic duct with an opening diameter of ≈ 300 μm. Detection of the donor-derived PKH 26 labelled SSCs after 14 hours of transplantation in the seminiferous tubule was made by fluorescence microscopy. Of 36 recipients, 33 showed colonization and spermatogenic cysts originating from PKH26 labelled, transplanted SSCs. However, it was not clear to Lacerda et al. whether all these 33 recipients were allogenics or some of them had regenerated testis from a fraction of endogenous testes that defied sterilization by busulfan treatment (see also

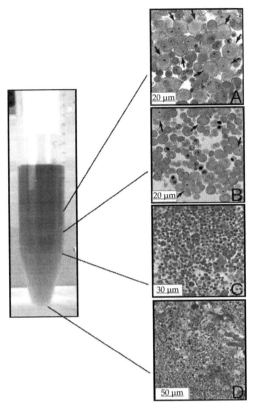

Fig. 43 Selection and isolation of germ cells from testis of mature *O. niloticus* using percoll gradient. Note the distinct separation of different cell types in the Falcon tube. In the upper 2 bands of A and B predominantly spermatogonial (stem) cells, as indicated by arrows, are separated. In the band C spermatocytes and spermatids are present and in the band D red blood cells and sperm are presents (from Lacerda et al., 2006).

Lowe and Larkin, 1975; Underwood,1986). These recipients were sacrificed 2 months after the SSC transplantation and were found to have developed new gonads. Lacerda et al. (2008) also claimed that they have generated three xenogenic *Cichla monoculus* using three recipient Nile tilapias.

On the other hand, Okutsu et al. (2006b) prepared spermatogonial cell suspension of testis excised from a dominant albino immature male carrying rainbow trout specific p*vasa-Gfp* sequence. Their experimental design of including a sterile triploid masu as a recipient resembles that of Yamaha et al. (2003), who used sterile diploid hybrid (goldfish x carp) as a recipient. Approximately, 18,000 cells containing 10,000 germ cells were intra-peritoneally microinjected into the recessively coloured alevins (32–35 dpf). The green colour of *Gfp* in the recipient was visually traced for the first 20 days and found that only the transplanted SSCs migrated to the genital ridges of the recipient, proliferated and colonized the gonadal anlagen (Fig. 44). The testicular SSCs differentiated into spermatozoa in the recipient males and fertile eggs in the recipient females. Surprisingly, the donor-derived SSCs also colonized ovaries of 37% of recipient females. The results obtained by Okutsu et al. (2006b) also gave evidence that the SSCs possess a high level of plasticity and sexual bipotency, even after the fish has reached sexual maturity and spermiated, and that SSCs have the partial

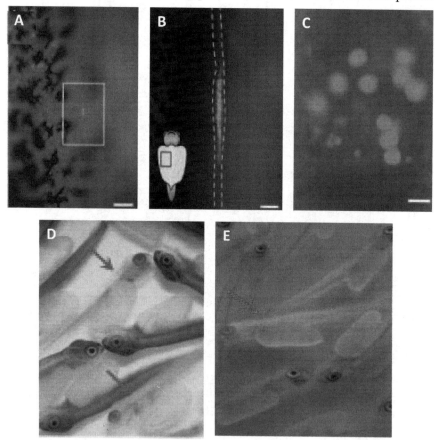

Fig. 44 Transplantation of donor *pvasa-Gfp* labelled SSCs into the recipient gonad of rainbow trout. (A) Fluorescent view of incorporation of donor SSCs into the recipient at 20 dpf, (B) Fluorescent view of proliferation of donor SSCs germ cells in a recipient testis at 7 months post-transplantation (pt) (C) Fluorescent view of donor SSCs–derived oocytes in a recipient ovary at 7 months pt (D) Donor derived F_1 albino progenies of a male recipient and (E) F_1 albino progenies from SSCs-derived female recipient (from Okutsu, T, Takeyuchi, Y, Takeyuchi, T, Yoshizaki, G, 2006b. Testicular germ cells can colonize sexually undifferentiated embryonic gonads and produce functional egg in fish. Proceedings of the National Academy of Sciences, USA, 103: 2725–2729).

Colour image of this figure appears in the color plate section at the end of the book.

functional competence of PGCs. Of the 26 male recipients which matured at age 1+, 13 (50% colonizing efficiency) were *Gfp*-positives. Among the F_1 progenies identified by PCR analysis for carrying the donor derived SSC, the transmission frequency was 5.46%. Of the 40 female recipients which matured at the age 2+, PCR analysis detected only 16 (40% colonizing effiency) as allogenics with green fluorescent germ cells and albino phenotype (Fig. 44). The transmission frequency of these allogenic females to F_1 progenies averaged to 2.14% with a wide range of 0.06% to 9.90%.

As indicated elsewhere, the number of mitotic divisions that a spermatogonial cell undergoes is seven times for salmonids, followed by two consecutive cycles of meiosis . Thus one founder spermatogonium can produce upto 512 spermatozoa. The mean number of incorporated SSCs in the trout of Okutsu et al. (2006b) was 4.6 ± 2.2. However, the donor-derived spermatozoa produced by an allogenic male were of the order of 20.1×10^7 and thereby implied that each spermatogonium underwent mitosis nearly 27 times instead of 7–8 times. Incidentally, allo-androgenic males of *Puntius conchonius* (Kirankumar and Pandian, 2004b) and *Hemigrammus caudovittatus* (David and Pandian, 2006b) also produced an order of greater number of spermatozoa (Table 26). Presumably, the number of mitosis undergone by a spermatogonial cell increased in these allo-androgenics too. It is not clear how induction of either androgenics or allogenics increased mitotic divisions in spermatogonial cells and produced an order of greater number of spermatozoa.

Okutsu et al. (2007) microinjected SSCs labelled with p*vasa-Gfp* construct drawn from the dominant orange coloured mutant heterozygous rainbow trout *Oncorhynchus mykiss* into the alevins of triploid sterile masu salmon *O. masou*. Incidentally, their experimental design included fertile diploid trout as a donor and a sterile triploid masu as the recipient and the design resembles that of Yamaha et al. (2003), who used fertile diploid goldfish and sterile diploid hybrid (goldfish x carp) as the donor and recipient, respectively. While the testes of the 2-year old triploid salmon remained immature and contained mostly spermatogonia, testes of the recipient salmon were mature and fertile. More than 84% of the recipient males produced milt. RAPD analysis showed that 50% progenies displayed the same fingerprint pattern as rainbow trout, indicating that the male triploid salmon produced donor-derived trout. At the age of 17 months, the recipient females bore donor specific green fluorescent vitellogenic oocytes, while the normal triploid had none. Between 2 or 3 years of age 10% of the recipients spawned fertile eggs. On fertilization of these eggs with milt of the recipient males, 90% of the eggs hatched. RFLP analysis of mitochondrial DNA of these alevins revealed that all the F_1 progenies carried trout's mitochondria. F_1 progenies matured and produced the trout F_2 progenies. Thus Okutsu et al. have established a surrogate broodstock technique for salmonids, in which SSCs can be transferred into sterile triploid xenogenic recipients to produce successive generations consisting of almost entirely donor-derived fish. They also confirmed that 45% trout SSCs frozen in cryomedium survived and could be used to xenogenic recipients.

Cell separation generally made by density-gradient centrifugation, as has been performed by Lacerda et al. (2006), may still suffer from contamination. To obtain pure SSCs, Yoshizaki's team employed two ingenious procedures. They developed (i) molecular markers *rtili, rt-scp3* and *rt-shippo* for spermatogonia, spermatocytes and spermatids of rainbow trout, respectively and (ii) the *vasa-Gfp* sequence present in the flow cytometrically (FCM) isolated *Gfp*-positive PGCs expressed abundantly in the gonads but its fluorescence began to diminish with the advancing spermatogonial process. Hence the FCM measured fluorescence intensity itself can be a powerful method to isolate cells of different stages of spermatogenesis. Based on *Gfp* intensities, tubicular cells were divided into five fractions. Of them *ritili* was used to isolate spermatogonia, which again contains A type and B type. The strongest *Gfp* fluorescence and the colonizing efficiency of the spermatogonial cells were used to identify and isolate testicular SSCs. The colonizing efficiency of the purified A type SSCs was 51 ± 9.0% producing colonies containing 10 ± 3.8 cells (Yano et al., 2008).

As the first step towards establishing a spermatogonial cell line for use in surrogate broodstock and cell mediated gene transfer systems, Shikina et al. (2008) optimized the culture conditions. They found that the spermatogonial survival and mitotic activity of rainbow trout improved in cultures containing Leibovitz's L15 medium (pH 7–8) supplemented with 10% fetal bovine serum at 10°C. Secondly, the elimination of testicular somatic cells promoted spermatogonial mitotic activity. Thirdly, besides insulin, trout embryonic extract and fibroblastic growth factor further improved the mitotic activity. The colonizing efficiency of 1 month old cultured SSCs was 52 ± 16.8%, against the efficiency of 78 ± 2.8% for the freshly purified SSCs.

In keeping view of their ultimate objective of using SSCs of the Pacific tuna *Thunnus orientalis* to mass produce the tuna seedlings using mackerel as a surrogate, Yoshizaki's team has isolated, cloned and characterized the *vasa* gene of the tuna (Nagasawa et al., 2009). Incidentally, the *vasa* gene is now known to express in the premiotic germ cells of zebrafish, rainbow trout, Nile tilapia, medaka, gibel carp, swampeel, goldfish, gilt-head sea bream and shiro-uo. Hence, the *vasa* gene is an excellent molecular marker for PGCs and SSCs of fishes.

The colonizing efficiency of SSCs ranges from 37 to 92% (Table 47) and is nearly 2–3 times higher than those reported for that of the PGCs (Table 44). Hence the SSC transplantation has another advantage over that of the PGCs. However, the transmission frequency remains almost equal in both

Table 47 Colonizing efficiency and transmission frequency of allogenics derived from transplanted SSCs.

Description	Data
I. Nile tilapia	
1. 10 x 10⁶ SSCs injected into sterilized adult male tilapia	
a) SSC/colonizing efficiency (%)	92
II. Rainbow trout	
2. 2.1 x 10⁴ SSCs into alevin, traced by *Gfp* marker & progeny testing	
a) SSC colonizing efficiency (%)	
i. for ♀	37
ii. for ♂	40
b) Sex ratio	0.46 ♀ : 0.54 ♂
c) Transmission frequency (%)	
i for ♀	2.1
ii for ♂	5.5
3. Transplantation of cryopresend SSCs	
a) Colonizing efficiency (%)	52
4. Transplantation of purified *ritili* specific A type SSCs	
a) Colonizing efficiency (%)	51–78*
5. Transplantation of trout SSCs to sterile masu	
a) Colonizing efficiency (%)	50

from: 1 Lacerda et al. (2006, 2008), 2. Okutsu et al. (2006b), 3. Shikina et al. (2008), 4. Yano et al. (2008), * see Shikina et al. (2008), 5. Okutsu et al. (2007)

the transplantation systems. Notably, it is not clear why there is such a wide difference in colonizing efficiency of 92% in diploid sterile adult tilapia and 50% in triploid sterile masu alevin.

8.8 Relevance of PGCs and SSCs

From their transplantation experiments involving sex specific PGCs into an opposite sex specific germ cell supporting somatic cells in the chimeric blastula of medaka, Shinomiya et al. (2002) concluded that sex is determined by the germ cell supporting cells rather than the PGCs. Table 48 briefly summarizes the gist of most important experimental observations relevant to the interplay between PGCs/SSCs on one side and the germ cell supporting somatic cells on the other. Incidentally, this summary also bridges the gap between academicians, who are in search of sex determining system/gene(s) and fishery biologists, who aim to generate chimeras and xenogenics. An incisive analysis of the results reported by different groups clearly indicates that the maternally derived PGCs and in turn, the PGC-derived SSCs channelize the development programme towards feminization and sexual dimorphism, and in the absence of PGCs or the PGCs-derived SSCs, or at the deficiency of PGCs as in Single PGC transplantation (SPT)-

Table 48 Results from experiments on transplantation of PGC or Morpholino nucleotide-treated embryos of fishes and suggested role played by PGCs and germ cells supporting somatic cells in sex determination in fishes.

Species used for experiment	Results	Observations
Danio rerio Hashimoto et al. (2004)	On surgical removal of PGC presumptive mRNA from unfertilized eggs, no PGC originated	Hence PGCs are maternally inherited cells. Do they carry XX genotype?
Oryzias latipes Shinomiya et al. (2002)	On transplantation female (XX) specific PGCs into blastula with germ cell supporting cells that were committed to become (XY) male, males were produced. Reciprocal chimeric transplantation confirmed the results	Germ cell supporting cells determine sex of progenies implying no role for PGCs in sex determination
Danio albolineatus (donor) *D. rerio* (recipient), Saito et al. (2008)	Development of endogenous PGCs was blocked by morpholino nucleotide (MO)	Only males with single but normal testis are produced
Danio rerio Slanchev et al. (2005)	MO-mediated blocking of endogenous PGCs	PGCs are dispensable for male development but are required for maintenance of gonad only
2n goldfish (donor); 2n sterile hybrid gold fish x carp (recipient) Yamaha et al. (2003)	On transplantation of PGC-'sandwich' into blastula, xenogenics containing donor-derived gonads produced at the ratio of 40% ♀♀ and 60 % ♂♂	In the presence of donor's PGCs surrounded by recipients germ cell supporting cells, allogenics produce males and females, implying a definite role for the PGCs
Pearl danio, goldfish, loach (donors) Saito et al. (2008)	MO-mediated blocking of endogenous PGCs, recipients received one PGC, but limited to only one cell each	Only male progenies with a single testis
Oryzias latipes Kurokawa et al. (2007)	MO-mediated knockdown of *Cxcr4*, the chemo-attractant (of PGCs) receptor gene	Germ cell supporting somatic cells are predisposed to develop towards male; hence confirm the results of Slanchev et al. and Saito et al. The presence PGCs is essential for sustenance of sexual dimorphism. Thereby implies that the maternally inherited PGCs induce development towards female

Triploid gynogenetic goldfish (donor); 2n bisexual crucian carp (recipient) Yamaha et al. (2001)	Induced germ line allogenic by transplantation of PGCs "sandwich". Male xenogenics were sterile, as in donor. Females were fertile but simultaneously produced haploid and triploids eggs.	Transplanted PGCs from triploid donor have overriding role in sex determination, and production of females
Oncorhynchus mykiss Takeuchi et al. (2003)	Transplantation of 10–15 PGCs from alevin to blastula produced allogenics at the ratio of 0.55 ♀: 0.45 ♂	Clearly implies a role for PGCs in sex determination
Oncorhynchus mykiss Okutsu et al. (2006b)	Transplantation of 10,000 SSCs into trout's alevin produced allogenics at the ratio of 0.46 ♀: 0.54 ♂	Clearly implies a role as the PGCs in sex determination
Oncorhynchus mykiss (donor); *O. masou* recipient Okutsu et al. (2007)	Transplantation of 10,000 SSCs from 2n diploid trout into alevin of triploid sterile masu produced xenogenic females and males of trout	Clearly implies the overriding (the germ cell supporting triploid sterile masu) role of SSCs in sex determination and production of ♀♀

the predisposed germ cell supporting somatic cells reverse the programme towards male development.

8.9 Surgical gonadectomy

Briefly, Yoshizaki and his group have demonstrated through a series of path-breaking contributions that the PGCs and SSCs can be transplanted

Table 49 Retention of bisexual potency by matured fishes, which underwent one or other gonadectomy.

Species	Reported observations	Reference
Betta splendens	65% surviving ovariectomized ♀♀ regenerated functional testes; crossed with normal ♀♀, the regenerated ♂♂ produced F$_1$ progenies, which sexually mature.	Lowe and Larkin (1975)
Oryzias latipes *O. latipes*	Treated with estrogen, ovariectomized ♀♀ failed to regenerate ovary Treated with estrogen, ovariectomized ♀♀ failed to regenerate ovary	Noble and Kumpf (1936) Okada and Yamashita (1994)
Carassius auratus	Treated with 11-Ketotestosterone, ovariectomized ♀♀ regenerated testis	Kobayashi et al. (1991)
Oncorhynchus mykiss	Gonadectomized neomales (X^{1}X^2) regenerated testis	Kersten et al. (2001)
O. nerka	Gonadetomized ♂♂ regenerated functional testis but on 'wrong site', i.e. adjacent to spermatic duct. Ovariectomized ♀♀ developed ovaries but with reduced number of eggs	Robertson (1961)
Oreochromis niloticus	67% of ovariectomized ♀♀ regenerated ovaries within 119 days	Akhtar (1984)
Ctenopharyngodon idella	Since Sonneman (1971) suspected the residual tissues adjacent to the removed gonad may regenerate as ovary or testis, the gonads including surrounding mesentery were removed from mature grass carp. Testes and ovaries were completely regenerated in 91 % ♂♂ and 38% ♀♀ gonadectomized grass carp	Underwood (1986)

into homologous and heterologous alevins, and allogenics and xenogenics can be successfully generated. From the point of sex determination, both the Japanese groups led by Dr. G. Yoshizaki and Dr. K. Arai have shown that the maternally derived PGCs and in turn, the PGC-derived SSCs have retained the bisexual potency even after sexual maturity and spermiation in the tested fishes, into which the PGCs/SSCs were transplanted. Incidentally, the results, involving sophisticated techniques and highly skilled experimental designs, confirm those obtained from conventional surgical experiments by zoologists during the period 1936 to 1994. The experimental observations listed in Table 49 clearly indicate that the bisexual potency has been retained in the tested sexually mature fishes. Notably, *Betta splendens*, *Carassius auratus* and *Oncorhynchus mykiss* regenerated the gonads in a direction opposite to the previous genetic sex, whereas *O. nerka*, *Oreochromis niloticus* and *Ctenopharyngodon idella* regenerated the gonads in the same direction, as that of the previous genetic sex. Furthering these classical experiments, especially from the point of understanding the mechanism of sex determination shall be rewarding. However, the transplantation experiments with the PGCs and SSCs have brought arguably unequivocal evidence that fishes have retained the bisexual potency even after sexual maturity and spermiation.

9

Sex Change and Hermaphroditism

In fishes, sex change is a unique but an intriguing phenomenon, especially from the point of genetic basis of sex determination. Instantaneously, this unique phenomenon poses a series of challenging questions regarding the presence of sex chromosome system, and sex determining genes. However, with the present understanding from the PGC/SSC transplantation studies that the bisexual potential is retained by fishes even after sexual maturity and spermiation, it may not be difficult to reconcile with the presence of 367 sex changing fish species. Consistently, these sex changers are considered as hermaphrodites and their presence is limited to tropical coral reefs. Nevertheless, simultaneous possession of ovarian and testicular tissues in gonads among cichlids (Peters, 1975) and sex changes in some cichlids (e.g. *Crenicara punctulata*, Zupanc, 1985) and anabantids (e.g. *Macropodus opercularis*, Koteeswaran and Pandian, 2011) are also known. Yet, the sex change in sequential and serial hermaphrodites is a natural genetic phenomenon but in cichlids and anabantids is perhaps a social pressure-induced phenomenon.

In hermaphroditism, both male and female reproductive function is expressed simultaneously or sequentially in a single individual. Being a strange phenomenon, hermaphroditism has evoked much interest and attention. The remarkable discovery of sex change induced by social factors (see Sadovy and Liu, 2008) accelerated the research output. The interesting contribution by Kuwamura et al. (1994) on serial sex change led to further research. Due to plasticity in sexuality and flexibility in the mating system (see Pandian, 2010), the detection of truly functional hermaphroditism in some 350–380 species has been indeed a task. Not surprisingly, the number of families, to which these hermaphroditic sex changers are included, has been changing from time to time; for instance, more than 34 families (Nakazona and Kuwamura, 1987), 23 families (Munday, 2001) and 27 families (Sadovy and Liu, 2008).

9.1 Extant of hermaphroditism

Whereas hermaphroditism is a rare phenomenon among higher vertebrates, the existence of several different patterns of hermaphroditism within fishes has evoked more excitement and curiosity. Three patterns of functional hermaphroditism have been recognized: simultaneous, sequential and serial (Fig. 45). Simultaneous hermaphrodite functions both as male and female at the same time or within a short span of time. A couple of species belonging to genus *Kryptolebias* (= *Rivulus*) are the only vertebrates, in which internal fertilization takes place. The externally gamete exchanging simultaneous hermaphroditism occurs in all species belonging to the genera *Serranus* and *Hypoplecturus* (see Pandian, 2010). The occurrence of this pattern of simultaneous hermaphroditism has also been recently confirmed in the following genera: *Displectrum, Pseudogrammus, Seraniculus,* (Serranidae), *Pagellus* and *Sparus* (Sparidae) (Sadovy and Liu, 2008).

Histological sections of the gonad of rivulus *Kryptolebias marmoratus*, the self-fertilizing hermaphrodite confirm the simultaneous presence of testicular and ovarian tissues (Fig. 46A). Harrington (1967) has reported the appearance of less than 5% primary males, when the eggs are incubated at 19°C or less. However, continuous exposure of post-hatchlings to temperatures below 20°C for 3–6 months fails to induce sex change (Harrington, 1971). Apparently, sex change can be induced only during the labile embryonic period. However, some hermaphrodites can be induced to develop into secondary males by rearing the hatchlings at 28°C. High water temperatures prevailing in the tropics may induce some hermaphrodites to change sex into secondary males (Fig. 46B); not surprisingly, the rivulus collections made by Tatarenkov et al. (2009) from the tropical Belize were comprised of 25% (secondary) males. Understandably, the sex differentiation is a very labile process in rivulus. From their cytogenetic studies, Sola et al. (1997b) also found 46 homomorphic chromosomes in the rivulus. The juveniles of the rivulus are amenable to feminization by hormone treatment; immersion of the juveniles in 1 ppb ethinyl estradiol for the first 28 days following hatching produced 100% females (unpubl. data; see Orlando et al., 2006). Hence the sex steroids, their receptors (e.g. estrogen receptor ER) and aromatase (Aro) play an important role in sex differentiation of the rivulus. In contrast to mammals, which have two *ERs* and a single aromatase gene, the fishes have three *ERs*, namely *ERα* and two isoforms of *ERβ*, *ERβ1* and

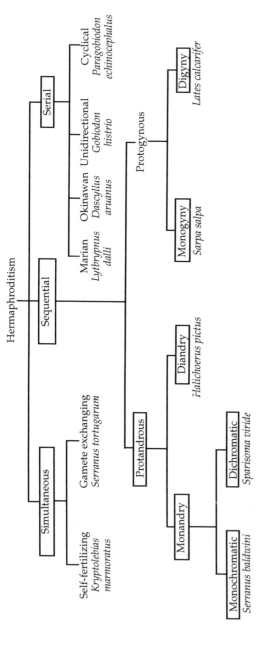

Fig. 45 Patterns of functional hermaphroditism in fishes.

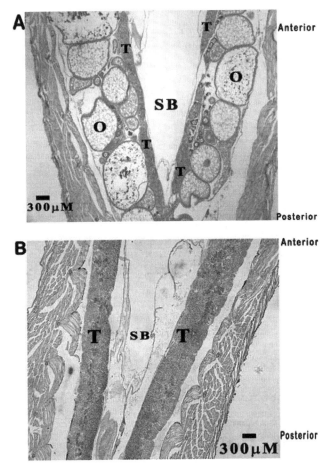

Fig. 46 Ventral sections of (A) hermaphrodite and (B) male rivulus *Kryptolebias marmoratus.*
T = testicular tissue, O = ovarian tissue, and (SB) swim bladder (from Orlando, EF, Katsu,
Y, Miyagawa, S, Iguchi, T, 2006. Cloning and differential expression of estrogen receptor and
aromatase genes in the self fertilizing hermaphrodite and male mangrove rivulus *Karyptolebias
marmoratus.* Journal of Molecular Endocrinology, 37: 353–365).

ERβ2 (Hawkins et al., 2000) and two isoforms of aromatase (Pellegrini et
al., 2005). The two isoforms *ERβ1* and *ERβ2* arose from a gene duplication
following the split between tetraploids and teleosts (Hawkins et al., 2000).
Orlando et al. (2006) have shown that the hermaphroditic rivulus have
stronger *ERα, ERβ, AroA* and *AroB* gene expression in their liver, brain and
gonad, as compared to that of males.

On the other hand, the largest adult individuals of gamete exchanging
hermaphrodites of *Serranus baldwini, S. fasciatus* (Fischer and Petersen, 1987)
and *Mycteroperca rosacea* (Erisman et al., 2008) may lose the female function
and become functional males. In fact the sex changed males of *S. fasciatus* and

S. baldwini maintain a harem, in which six to eight hermaphrodites are held as 'female' members of the harem (Petersen, 1987). Despite much interest and curiosity on these simultaneous hermaphrodites, no information is yet available on their cytogenetics and sex chromosome system.

Sequential hermaphroditism may be either protogynous or protandrous; in the former, a functional female changes sex to a male and the course of sex change is in the opposite direction, i.e. male to female in the latter; however, the sex changing event is limited to only once in the life time among the sequential hermaphrodites. On the other hand, it may occur more than once in serial hermaphrodites (Fig. 45). The protogynous species can be monandric, wherein all the males, called secondary males, arise only from functional females. In diandrics, there are primary males and secondary males; the primary males do not pass through the female phase. A similar dichotomy has also been recognized in the protandrous species and the corresponding terms, monogyny with only secondary females and digyny with primary females and secondary females. The monandrics are further divided into the monochromatics with all the matured females displaying the same dull body colour as the juveniles and the dichromatics in which the female changes her first body colour to the bright terminal one with increasing body size/age and as they approach transition to males (Pandian, 2010).

The serial hermaphrodite simultaneously maintains both ovarian and testicular tissues in its gonads but functions either as a male or a female. There is a progressive increase in the number of sex changing event from almost once in solitary Marian hermaphrodites to more than once in cyclical hermaphrodites. Depending upon the gonadal allocation to ovarian and testicular tissue in the gonads, there can be 'pure' males, male-biased hermaphrodites, female-biased hermaphrodites and pure females among the Marian hermaphrodites, which are solitary, and reported from four species belonging to the genus *Lythrypnus* (St Mary, 1998). Okinawan hermaphrodites are colonial; a colony, dominated by no master or mistress, consists of females, female-active hermaphrodites, and male-active hermaphrodites; a male may be present or absent in a colony. This hermaphroditism is represented by two species one each belonging to *Gobiodon* (Cobiidae, Cole and Hoese, 2001) and *Dascyllus* (Pomacentridae, Cole, 2002).

At present *Gobiodon histrio* is the only species representing the unidirectionally sex changing hermaphrodite; being a protogynous hermaphrodite, sex change is inducible in individuals from female to male and male to female, i.e. protogynous and protandrous sex changes within one species (Munday et al., 1998). Cyclical or bidirectional sex change is known to occur and can be induced in the same individual. The cyclically sex changing hermaphrodites are more prevalent than it was appreciated earlier. It is now confirmed to occur in 14 species belonging to genera *Pseudochromis* (Pseudochromidae), *Centropyge* (Pomacanthidae), *Cirrhitichthys* (Cirrhitidae), *Halichoeres, Labroides, Pseudolabrus* (Labridae), *Bryaninops, Gobiodon, Lythrypnus, Paragobiodon,* and *Trimma* (Gobiidae). In *Lythrypnus* spp and *Gobiodon histrio,* not much is known about the sex chromosome system.

9.2 Sex change in gonochores

If sex change is to increase the life time reproductive success of a fish, then there is no reason why the sex changing phenomenon is not extended to gonochores? A couple of species, in which sex change occurs, merits description (Oldfield, 2005). As in protogynous labrids, sequential hermaphroditism is typical of the checkerboard cichlid *Crenicara punctulata.* Using the magazine publications of D Ohm, Zupanc (1985) described the sex changing process in this cichlid. All checkerboard individuals raised in isolation pass through a female phase, but then change into males at the age 7–10 months. But in group-rearing, some 2-years old females differentiate into sub-males. These sub-males have the male's body colouration but are not strongly territorial. When the α-male is removed from the group, the sub-male begins to reproduce with the remaining females of the group. In a natural territory, the dominant α-female, on removal of the territorial male, changes sex and begins to mate normally with the remaining females. Carruth (2000) has undertaken proper experiments and confirmed the descriptions of Zupanc.

The sex changes in the adult Chinese Paradise fish *Macropodus opercularis* are more dramatic. Describing the social control of sex differentiation and sex change, Francis (1984) found that the male ratio decreased with increasing number of individuals in a stock. In our laboratory, the paradise fish matures at the age of 3 months and lives for a period of longer than 2 years. A female completes successive spawning cycles, each cycle within a period of 18–20 days and a male completes the bubble nest building or spermiation cycle once every 4–5 days. Reared individually, an immature juvenile displays a greater tendency to mature as a male. Experiments, facilitating visibility to each other but without allowing pheromonal interaction, induces sex change to males in three females out of 10. A male could also be induced to change his sex to female, when reared alone. Thus the paradise fish is capable of bidirectional sex change. Sex change is completed within 30 days, irrespective of the direction of sex change. Rearing in groups of 4, 9, 14, 19, 24, 29 and 34 with a single male each results in the appearance of 2–3 males, beyond the so called critical minimum density of 19 females per male in a tank. A sex changed female progressively increases her clutch size with successive spawning and almost achieves a clutch size with equal number of eggs, as that of a normal female in her fourth spawning cycle following sex change (Fig. 47). The trend obtained for hatching success of the sex changed female also runs parallel to that of fecundity versus spawning cycle (Koteeswaran and Pandian, 2011).

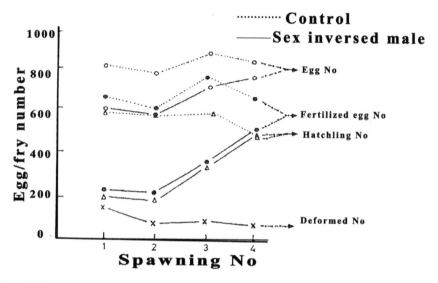

Fig. 47 Progressive compensation of fecundity and hatching by the sex changed female *Macropodus opercularis* during her successive spawning cycles (from Koteeswaran and Pandian, 2011).

9.3 Sex allocation theory

This theory offers an explanation for sex change in terms of which species should do it and in what direction it should be (Charnov, 1982). The general perception is that sex change is favoured, when (i) the reproductive performance of an individual is closely related to its size rather than age, as in protogynous hermaphrodites (see Pandian, 2010) and (ii) the relationship is different for the sexes. In this case, selection may favour genes that cause an individual to first be the sex, whose reproductive performance increases more slowly with age/body size and then changes to the other sex later (Allsop and West, 2004). The theory has made three predictions: 1. The sex ratio is biased towards the first sex to which individuals attain reproductive maturity. 2. It is less biased in species, in which some individuals precociously mature as the second sex, and 3. Protogynous hermaphrodites display a greater deviation from a sex ratio of equality than protandrous species. While the theory has given due consideration to the precociously sex changing individuals, it has not taken into consideration a certain percentage of females in protogynous species (e.g. *Pagrus pagrus*, Kokokris et al., 1999) and males in protandrous species (e.g. *Diplodus vulgaris*, Pajuelo et al., 2006), who may not change sex during the entire life span, a feature not uncommon among sequential hermaphroditic fishes.

Considering the sex ratio of 121 species spanning from annelid worms to fishes (76 species), Allsop and West (2004) did not find support for prediction 2, which states that sex ratio should be less biased in species, in which some individuals precociously mature as the second sex, especially with reference to protandrous species. However, the following merit consideration, especially from the point of fishes: 1. In comparison to the wide spread occurrence of confirmed protogyny in about 140 species belonging to 50 genera and 15 families, the incidence of confirmed protandry is limited to 25 species belonging to 17 genera and 6 families (Sadovy and Liu, 2008). 2. Among these limited number of protandrous species, only a few species have received attention, especially from the point of sex change and sex ratio. 3. As much as there are precociously maturing males, e.g. *Lates calcarifer* (see Pandian, 2010), there are also females, who may not change sex during the entire life time, e.g. *Diplodus vulgaris* (Pajuelo et al., 2006) and 4. Unlike the protogynous harems each consisting of one master male, and a few females, the protandrous harems each consists of one mistress female, one male and many subadults, who are inhibited to remain sexually immature (e.g. *Amphiprion alkallopsis*, Fricke and Fricke, 1977). Hence there is an inherent constraint in providing data on sex ratio of these harem owners (see Table 50). Apparently, these have not been taken into consideration by Allsop and West (2004). From the available data presented in Table 50, there is adequate supporting evidence for the correctness and acceptability of all three predictions of sex allocation theory. The scarid *Leptoscarus vaigiensis* is known to have become secondarily a gonochore but it has not 'forgotten' its original dichromatism, though it does not change sex. As sex change can be induced in some cichlids (e.g. *Crenicara puntulata*, Zupanc, 1985) and may be inducible in haremic gonochore species, some of data are also included in Table 50. These data for the gonochoric haremic species also confirm with those of sex changing hermaphrodites.

9.4 Sex changing age

Sex changing is considered as advantageous in protogynous species, in which size-reproductive success relationship differs between sexes so that at smaller size, it is effective to reproduce as a female and to facilitate more effective reproductive performance in a larger male. Then the female-biased sex ratio and reduced sperm competition are evolutionary consequences. In the female-biased species, females exhibit a relatively low reproductive success, as compared to the few individuals of the second sex (Chopelet et al., 2009). With increasing fishing pressure, the long living large serranids

Table 50 Sex ratio reported for different hermaphrodites and haremic and dichromatic gonochores (from Pandian, 2010).

Hermaphrodites/gonochores	Sex ratio	
I a Protogynous hermaphrodites 1. Monandric monochromatic	♂ ratio 0.31	♀♀/♂ 2.23
2. Monandric dichromatic	0.30	2.57
3. Diandric dichromatic	0.23	3.34
4. Haremic	0.22	3.55
I b Gonochores	♂ ratio	♀♀/♂
1. Dichromatic		
Leptoscarus vaigiensis	0.29	>5.00
*Macropodus opercularis**	0.16	3.00
2. Haremic		
Acanthurids, Ostraciids, Labrids, Pomacanthids, Cichlids	<0.25	>3.00
II Protandrous hermaphrodites	♀ratio	♂♂/♀
1. Monogynic *"Diplodus"* model	0.50	1.00
2. Monogynic *"Amphiprion"* model	0.31	2.23
3. Digynic *"Lates"* model	0.50	1.00
4. Digynic *"Gonostoma"* model	0.20	3.80
III Marian hermaphrodites	♀ratio	♂♂/♀
1. *Lythrypnus* spp	0.67	4.93

*Not dichromatic, Koteeswaran and Pandian (2011)

provide a good example; for every male, there are as many as 15 females in the black grouper *Mycteroperca bonaci* in Florida, 30 in Cuba and 77 in Mexico (see Pandian, 2010). Hence the expected consequences are: 1. Reduction in male ratio and 2. Reduction in gonadal allocation/investment. From an analysis, Molloy et al. (2007) came to the conclusion that the male ratio for the protogynous species can be 0.29, which is closer to that independently arrived at by Pandian (2010, see also Table 50). Using gonado somatic index (GSI) value of 14 protogynous species, Molloy et al. also estimated an average GSI value for the protogynous species as 0.006 for males and 0.032 for females (Fig. 48). The corresponding value is 0.03 for males and 0.057 for female gonochores. According to Molloy et al., the sex changing males seem to invest just 20% of that of the gonochores, while the females about 50% of that of the gonochores.

In a limited sense, the sex change may be defined as the proliferation of gonadal tissue of secondary sex and simultaneous degeneration of gonadal tissue of primary sex (Frisch, 2004). But the entire sequence of chronological events requires a longer duration; for instance, on removal of the territorial male from a harem of the bucktooth parrotfish *Sparisoma radians*, the transitional female assumed the male's body colour on the 9th day, commenced patrolling of harem on the 7th–14th day but successfully

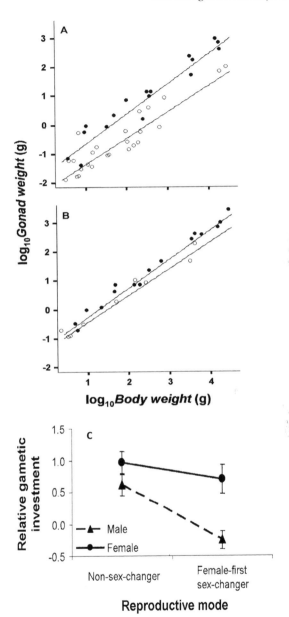

Fig. 48 Relationship between body weight and gonad weight for (A) males and (B) females of non sex-changing gonochoric (filled circles) and protogynic sex changing (open circles) fishes. Lines represent the best fit linear regression for each sex of each reproductive mode. (C) Relative gametic investment as a function of reproductive modes in sex changing and gonochoric fishes (from Molloy et al., 2007).

fertilized eggs only after the 12th–18th day (Munoz and Warner, 2003), suggesting the involvement of a cascade of genes each with a specific role to play in the entire sequence of masculinization of a female. This duration of 18 days required for the sexual transition provides a unique opportunity to investigate the mechanism of controlling the sex change and redetermination. Within an individual under experimental conditions, the duration of sex change spans from a minimum of 6 days in *Centropyge fisheri* to a maximum of 63 days in *Amphiprion alkallopsis* (Table 51). However, *A. alkallopsis* requires just 23 days under field conditions (Fricke and Fricke, 1977). With the respective longest transition durations, the protogynous *G. watanabei* and protandrous *A. alkallopsis* may serve as ideal models to undertake research for a better understanding of the mechanism of sex redetermination. Besides one or another morphological feature may also serve as a biomarker of the status of transition; for instance, the elongation of the second ventral fin ray in the peacock wrasse *Cirrhilabrus temmencki* (Kobayashi and Suzuki, 1990) and the shape of genital papilla in the Caribbean reef goby *Coryphopterus personatus* (Cole and Robertson, 1988). In the Japanese wrasse *Pseudolabrus sieboldi*, the colour of the analfin dramatically changes from yellow to red and is a useful biomarker showing the status of sex change (Ohta et al., 2003).

On the other hand, a longer duration is required even for 50% females of a population to change sex in the natural habitats (Table 51). The duration required is a minimum of about 4 months in *Thalassoma bifasciatum* and the maximum of longer than 36 months in the dusky grouper *Ephenephelus marginatus*. Not surprisingly, Chinese scientists have chosen *E. marginatus* to study the molecular mechanism underlying sex change and sex determining mechanism (Zhou and Gui, 2008). Interestingly, Allsop and West (2003) predicted that (i) fishes change sex when they are 80% of their maximum body size and (ii) 2.5 times their age at maturity. They also claim that the sex changing event occurring in 52 fish species supports these predictions, of course with 60 and 25 fold differences across species in maximum size and age at maturity. Incidentally, the sex change occurs already at 50% of the maximum body size and 2 times their age at maturity in *Hypoplectrodes maccullochi* (Webb and Kingsford, 1992). In *Amphiprion percula,* the sex change occurs already at 43% of the maximum body size (Madhu and Madhu, 2007). In view of the fact that the half banded sea perch and the clown anemonefish are not included in the development of the dimensionless theory by Allsop and West (2003) to predict the age at which sex change should occur, it is suggested that more species from the 360 and odd sex changing fishes are included in developing a theory to predict more precisely the age of sex change in fishes.

Table 51 Duration required for completion of sex change in fishes at individual and population levels (references from Pandian, 2010).

Species	Required duration
I Experimental (gonochore) at individual level (days)	
Macropodus opercularis[1]	30
II Experimental (hermaphrodites) at individual level (days)	
Bodianus rufus	7–10
Labroides dimidiatus	17
Halichoeres melanurus	14–21
Sparisoma radians	12–18
Genicanthus watanabei	25
Apolemichthys trimaculatus	25
Centropyge vroliki	10–16
C. fisheri	6
C. acanthrops	8
C. bicolor	<20
Coryphopterus personatus	20
C. glaucofraenatus	20
Halichoeres trimaculatus	38–42
Lythrypnus zebra	14
Amphiprion alkallopsis	23/63
III Natural at population level (months)	
Thalassoma bifasciatum[2]	≈ 4
Coris julis[3]	≈ 8
Pagrus pagrus[4]	≈ 20
Mycteroperca bonaci[5]	24
Hypoplectrodes maccullochi[6]	24
Epinephelus tauvina[7]	≈ 18
E. marginatus[8]	≈ 36

1. Koteeswaran and Pandian (2011), 2. Shapiro and Rasotto (1993), 3. Lejuene (1987), 4. Kokokris et al. (1999), 5. Crabtree and Bullock (1998), 6. Webb and Kingsford (1992), 7. see Zhou and Gui, (2008) 8. Kuwamura et al. (2007).

9.5 Genetic or environmental?

Coral colonies are a limited resource and only corals above a certain size can support a breeding pair of fishes (Hobbs and Munday, 2004). In sex changing hermaphroditic gobies, which obligately dwell in corals, an unpredictable patchy resource, sexual maturation and direction of sex differentiation may be under environmental social control (Charnov and Bull, 1989). Social interactions may suppress/delay or induce/advance sexual maturation/ sex change (Hoffman et al., 1999). By manipulation of pair combinations, Hobbs et al. (2004) demonstrated that in the presence of a sexually mature partner, most (91%) of the paired juveniles of the coral-inhabiting goby *Gobiodon erythrospilus* in smaller size classes were mature (Fig. 49), while more than half (48%) of the solitary gobies continue to remain juveniles. Following a period of 42-day stay with either a mature male or a mature female, the juveniles matured into a female or a male (Fig. 49). Thus Hobbs et al. (2004) showed that the presence of a mature partner within a coral and its sex induce sexual maturation and direction of sex differentiation. This finding also confirms the first report on similar sexual maturation into male or female, when reared with a larger female or male in free-living (but not space-limited, as in *G. erythrospilus*) serranid *Cephalopholis boenak* (Liu and Sadovy, 2004a). Similar direct differentiation to the final sex, i.e. female has also been observed in protandrous anemonefishes *Amphiprion bicinctus* (von Brandt, 1979) and *Premnas biaculeatus* (Wood, 1986). Perhaps PCR analysis of the genomic DNA of the juvenile and the social pressure-induced males and females could have conclusively shown whether the sex in these juveniles is genetically determined or not and what are the true genetic sexes of these social pressure induced males and females. For PCR analysis has revealed that the unexpected female progenies sired by the supermales (Y^1Y^2) and androgenic males (Y^2Y^2) of the rosy barb *Puntius conchonius* are indeed genetic males but functions as phenotypic females (Kirankumar et al., 2003). This finding has also been confirmed in *Hemigrammus caudovittatus* (David and Pandian, 2006b), Hence it is likely that the sex of the goby *G. erythrospilus* is genetically determined and that the goby becomes a phenotypic male or female, depending on the sex of its respective partner in the coral. PCR analysis to determine the true genetic sex of this social pressure- induced sexually maturing and sex differentiating fishes may prove to be a real path-breaking research.

Incidentally, Liu and Sadovy (2004a) also induced sex change in *Cephalopholis boenak* by maintaining one mixed group of one male plus two females and paired group one male and one female. Following the removal of males in these groups, sex change did not occur in the entire paired group and but it did occur in the entire mixed group, suggesting the need for the presence of a smaller female to induce sex change. This observation is similar to that of Ross et al. (1983), who found that it is sex ratio rather than body size that induces sex change. Hence, a brief discussion may be relevant to know the role of body size on sexual maturation and sex change in the context of genetic and/or environmental role in sex determination. Both in haremic and non- haremic hermaphrodites and gonochores as well, body size seems to play a role in sexual maturation and sex change in some fishes; for instance, in gonochoric cichlid *Cichlasoma citrenellum* larger and smaller juveniles differentiate into males and females, respectively (Francis and Barlow, 1993). In the haremic parrotfish *Sparisoma radians*, it is the largest female that changes sex to male following the removal of territorial

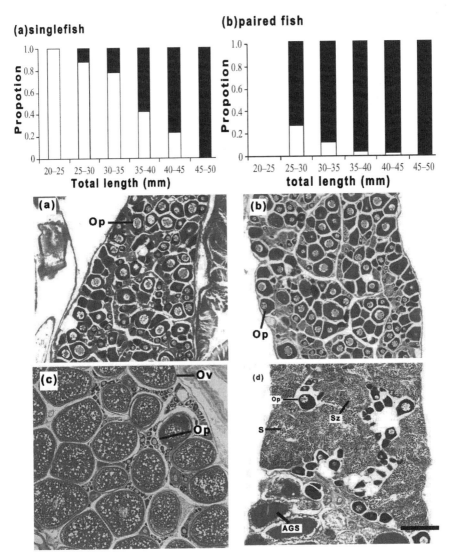

Fig. 49 Upper panel: The proportion of immature (open square) and mature (closed square) individuals in each size class for (a) single fish and (b) paired *Gobiodon erythrospilus*. Lower panels: Longitudinal section from (a) a typical juvenile (immature female) prior to experimental manipulation, (b) a juvenile (immature female) after remaining as a single for 42 days, (c) a juvenile paired with adult male for 42 days and (d) a juvenile paired with an adult female for 42 days. Op = previtellogenic oocytes, Ov = vitellogenic oocytes, S = spermatocytes, Sz = spermatozoa, AGS = accessory gonad structure. Scale bar = 0.1 mm (With kind permission by The Royal Society, London; from Hobbs, J-PA, Munday, PL, Jones, GP, 2004. Social induction of maturation and sex determination in a coral reef fish. Proc R Soc, London, 271B: 2109–2114).

Colour image of this figure appears in the color plate section at the end of the book.

male (Munoz and Warner, 2003). On the other hand, there are others, in which body size seems to play no significant role in sex change; even the last remaining female in a haremic *Pseudoanthias* (= *Anthias*) *squamipinnis* changed sex following serial male removal (Fishelson, 1970). Irrespective of body size, isolated females of the reef goby *Coryphopterus personatus* (Cole and Robertson, 1988) changed sex. These examples clearly indicate the less significant role of size in induction of sex change. Incidentally, there are many hypotheses that explain either the role of body size or sex ratio in induction of sex change in haremic hermaphrodites (see Pandian, 2010) but these hypotheses are inadequate to explain the role of body size in induction of sexual maturation and sex change in non-haremic hermaphrodites and others. Lawrence et al. (2008) obtained a strain of the zebrafish *Danio rerio* (from Tubingen, Germany) that were originally wild catch from India; placing single males and females in mating cages, they found that a higher proportion (76%) of F_1 female progenies among the high ration fed, fast growing groups than in low ration-fed, slow-growing groups (51%). Interestingly, they also noted that cross breeding resulted in the production of 84–93% females but only 45% females in the pure-bred groups. In the roach *Rutilus rutilus*, Paull et al. (2009) found that there were fewer female progenies (19%) among the low-ration-fed, slow growing groups than among the high ration-fed, fast growing groups (36%). The abundance of females among the high ration fed fast growing fishes may be due to the fact that fecundity in females is limited by body size (e.g. Naddafi et al., 2005; see also Pandian, 2010) and that the males are able to reproduce at a significantly smaller body size than females (e.g. Lawrence et al., 2008). Hence these studies clearly indicate the importance of body growth on sex differentiation in gonochoric fishes. Clearly, the area requires research especially from endocrinologists, who are pursuing endocrine mechanism of sex change in hermaphrodites.

In diandric species, secondary males are derived from females following sex change. But the primary males are hatched as males and are considered not to undergo sex change during their life time (Reinboth, 1970). Primary males and secondary males are readily distinguishable by size as well as by their respective dull and bright body colours. Anatomically and histologically also, their testes and gonadal systems are distinctly different (Lo Nostro and Guerrero, 1996). Exceptionally, the testis of the primary of males of the rockcod *Cephalopholis boenak* develops from a bisexual gonad with ovarian lumen, known to be a relic of the female (Liu and Sadovy, 2004b). Consequently, the testicular structure of primary males is similar to that of secondary males. Nevertheless, differentiation of the testis of primary males in other diandrics including the bluehead wrasse *Thalossoma bifasciatum* occurs prior to the appearance of any distinctive ovarian features in the gonad (Shapiro and Rasotto, 1993). These striking anatomical differences have been presumed to be the result of genetic determination. An alternative consideration is that individuals possess environmentally sensitive development reaction norms (Suzuki and Nijhout, 2006) that determine their capacity to become either primary males or females (Munday et al., 2006a).

From an interesting field survey in the Caribbean Island of St Croix, Munday et al. (2006b) found only 5 to 14% primary males of the bluehead wrasse *Thalassoma bifasciatum* in the reefs with small populations of 44 to 55 individuals during May 2004 but 13–58% primary males in the reefs with larger populations of 105–137 individuals during September 2004. Only 6% of the juveniles matured into primary males, when reared in isolation, irrespective whether juveniles were brought from the reefs with larger or smaller populations. These observations suggest that the bluehead wrasse may differentiate into females or primary males, depending upon the female population density of the reef, in which they happened to settle, i.e. the social pressure-induced direction of sexual maturation, as has been reported for the obligate coral dwelling monogamous goby, *Gobiodon erythrospilus* with limited home range and powers of mobility. Contrastingly, the wrasse maintains a large harem with many female members and the males are both pair-spawners and group spawners (Marconato et al., 1997); hence the wrasse is expected to have more males, especially primary males. Incidentally, even in the rockcod *Cephalopholis boenak*, in which a male maintains a small harem with one or two females and only pair-spawns, social pressure-induced sexual differentiation of larger individual is in the direction of males and equal sex ratio of 1♀ : 1♂ is maintained in most cases (Liu and Sadovy, 2004a). While these observations may go against the conclusion of Munday et al. (2006b), it must be stated that all these interesting observations reported by Hobbs et al. (2004), Liu and Sadovy (2004a) and Munday et al. (2006b) are clearly the social pressure- induced sex (maturation) differentiation rather than sex change.

In fact, the first discovery of sex change in primary males of *Halichoeres trimaculatus* was reported by Kuwamura et al. (2007) from their field-cum-aquarium observations. From marking and recapture of this diandric three spot wrasse, one primary (= initial phase IP) male of 9.6 mm size was found in the summer of 2002 to participate in group-spawning and to streak and sneak, typical of a primary male. Having grown to 13.6 cm by the spring of 2003, it was found to have a swollen belly with urinogenital papillae, typical of a female and to pair spawn with a territorial male (TP). Thus, the first discovery of a primary male changing sex to female was made on this single individual in the field. In the aquarium, five primary males (4.8 cm –10.4 cm) underwent protandrous sex change to become functional females, of which the largest female reverted to protogynous sex change and became a primary male. Incidentally, this three spot wrasse is a cyclical sex changer, i.e. female changes sex to male and then revert back to female. Hence the primary male is also a cyclical sex changer. Incidentally, sex change in primary males occurred only in the tank with a few TP males and females larger than the TP males. Such protandrous sex change of primary males did not occur in other tanks with a similar sex combination. Hence the precise social situation that may induce protandrous sex change in primary males remains to be elucidated.

Search for detection of sex chromosomes in sex changing hermaphrodites has however, not yielded any success, though the search has been very limited. Undertaking chromosome analysis, Ruis-Carus (2002) has found no difference in chromosome number or morphology between female and sex changed males in the monandric grouper *Epinephelus guttatus* and diandric protogyny *Thalassoma bifasciatum*. Nevertheless, there are strong evidences for the presence of sex determining genes and retention of bisexual potency in sex changing hermaphrodites. The Chinese investigations have clearly shown the involvement of several genes in sex determination in groupers (Zhou and Gui, 2008). In the diandric protogynous hermaphrodite *Synbranchus marmoratus*, a single gonium generates germ cells of both sexes (Lo Nostro et al., 2003). In unidirectionally sex changing hermaphroditic goby *Gobiodon erythrospilus*, the sex changed individuals exhibit little or no remnant tissue of the alternative sex (PL Munday pers. commun. to Sadovy and Liu, 2008). Hence the genetic determination of sex and sex change may remain unquestionable, until such a time that the genetic sex of the juveniles of such species, in which the environmental role of sex determination is claimed, and is proved by PCR analysis of their true genetic sex.

10

Unisexualism and Reproduction

Approximately 80 distinct unisexual biotypes of fishes, amphibians and reptiles are presently recognized (Alves et al., 2001). However, only eight unisexual fish species are known (Table 52). Unlike amphibians and reptiles, all the unisexual fishes are readily amenable for rearing and experimentation. In fact Hubbs and Hubbs (1932), the first discoverers of unisexual vertebrate, have reared and experimented with the Amazon molly *Poecilia formosa* for 15 years. Not surprisingly, a relatively large volume of literature is available on the unisexual fishes (Pandian, 2010).

Essentially, all the unisexual fishes are hybrids between two or more biparentally reproducing sexual species (Table 53). They have arisen from a particular combination of genomes that have disrupted oogenic meiosis to produce unreduced diploid, triploid or tetraploid eggs without having undergone recombination and reduction in ploidy. Assuming such hybrids are viable and fertile, an 'oogenic rescue mechanism' is rapidly fixed, when demographic advantage also arise with female reproduction. Consequently, the most widely prevalent mode of reproduction in these unisexuals is gynogenesis (Fig. 1). Gynogenesis is strictly a clonal mode of reproduction that faithfully replicates the maternal genotype. Unlike in parthenogenesis, the gynogenesis requires sperm from one or more bisexually reproducing ancestral parental species to initiate embryogenesis in their unreduced eggs. This has two implications: 1. More than the meiogynogenic gonochores, unisexualism provides an opportunity to investigate the independent role of sex determining gene(s) of a female, and 2. The indispensable sperm dependence from sexual ancestral sympatric population prohibits the unisexuals to occupy new niches, which are markedly different from the niches of the sexual parent. Thus the geographical distribution of the unisexuals is effectively limited to certain isolated niches alone (Table 52). Incidentally, different reproductive strategies of the unisexuals provide avenues to avoid the Muller's ratchet or mutational meltdown by adding freshly recombined genetic material (Bogart et al., 2007).

Table 52 Polyploidization and reproductive modes in unisexual fishes (from Pandian, 2010; updated).

Unisexual	Sexual pattern	Mode of reproduction	Sperm source
Menidia clarkhubbsi Athernidae Capano Bay, Texas	2n ♀ only	Gynogenesis	*M. beryllina* *M. penninsulae*
Fundulus sp Fundulidae Nova Scotia, Canada	2n ♀ only but 3n rarely	Gynogenesis?	*F. heteroclitus* *F. diaphanus*
Poeciliopsis spp Poeciliidae, Mexico	2n ♀ 3n ♀	Hybridogenesis Gynogenesis	*P. lucida, P. monacha* *P. occidentalis, P. infans*
Poecilia formosa Poeciliidae Texas, north east Mexico	2n ♀, ♂ extremely rare, 3 ♀, ♂ rare	Gynogenesis Gynogenesis	*P. mexicana* *P. latipinna*
Phoxinus eos neogaeus Cyprinidae East Inlet pond, New Hampshire	2n ♀ only 3 ♀; 0.5% sterile ♂ 3 ♀ clonal; non clonal	Gynogenesis Variant in hybridogenetic direction	*P. eos, P. neogaeus*
Carassius auratus Cyprinidae *gibelio*, north China *langsdorfi*, Japan *sugu* south China	2n ♀, 50% sterile ♂ 3n ♀, 50% sterile ♂ 4n ♀, 50% sterile ♂	Gynogenesis	*Cyprinus carpio*, other cyprinids
Squalius alburnoides Cyprinidae West Spain, Iberia	2n ♀, ♂ 3 ♀ only 4n ♀, ♂	Gynogenesis Hybridogenesis Gonochorism	*Squalius pyrenaicus* *S. carolitertii*
Cobitis granoei taenia Cobitidae Moscow River.	3n ♀ only 4n ♀ only 4n ♀, ♂	Gynogenesis Gynogenesis Gonochorism	*C. granoei, C. taenia*

10.1 Mutational meltdown

In most organisms, the deleterious mutation arises at a high rate of 1% per diploid per generation, resulting in 1% reduction in fitness per generation. Hence the accumulation of deleterious mutations is expected to result in an environmental reduction in population size, as in unisexual fishes, especially *Menidia clarkhubbsi* (Table 53), and that accelerates the chance of further accumulation of future mutations. This synergistic interaction between reduction in population size and accelerated accumulation of mutation

Table 53 Hybrid origin and reproductive strategies of unisexual fishes to escape from extinction.

Unisexual species	Hybrid origin	Proportion in sympatric population (%)*	Reproductive strategies
Menidia clarkhubbsi Echelle et al. (1989)	*M. beryllnia* x *M. penninsulae-* like ancestor	2.8	All females; gynogenesis only; no male so far found
Fundulus sp Dawley (1992)	*F. heteroclitus* x *F. diphanus*	7.9	All females; gynogenesis only; no male so far found; 3n with *diphanus* genome addition
Poeciliopsis spp Schultz (1977)	*P. monacha* ♀ x *P. lucida* ♂	95.0	All females; no male so far found; (i) 2n reproduce by hybridogenesis (ii) 3n has gained genome addition but reproduces by gynogenesis
Poecilia formosa Balsano et al. (1985)	*P. mexicana latipinna* ♂ x *P. latipinna* ♀	32.0	Though reproduce by gynogenesis only (i) 3n and 4n have gained additional genome (ii) 2n incorporate host leaked subgenome and β chromosome to increase diversity
Phoxinus eos neogaeus Dawley and Goddard (1988) Elder and Schlosser (1995)	*P. eos* x *P. neogaeus*	59.0	Though reproduce mostly by gynogenesis (i) high incidence (50%) of fertilization by donor sperm (ii) consequent genome addition to 3n and 2n–3n mosaics (iii) Hybridogenesis?
Carassius auratus Wu et al. (2003)	*C. auratus* ♀ x *Cyprinus carpio* ♂	13–49	Though reproduce only by gynogenesis has gained genome addition from one of 4 cyprinid species
Squalius alburnoides Ribeiro et al. (2003)	*Squalius pyrenaicus* ♀ x *Anaecypris hispanica*-like ♂ancestor	57.0	(i) Genome addition to 3n and 4n (ii) 3n reproduce by hybridogenesis (iii) Hybrid ♂ produce clonal 2n sperm and haploid sperm
Cobitis granoei taenia Vasil'yev et al. (1990)	*C. granoei* x *C. taenia*	34.0	Genome addition to 3n and 4n unisexuals

*For references see Pandian (2010); however, for data on production of unisexuals in sympatric population, references are given along with species names may theoretically lead to the extinction of the unisexual fishes within 10,000 to 100,000 generations and the process is known as mutational meltdown (Lynch et al., 1993). However, the addition of freshly recombined genetic material through incorporation of subgenomic amounts of DNA (Schartl et al., 1995a), B chromosomes (Lamatsch et al., 2004; Nanda et al., 2007) and/ or genome addition to triploid (however see also Lamatsch et al., 2009) and tetraploid (Lampert et al., 2008) enabled the unisexuals like *Poecilia formosa* to avoid the Muller's ratchet for 840,000 generations over a period of 280, 000 years (Lampert and Schartl, 2008).

Whereas oogenesis in females is largely complete at hatching, spermatogonia continue to divide throughout the life time of males. Consequently, males show a higher (1–10 times) mutation load than females. Hence evolution is regarded as 'male-driven'. In contrast to gynogenic, hybridogenic (in line with gynogenic, androgenic, xenogenic, transgenic) females discard the paternal genome, produce haploid (e.g. *Poeciliopsis monacha lucida*) or diploid (e.g. *Squalius alburnoides*) eggs and hybridity in their progenies is restored by backcrossing with males of parental species. In this inclusion of paternal genome, they resemble true bisexual species (Fig. 1). Thus, the genome of this hybridization is hemiclonal, consisting of a clonally inherited maternal part and the newly acquired paternal part with no recombination between them. Hence the hybridogenics transmit maternally inherited mutations alone. This kind of 'Montecarlo' simultation facilitates the hybridogenics to carry a lower mutational load and thereby reduce the speed of entry into the Muller's ratchet (Som et al., 2007; Loewe and Lamatsch, 2008). For instance, hybridogenesis has perpetuated *P. monacha lucida* for 100,000 generations (Quattro et al., 1992) over a maximum period of 270,000 years (Mateos et al., 2002).

Strategies adopted by unisexual fishes to overcome the constraints of mutational meltdown and/or Muller's ratchet are: 1. Polyploidization (Beukeboom, 2007), i.e. paternal introgression eventually leading to triploidization in all unisexual fishes except in *Menidia clarkhubbsi* and to tetraploidization in *P. formosa* (Lampert et al., 2008), *Carassius auratus, Squalius alburnoides* and *Cobitis granoei-taenia*. 2. Hybridogenesis, which occurs in three unisexual fishes, as in diploid *P. monacha lucida* and triploid *S. alburnoides* and perhaps *P. eos neogaeus*. 3. Incorporation of paternally leaked DNA, and B chromosomes, as in *P. formosa* and 4. Simultaneous sexual and hybridogenic/gynogenic reproduction, as in *C. auratus, S. alburnoides* and *C.granoei taenia*.

The majority (> 60%) of vertebrate unisexual biotypes are polyploids. Most of them, however, exihibit low mtDNA diversity and little sequence divergence from their closest sexual relatives (Avise et al., 1992), although some notable exceptions in fishes are *P. eos neogaeus* (Goddard and Dowling, 1989) and *P. monacha lucida* (Quattro et al., 1991). Though associated with unisexuals, it is not clear whether polyploidy is the cause or the consequence (Mable, 2004). The most obvious advantage of polyploidization is the gross increase in DNA content. Gene duplication results in gene redundancy and heterosis (Comai, 2005). With redundant availability of genes, evolution may assign new functions. Disadvantages of polyploidy are: (i) the disrupting effects of nuclear and cell enlargement and (ii) increased probability of mitosis and meiosis to produce aneuploid cells as well as unbalanced gene expression in polyploids (Pandian and Koteeswaran, 1998, Comai, 2005).

The possibility of 'paternal leakage' of subgenomic amounts of DNA into the gynogenic 'recipient' *Poecilia formosa* was originally suspected by Rasch and Balasano (1989) but proved by Schartl et al. (1995a), who demonstrated the incorporation of subgenomic amounts of DNA from the bisexual species to the gynogenic fish *P. formosa*. Supernumerary or B chromosomes are unstable, dispensable constituents of the genome of many fishes (Carvalho et al., 2008). Their function may range from deleterious to neutral or even beneficial. The high incidence of B chromosomes both in the laboratory-reared and wild caught *P. formosa* individuals from Rio Purificacion, Rio Soto la Mariana River system, Mexico (Lamatsch et al., 2004) suggests that they are beneficial. Nanda et al. (2007) have traced the origin of these B chromosomes comprising one to three microchromosomes in the gynogenic *P. formosa* from its bisexual sperm donor *P. mexicana*. These microchromosomes have centromeric heterochromatin but usually only one has a telomere. They are traced to be stably inherited over eight generations (Fig. 50). In some cases, the microchromosomes carry functional gene lending support to the hypothesis that B chromosomes in *P. formosa* may increase genetic diversity and to some degree counteract the genetic decay of the clonal lineage in this gynogenic fish.

10.2 Exploration of reproductive modes

Having arisen as hybrids on multiple occasions, each unisexual fish species has explored its own uncommon mode of reproduction. Gynogenesis is the chosen mode of reproduction by all the eight unisexual fishes but to escape from the consequent clonal uniformity, each one has added one or another component to the gynogenesis. Thus unisexualism has 'opened' many new avenues with complicated reproductive pathways; some of these pathways are illustrated in Fig. 51. The Atheriniformes *Menidia clarkhubbsi* (Atherinidae) has adopted gynogenesis as the only mode of reproduction. Consequently, it has the smallest share of 2.8% in its sympatric population complex (Table 53) and according to Lynch et al. (1993), it may be on the verge of extinction and requires suitable conservation measures. The simple addition of a genome seems to have almost trebled the population share of the Cyprinodontiformes *Fundulus diaphanus heteroclitus* (tentative name)

Pedigree III/9 Pedigree III/4

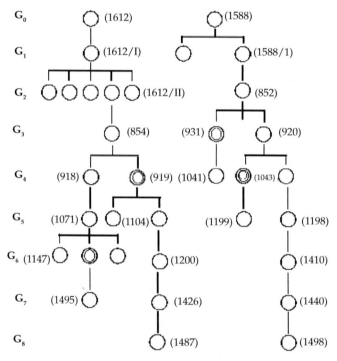

Fig. 50 Upper panel: Metaphase spread of of *Poecilia formosa* with two microchromosomes. Arrowhead indicates the the stable microchromosome and the arrow new microchromosome. Lower panel: Pedigrees of Amazon mollies studied for inheritance of microchromosomes. Open circle: wild type pigmented; filled circle: spotted and double circle: molly with additional microchromosomes. Numbers in parantheses are specimen code. G_0–G_8 generation number (from Nanda et al., 2007).

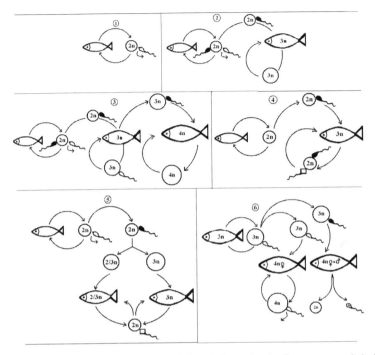

Fig. 51 Reproductive modes in unisexual fishes. 1. Reproduction by gynogenesis in *Menidia clarkhubbsi*, 2. Reprodution by hybridogenesis in diploids and gynogenesis in triploid *Poecilia formosa*, 3. Reproduction by gynogenesis but with genome addition to triploids and tetraploids in *Carassius auratus*, 4. Reproduction by gynogenesis in diploids and hybridogenesis in *Poeciliopsis monacha lucida*, 5. Reproduction by gynogenesis in diploids and hybridogenesis producing diploid-triploid mosaics and triploids in *Phoxinus eos neogaeus* and 6. Reproduction by gynogenesis in diploids and triploids and gynogenesis with genome addition in *Cobitis granoei taenia*.

(Fundulidae). The other two cyprinodontids *Poecilia formosa* and *Poeciliopsis monacha lucida* (Poeciliidae) have a share of more than 32% in their respective sympatric population complexes. Diploids of *P. monacha lucida* reproduce by hybridogenesis, while the triploids by gynogenesis. Diploids of *P. formosa* incorporate paternally leaked DNA and B chromosomes. Besides, diploids of *P. formosa* continuously generate triploids at 1.5% frequency (Balsano et al., 1989) and the triploids generate tetraploids at 0.15% frequency (Lampert et al., 2008). Consequently, triploids constitute nearly two times (21%) the population size of the diploids (11% , Balsano et al., 1985). With genome addition and hybridogenesis, the cyprinid unisexuals consistently produce 0.4 to 50% sterile or fertile males and ensure more than 50% share in their respective sympatric population complexes (Table 53). Two points need to be mentioned. Firstly hybridogenesis takes place in diploid poeciliids but in triploid cyprinidontiformes. The triploids are generated at low frequencies of 1.5 to 6.0% in poeciliids but at the high frequency of more than 50% in the cyprinid like *P. eos neogaeus* (Dawley and Goddard, 1988). However, hybridogenesis in either case provides an opportunity to test a new genome at every generation. Whereas most unisexual clones remain frozen and isolated, *P. eos neogaeus* has chosen to link them together and introduces clonal diversity by discovering 'mosaic' forms. The same is true of other cypriniformids except for the production of 'mosaics'.

10.3 *Squalius alburnoides*

This endemic Iberian minnow is not truly a unisexual, which is characterized by all females producing unreduced eggs. Nevertheless, its genome is clonally transmitted either by gynogenesis or karyogamy, although its reproduction frequently involves recombination and meiotic reduction (Alves et al., 2001; Gromicho et al., 2006). A host of features like the multiple ploidy levels, rare production of males (0.4% in triploids, Ribeiro et al., 2003), incorporation of paternal genome (Alves et al., 1996) and recombination (Alves et al., 1997) makes *S. alburnoides* a unique system, with unusual reproductive modes in its various forms and lineages to compensate for the disadvantages of gynogenesis/hybridogenesis and to improve survival of its unisexual populations.

The *Squalius alburnoides* complex comprises of diploid (2n = 50), triploid (3n = 75) and tetraploid (3n = 100) females and males with genomic composition that arose by hybridization between the sympatric bisexual *S. pyrenaicus* female and *Anaecypris hispanica*–like male ancestor (Alves et al., 2004; Robalo et al. 2006).Though *S. carolitertii*-like female ancestry was considered (e.g. Carmona et al., 1997), sequence analysis of introns from two nuclear genes, i.e. aldolase B and triose phosphate isomerase B, has shown that the paternal contributor is the *A. hispanica*-like ancestor (see Alves et al., 2001). Thanks to the consistent endeavour by Portuguese ichthyologists, perhaps the most diverse modes of reproduction of *S. alburnoides* involving hybridizations, polyploidization, gynogenesis and hybridogenesis have been unravelled.

The *Squalius alburnoides* complex appears to have a single origin some 700, 000 years ago from the Tagus/Guadiana River basin (Fig. 52; Sousa-Santos et al., 2007a). It seems to have dispersed to the northern Iberian rivers through Tejo River basin, where *S. caroliterlii* has been the most likely sexual host for the unisexual complex (Alves et al., 1997). The complex has colonized the northern Douro River basin over 100,000 years ago (Sousa-Santos et al., 2007a). During this period of dispersal of the complex, hybridizations within the complex have apparently 'opened' many novel reproductive pathways modifying both oogenesis and spermatogenesis; some of these modifications are listed in Table 54. Consequently,

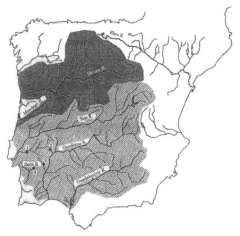

Fig. 52 Map of the Iberian Peninsula showing the distribution of the river basins.

Table 54 Reproductive characteristics of the Iberian minnow *Squalius alburnoides*, C = *Carolitertii* genome; P = *Pyrenaicus* genome and A = genome of unknown ancestor.

Ploidy (n)	Sex	Genome constitution	Reproductive characteristics	Ploidy ♀ gamete (n)	Reference
2	♀	CA	Hybridogenesis	n	Carmona et al. (1997)
2	♀	PA	Unreduced sperm	2	Alves et al. (1998)
2	♂	PA	Unreduced fertile sperm	2	Alves et al. (1999)
2	♂	AA	Meiosis	n	Alves et al. (1998)
3	♀	CAA	Hybridogenesis	2	Carmona et al. (1997)
3	♀	PAA	Meiotic hybridogenesis	n, 2n	Alves et al. (1998)
3	♂	PAA, PPA	Fertility ?	3n	Alves et al. (1999)
4	♂	PPAA	Meiosis	2n	Alves et al. (1999)
3*	♀	PPA, A	Clonal, meiosis	3n, n	Alves et al. (2004)

*Simultaneously producing large (PPA) and small (A) eggs

S. alburnoides population in any Iberian river system comprises of different proportions of males and females bearing different levels of ploidy (Table 55). Notably, tetraploids, which are totally absent in the southern Guadiana River basin, do appear, though in smaller proportions, in the northern river basins. Remarkably, a recent flow cytometric study undertaken on these minnow populations from Lodeiro and Paiva rivers of the Douro basin has shown the dominant presence of tetraploids with no sex ratio bias. These sympatric populations consist of 87% *S. alburnoides* and 17% *S. carolitertii*, the sexual host of the unisexual in these river basins. More than 85% and 97% of *S. alburnoides* population are tetraploids in the Lodeiro and Paiva Rivers, respectively. Combined analysis of six microsatellite loci and the measurement of DNA content of erythrocytes and sperm of the tetraploid

Table 55 Distribution of diploid, triploid and tetraploid males and females of *Squalius alburnoides* in the Iberian River basins. Values represent the minimum and maximum relative frequencies. 2nNH ♂ represents nuclear non-hybrid male (from: Collares-Pereira, 1985; Alves et al., 1997; Carmona et al., 1997; Martins et al. 1998).

Ploidy and sex	Sex ratio (%) in the River basins				
	Guadiana	Sado	Tejo	Mondego	Douro
2n ♀	0–35	36–77	0–15	0–10	0–4
2n ♂	–	–	0–23	5–10	10–14
3n ♀	11–88	19–70	50–100	80–90	36–90
3n ♂	0–5	0–4	0–22	–	0–1
4n ♀	–	0–2	0–10	–	0–1
4n ♂	–	0–2	0–14	0–5	0–1
2n NH ♂	8–89	0–48	0–16	–	–

S. alburnoides has also shown that the tetraploid minnow carries CCAA, in which A represents the genome of *Anaecypris hispanica*–like ancestor of *S. alburnoides* and C, the *carolitertii* genome. Apparently, the male and female tetraploid *S. alburnoides* in these river basins produce reduced diploid eggs and reduced diploid sperm (Cunha et al., 2008).

The diploid males of *S. alburnoides* are comprised of two distinct genotypes: 1. The non-nuclear hybrid genotype from the southern river basins (Table 55) possesses the *S. pyrenaicus*–like mitochondria, undergoes normal Mendelian meiosis and produces haploid sperm, and 2. The hybrid genotype produces unreduced clonal sperm, a very rare event among vertebrate hybrids (Alves et al., 1999). Diploid female clonally transmits hybrid genome to egg.

The presence of fertile triploid male is also another very rare feature among vertebrate hybrids (Sousa-Santos et al., 2007b). The triploid male also produces unreduced 3n sperm. Diploid and triploid females of the northern river basins reproduce by hybridogenesis, where the *Carolitertii* genome is discarded during oogenesis and replaced in each generation (Carmona et al., 1997). Triploid females of the southern Tejo and Guadiana River basins undergo a modified hybridogenesis, in which *Pyrenaicus* (P) genome is discarded in each generation (Alves et al., 1998). Meiosis in the remaining AA^1 genome involves random segregation and recombination between the homospecific genome and genetically distinct haploid A^1 and diploid AA^1 eggs are produced (Fig. 53); this modified meiosis is named meiotic hybridogenesis (Alves et al., 1998). Unlike in normal hybridogenesis, a sperm genome that is incorporated in the diploid egg may remain in the hybrid lineage for more than one generation; the sperm genome incorporated into a haploid egg is clonally transmitted by the diploid female to its egg. The eggs produced by these diploids and triploids require sperm from host species to initiate embryogenesis. However, 3% of eggs produced

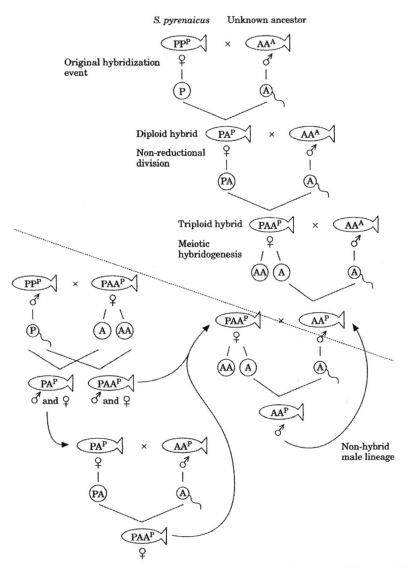

Fig. 53 *S. alburnoides*: Suggested reproductive pathways of all-male non-hybrid lineage within hybrid complex. Above dash line: mechanisms of the origin of the hybrid complex. Below dash line: perpetuating mechanism of the complex. P and A represents the genome of *S. pyrenaicus* and the other ancestor, respectively. The main letters indicate the nuclear genome and the prime ones the mitochondrial genome (With kind permission of The Fisheries Society of the British Isles/Wiley Blackwell; from Alves, MJ, Collares-Pereira, MJ, Dowling, TE, Coelho, MM, 2002. The genetics of maintenance of an all-male lineage in the *Squalius alburnoides* complex. J Fish Biol, 60: 649–662).

by diploid females develop directly without the need for the trigger by the sperm of the host male, as evidenced by the absence of fingerprinting bands of the male parent. Thus "reproduction by the different types of diploid and triploid has introduced high genetic diversity into the hybrid populations of *S. alburnoides* complex and allows purging of deleterious genes and incorporation of beneficial mutations in the same genome, characteristics believed to be major advantages of sexual reproduction" (Alves et al., 1998).

Restoration of an even ploidy in tetraploids may allow *Squalius alburnoides* to return to normal meiosis and serve as a stepping stone to biparental reproduction. Although the natural occurrence of tetraploid minnow in the northern Iberian river basins has been reported (see Table 55; Cunha et al., 2008), studies on experimental crossings have so far confirmed only two routes, through which tetraploids are generated: 1. Crosses between the all male tetraploids with PAAA genome, in which P represents *S. pyrenaicus*, the female ancestor of *S. alburnoides* and dominant female tetraploid with PPAA genome produce (Alves et al., 1999) and 2. Crosses between 4n ♀ (CCAA) and 4n ♂ (CCAA) produce tetraploidy progenies with normal sex ratio (Cunha et al., 2008).

10.4 Males in unisexuals

Within unisexual fish species, males do occur very rarely in poeciliids but almost regularly in Cypriniformes. Natural occurrence of the male is extremely rare in *Poecilia formosa* (Hubbs et al., 1959). In a microchromosomes-carrying laboratory stock, Lamatsch et al. (2000) recorded the appearance of eight males of triploid *P. formosa*. Hence the males found by Hubbs et al. (1959) might have also been triploids. Experimental studies of Lamatsch et al. (2000) have shown that being triploids, the males could produce only aneuploid sperm, as evidenced by the fact that they have induced embryogenesis alone in unreduced eggs of the molly. Incidentally, the sex determination process in the unisexuals is becoming apparent. Like *P. sphenops*, *P. formosa* is likely to be a female heterogametic and produce unreduced clonal eggs carrying W^1Z^2 genotype. *P. formosa* borrows sperm from the female heterogametic *P. latipinna* (Sola et al., 1992); hence *P. formosa* triploids and tetraploids may carry $W^1Z^1Z^2$ and $W^1Z^1Z^2Z^2$ genotypes, respectively. Thus the presence of a single W^1 chromosome feminizes *P. formosa* despite the presence of $Z^1Z^2Z^2$ chromosomes. The gamety of *P. mexicana*, whose *limantouri* race is a predominant contributor of the sperm to 3n *Poeciliopsis* complex, is likely to be a female heterogametic. Likewise, *Poeciliopsis monacha*, *P. lucida* and *P. occidentalis* are likely to be female heterogametic, as all the triploids of the *Poeciliopsis* complex are females. Incidentally with the addition of one or two genomes to the clonal eggs

(X^1X^1) of the unisexual *Carassius auratus* (Wu et al., 2003) by the inclusion of sperm from the male heterogametic host cyprinids (indicated by italic alphabets), the triploid and tetraploid *C. auratus* X^1X^1 X^2 and X^1X^1 X^2X^2 genotypes in fertile females but X^1X^1 Y^2 and X^1X^1 X^2Y^2/X^1X^1 Y^2Y^2 genotypes in sterile triploid and tetraploid males, respectively. Rarely, the diploid X^1Y^2 hybrid, which normally remains sterile, may produce clonal sperm bearing X^2Y^2 genotypes.

In the unisexual lineages, genetic information specifying the other sex has apparently become dispensable. In an interesting experiment, Schartl et al. (1991) investigated whether these dispensed genes specifying the other sex are retained or lost. Following masculinizing hormone treatment, *Poecilia formosa* developed into a typical male with masculine stream line body stinted with xanthophore pigment and perfect gonopodium. Its anatomical and behaviour features also proved to be typically male. These developments revealed that structural genes specifying the male remained functional, though they were not expressed. Amazingly, the Amazon molly has not only retained the structural genes of male as functional but also expressed even after 280,000 years and after the passage of 840,000 generations (see Lampert and Schartl, 2008). Therefore unisexuals like *P. formosa* have also retained the bisexual potency for over 280,000 years (see also Schartl et al., 1995b).

Among the cyprinids, *Phoxinus eos neogaeus* produce 0.5% males, who could not father any progeny (Goddard and Dawley, 1990). The hybrid diploids triploid and tetraploid *Carassius auratus* produce 50% males but they are completely sterile (Wu et al., 2003). This is also true of another Cypriniformes *Cobitis granoei taenia* (Cobitidae), in which the tetraploids suffer in a yet to be known cytological problem, and remain sterile (Vasil'yev et al., 1990). Apparently, the sperm donors in these cyprinids and cobitids are male heterogametics.

On the other hand, the diploid, triploid and tetraploid males of the cyprinid minnow *Squalius alburnoides*, though in smaller proportions, do occur in most Iberian river basins (Table 55). Among the diploids, there are two distinct genotypes, the non-nuclear hybrid male produces (PA) clonal sperm, while the other haploid sperm. Triploids too produce unreduced clonal sperm. In the northern river basins, the tetraploids produce reduced diploid sperm. Alves et al. (1999), Gromicho et al. (2004) and Cunha et al. (2008) have successfully made experimental crossings and some of their results are briefly summarized hereunder:

1. 2n PA ♀ x 2n PA ♂ = 4n PPAA ♂ progenies
2. 2n PA ♀ x 2n AA ♂ = 3n PAA ♀ progenies only
3. 3n PAA ♀ x 3n PPA ♂ = 3n PAA progenies both sexes
4. 3n PAA ♀ x 2n* PP ♂ = 2n PA progenies of both sexes
5. 3n PAA ♀ x 2n AA ♂ = 2n AA all male progenies

6. 4n PPAA ♀ x 2n* PP ♂ = 3n PPA progenies
7. 4n PPAA ♀ x 4n PPAA ♂ = 4n progenies
8. 4n CCAA ♀ x 4n** CCAA ♂ = 4n CCAA progenies of both genes

* represents *pyrenaicus* genome, ** represents *carolitertii* genome

If female heterogametic (ZW/ZZ) mechanism of sex determination is to be applied, all the tetraploid progenies of the cross between 2n PA ♀ x 2n PA ♂ should have all been females, but they were all males. However, the male heterogametic (XX/XY) sex determining mechanism can not explain the presence of females in the cross between 3n PAA ♀ and 2n PP ♂. Hence, the mechanism of sex determination in *S. alburnoides* remains a puzzle and may require an intense molecular level search, for which a beginning is being made (see Pala et al., 2009).

11

Genetic Sex Determination

Sex determination, a delicate process, is one of the earliest and most basic 'decisions' made by a developing embryo. The decision whether to become male or female is conveyed very early in embryonic development by a primary gene(s); this signal is transmitted through a cascade of genes, which ultimately completes the production of distinct sexes. It is therefore not unreasonable to expect the genetic mechanism of sex determination to be among the most conserved developmental process. The entire sequence of the sex determination system commencing from the primary sex determining gene(s) down the genetic cascade completing the sexualization process seems to be very stable in some fishes but not in others. For instance, cyprinids like the male heterogametic crucian carp *Carassius auratus*, irrespective of being fertilized by the homologous sperm or by the male heterogametic sperm of the common carp *Cyprinus carpio*, sustains the expected sex ratio of 50% females and 50% males among their (F_1-F_{16}) progenies . Nor do the other cyprinids like *Puntius gonionotus*, on induction of gynogenesis, display any departure from the expected 100% female ratio (see Table 19). Likewise polyploidization of these hybrid progenies to 3n or 4n too does not alter the normal ratio. This observation holds good for the cobitid loach *Misgurnus anguillicaudatus* also. On the other hand, the departure from the expected sex ratio is ubiquitous within a single pair or when hybridized with male or female heterogametic cichlids . Apparently, the sex determination system is stably fixed in the cyprinids like the crucian carp but not in the cichlids. In fact the stability of the sex determination system appears to be continuum from the stably fixed one to the most labile one (see Baroiller et al., 2009a,b). In some fishes, one or more genes in the cascade seem to be influenced by environmental factors, rendering the sex determining system a labile one.

Unlike higher vertebrates, fishes display diverse mechanisms of sex determination (Sandra and Norma, 2009), sex chromosome systems, sex determining genes and operation of different cascades of genes in sex differentiation. In higher vertebrates, sex is determined by a single monogenic (XX/XY or ZZ/ZW chromosomal) system, i.e. sex is determined by a gene located on a certain chromosome and genes located in other chromosomes have little effect on it. But in fishes sex seems to be determined by one or two of many mechanisms including monogenic, polygenic and environmental sex determination. For example, both monogenic (ZZ/ZW) and polygenic determination systems have been described in *Xiphophorus helleri* (Woolcock et al., 2006). In *Oreochromis niloticus*, Baroiller et al. (1999) have shown the occurrence of both monogenic (XX/XY) and environmental (thermal) systems of sex determination. Instead of two chromosome systems, sex is determined by three chromosome systems in *X. maculatus* (Kallman, 1984). As indicated elsewhere, the mechanism of sex determination is rarely conserved in fishes and even within a genus (e.g. *Poecilia*: male heterogamety in *P. reticulata*, Nanda et al., 1990; female heterogamety in *P. sphenops*, George and Pandian, 1995) or in some cases even within a species (e.g. *Gambusia affinis*, Black and Howell, 1979). As a result, closely related species often determine sex by different (male heterogametic or female heterogametic) systems and different genes (e.g. the mammalisn *SRY-Sox* gene?, as in *Cyprinus carpio*, Du et al., 2007 or *Oryzias latipes Dmy* gene, as in *Oryzias latipes* and *O. curvinotus* Matsuda et al., 2002; Nanda et al., 2002). The master gene responsible for sex determination has been discovered in the medaka fish *O. latipes* (Matsuda et al., 2002; Nanda et al., 2002). But it is absent in other fishes, especially even in closely related species (Kondo et al., 2003). Hence there is no common sex determining gene in fishes.

Ferguson-Smith (2007) considered that the same cascade of genes operates in differentiation of gonads in all vertebrates. But in the *differentiated gonochoristic fishes*, the undifferentiated gonad directly develops into ovary or testis (e.g. *Oncorhynchus kisutch*, Piferrer and Donaldson, 1989). On the other hand, all the *undifferentiated gonochoristic fishes* have to pass through an ovarian stage, designated as 'juvenile hermaphroditism', prior to final differentiation into testis or ovary (e.g. *O. masou*, Nakamura et al., 1993). Hence, within the same genus *Oncorhynchus*, the cascade of genes responsible for gonadal differential is likely to be diverse. In fact, in *the third group of fishes*, juveniles possess a bipotential intersexual gonad, which subsequently develops into ovary or testis (e.g. *Gramma loreto*, Asoh and Shapiro, 1997). Besides there are also a host of simultaneous and sequential hermaphrodites; the latter changes to a second opposite sex after reproducing as female or male during their first sex phase. Molecular biologists are yet to look at the operation of cascade of genes in this group of fishes, especially in the sequential hermaphrodites during the sex changing phase (see Siegfried, 2010), although a beginning has been made (Zhou and Gui, 2008).

11.1 Molecular markers

In general markers, especially the sex-linked ones, provide an access to detect sex, sex chromosomes and sex determining genes. Classical studies examining sex linkages and phenotypic markers revealed the presence of XY system in fishes like *O. latipes, P. reticulata, Betta splendens* (Yamamoto, 1969) and likewise skin pigmentation specific markers have identified the dominance of Y chromosomes and the presence of YY chromosomes (Table 56). Protein markers are also useful in identifying sex linked mode of inheritance in fishes; for instance, the GPI-B locus located approximately at 16 map unit away from sex determining locus in *Ictalurus punctatus* is described by Liu et al. (1996). However, molecular markers may identify sex from the larval stage of fishes.

To begin with a few molecular markers were described to identify genetic sex of a fish species. For instance, the sex specific molecular markers have identified the true male genetic sex in a few unexpected phenotypic females sired by supermales (Y^1Y^2) and androgenics (Y^2Y^2) of *Puntius conchonius* and *Hemigrammus caudovittatus* (Table 56). However, sex specific probes in *Salvelinus alpinus* and microsatellite linkage analysis in *Salmo salar* identified the sex chromosomes in these fishes. Further studies on genome mapping of the sex determining region of the sex chromosomes with several different molecular markers have identified the location of the sex determining gene, SEX on the metacentric arm of chromosome 2 in *S. salar*.

However, the genome sequencing in *Danio rerio, Takifugu rubripes* and *Tetraodon nigroviridis* has neither facilitated the identification of sex chromosomes nor the detection of the master sex determining gene (Volff, 2005). No sex linked molecular markers (e.g. *D. rerio*, Wallace and Wallace, 2003, *Dicentrarchus labrax*, Piferrer et al., 2005) and no sex chromosomes have been identified so far in many fishes. From their electron microscopic studies, Mestriner et al. (1995) could not find any evidence for the atypical pairing that might characterize an XY pair in *Leporinus lacustris*, a close relative of *L. elongatus*, which is one of the rare teleost fishes claimed to have markedly differentiated sex chromosomes (Galetti and Foresti, 1986, see also Baroiller et al., 1996). A total of 310 randomly amplified polymerase DNA primers were used to screen 4,146 bands but they were present in both sexes of the beluga *Huso huso*. Hence Keyvanshokooh et al. (2007) concluded that either the sex specific DNA in the beluga is composed of sequences not complimentary to the 310 random decamer primers or sex chromosomes

Table 56 Representative examples of molecular markers to identify sex, sex chromosome and sex determining locus in selected fishes.

Species and Reference	Remarks
Xiphophorus maculatus Coughlan et al. (1999)	PCR based genomic DNA analysis identified the presence of 323bp fragment in XX and YZ females and 323 bp plus high molecular fragment in XY and YY males
Puntius conchonius Kirankumar et al. (2003)	*SRY* specific primer identified male specific 588, 333 and 200 bp fragments in males but only 200 bp fragment in females
Hemigrammus caudovittatus David and Pandian (2006b)	DMRT-1 specific primer identified 237 and 300 bp fragments in males but a 200 bp fragment in females
Takifugu rubripes Cui et al. (2006)	Sex linked anonymous marker sequence suggested female heterogamety
Clarias gariepinus Kovacs et al. (2000)	Sex linked RAPD marker *CgaY1* (2.6 kb) present in males and produced faint male specific band at low stringency in *Heterobranchus longifilis*. *CgaY2* (458 bp) produced similar hybridization pattern in both sexes of *C. gariepinus, C. macrocephalus* and *H. longifilis*
Oncorhynchus spp Du et al. (1993), Nakayama et al. (1998), Devlin et al. (2001)	Sex linked *GH* pseudogene identified in *O. keta, O. gorbuscha* and *O. kisutch*, but it is absent in *O. nerka, O. mykiss and O. clarki*
O. masou Zhang et al. (2001)	PCR based identification of growth hormone pseudogene *GHψ* in 93–98 % of males and in a few females, whose *GH-ψ* fragment was identical to that of males and shared 95 % homology with that in male *O. tshawytscha*
O. tshawytscha Nagler et al. (2004)	Real time quantitative PCR analysis detected the presence of a single copy of growth hormone pseudogene *GH*p in males but its absence in females
S. namaycush Stein et al. (2002)	Detection of sex linked microsatellite locus near the telomere of Y chromosome
Xiphophorus helleri Kazianis et al. (2004), Woolcock et al. (2006)	Identification of sex chromosomes by several molecular markers and mapping of sex determining region by linkage group 24
Salvelinus alpinus Woram et al. (2003)	Sex specific probes facilitate identification of sex chromosomes by FISH. Genome mapping identified 12 sex linked markers, 9 microsatellite loci and one functional gene, the SEX located at the distal end of linkage group AC4.
Salmo salar Artieri et al. (2006)	Microsatellite linkage analysis identified the sex determining locus 'SEX' on the metacentric arm of chromosome 2
Poecilia reticulata Kirpichnikov (1981), Khoo et al. (1999a, b, c) Lindholm et al. (2004), Nanda et al. (1992), Tripathy et al. (2009)	Traits like Black caudal peduncle (Bcp) and Red tail (*Rdt*) markers linked to the dominant T chromosome. Traits like *Macropodus, Aramatus* and *Pauper* pigmentation linked to YY chromosomes. Using (GATA)$_4$ and (GACA)$_4$ probes, sex chromosomes were identified. Genetic linkage map of the genome using 790 single nucleotide polymorphism markers

are weakly differentiated. Despite AFLP analysis of a total of 12, 815 loci produced by 256 primer combinations, Gao et al. (2010) could not detect a sex specific marker in *Lepomis macrochirus*. Testing five loci for linkage with the Y chromosome of *Oncorhynchus mykiss*, Alfaqih et al. (2009) excluded a role for *Dmrt*1 and *Sox*6 as candidate genes in sex determination. Despite fishes providing diversely rich material for research, impediments described above seem to have slowed down our understanding of the molecular mechanism of sex determination in fishes. Consequently, our knowledge about the sex determination gene is scarce in this species-rich group.

11.2 Sex chromosomes

The Y chromosome linkage maps of four salmonids *Salvelinus alpinus*, *Salmo salar*, *S. trutta* and *Oncorhynchus mykiss* reveal the telomeric placement of the sex determining region involving a smaller fraction of chromosome in the first three species but an intercalary position for *O. mykiss* (see Fig.1 of Woram et al., 2003). In each species, the Y specific region is very short and the major male sex determining gene is located on different linkage groups in these salmonids, indicating that different Y chromosomes have independently evolved in each species (Woram et al., 2003). The lack of conservation of sex linkages between species can be explained by (i) the translocation of the small chromosome arm containing the sex determining gene into an autosome resulting in the formation of a new 'Y' chromosome, (ii) the transposition of sex determining gene onto an autosome and (iii) the activation of unlinked master sex determining genes in different species (Woram et al., 2003). Sex determining regions are apparently very unstable in fishes like salmonids (Volf and Schartl, 2001). Likewise the sex determining region of the chromosome of the platyfish displays a high level of genomic instability characterized by frequent transposition, duplications and deletions (Volff et al., 2003b). Creation of a neo-sex chromosome may induce sex-determining genes and genes involved in mate choice. This may occur, for example, through transplantation of the master sex-determining gene from the sex chromosome onto autosomes, as in salmonids (Woram et al., 2003). Another possibility is the creation of a novel master sex determining gene on an autosome (Almeida-Toledo et al., 2000; Nanda et al., 2002). For instance, *Oryzias latipes*, like mammals has an XX-XY sex determination system. The X and Y chromosomes of medaka are morphologically indistinguishable (Matsuda et al., 1998). Such homomorphic sex chromosomes in a heterogametic system indicate that the gonochores are at an early stage of evolution (Ohno, 1967). Finally, the frequent switching between different sex determination systems and the rapid evolution of sex chromosomes might also be linked to the formation of new species.

Consequent to the possibility of creation of new sex-chromosomes, some fishes have more than one pair of sex chromosomes. The platyfishes seem to provide a wide range of sex chromosome combinations; for instance, there are three different chromosomes in *Xiphophorus maculatus.* Two hypotheses have been proposed to explain the situation. The first one proposes the distribution of male determining genes over the three different types of sex chromosomes (Kallman, 1984) and the second one conceives a dosage-dependent mechanism based on one gene being present in different copy numbers on the three types of sex chromosomes (Volff and Schartl, 2001). In *X. nezahualcotyl* and *X. milleri,* most males bear XY chromosomes and most females XX chromosomes. However, a second Y[1] chromosome is also present, which together with an autosomal modifier is responsible for the presence of XY[1] females (Kallman, 1984).

11.3 Sexonomics

Coined by Volff et al. (2007), the term sexonomics is now defined as the sex determination specific component of genomics. In most fishes, primary sex determination is genetically determined (Valenzuela et al., 2006). Nevertheless sex differentiation of fishes is remarkably plastic and is determined by both genetic and environmental factors in some fishes. The focus of this description is limited to primary sex determination and the genetic and environmental role of sexualization of fishes will be described elsewhere (Pandian, 2011).

The sex determination process varies widely among vertebrate taxa but no group offers as much diversity for the study of evolution of sex determination as teleost fishes (Pala et al., 2009). However, the molecular and evolutionary mechanisms driving sex determination and its variability in teleost fishes are poorly understood (Volff, 2005). Hitherto only two vertebrate sex determination genes have been identified; they are the mammalian *SRY* gene and the (*Oryzias latipes*) medaka's *DMY* gene. *SRY* encodes a transcription factor containing high mobility group (HMG) box domain, whereas *DMY* encodes a transcription factor similar to the *double-sex* and *mab-3* related transcription factor (*dmrt1*). In several fishes, the sex determination regions have been localized to chromosomes using either sex-linked traits (e.g. *Salvelinus namaycush,* Stein et al., 2002, see Table 56), quantitative trait loci (QTL) (e.g. *Oreochromis* spp, Shirak et al., 2006) or cytogenetics (e.g. *Salmo salar,* Artieri et al., 2006). Candidate sex determination genes have been localized within the genome in some fishes, although these are often not located within the sex determination region (e.g. *Xiphophorus maculatus,* Veith et al., 2003; *Oreochromis* spp, Shirak et al., 2006); a candidate gene *amh* (anti Mullurian hormone) has been mapped within an autosomal sex determining QTL (Shirak et al., 2006).

In a remarkable description, Schartl (2004) reconstructed the cytological events and evolutionary history of the Y chromosome and the sex determining gene in the medaka *Oryzias latipes*. Based on a precise linkage map of the sex chromosome, the placement of the sex determining locus was identified at the tip of linkage group 9 that contains the *dmrt*1 gene. This gene encodes a putative transcription factor with an intertwined *2n* finger DNA binding domain. The chromosomal segment containing *dmrt*1 became duplicated (Table 57) some 10 MYa (Zhang, 2004) and the duplicated version was inserted into another chromosome (Nanda et al., 2002). At the new location, all the other genes except the new copy of *dmrt*1 became activated and repetitive DNA as well as other transposable elements accumulated around *Dmy* (Matsuda et al., 2002). This process created a chromosomal region of 260 kbp on the Y that is absent in X. Inhibition of pairing and cross over led to recombinational isolation of the sex determining gene. The suppression of recombination established the integrity of the Y as the male determining sex chromosome (Schartl, 2004). However, the *Dmy* gene is found only in two closely related species, i.e. *O. latipes* and *O. curvinotus* of the genus *Oryzias* so far examined (Kondo et al., 2003; Matsuda et al., 2003). Hence the *Dmy* is not the universal master sex determining gene in teleost fishes (Volff et al., 2003a).

*Dmrt*1 belongs to a family of transcription factors related to sexual regulators and contains a DNA binding motif called DM domain (Zarkower, 2001). In mammals, DMRT1 is autosomal and involved in the male cascade downstream of SRY (Raymond et al., 1998). Only one characteristic amino acid distinguishes all *DMY* sequences from all DMRT1 sequences, suggesting that a single amino acid change of Ser in DMRT1 to Thr in *Dmy* may largely be responsible for the establishment of *Dmy* as the male sex determining master gene in medaka (Zhang, 2004). Whereas *Dmy*, designated as *Dmrt1by* in medaka, is expressed during early embryonic development prior to the formation of the testis, the autosomal *dmrt*1 is expressed predominantly in somatic cells of the testis. The *Dmrt1bY* may regulate proliferation of primordial germ cells (PGCs) and differentiation during early gonadal differentiation but *dmrt*1 is involved in spermatogonial differentiation (Kobayashi et al., 2004b). The critical role of *dmrt*1 in testicular differentiation has also been confirmed in hormonally sex reversed fishes. In XY fish treated with estrogen, *dmrt*1 is not expressed, but it is expressed in XX males (Guan et al., 2000; Kobayashi et al., 2003b). In zebrafish, an undifferentiated gonochore, all individuals pass through juvenile hermaphroditism prior to differentiation into male or female; the dmrt1 is expressed in the undifferentiated gonad in both sexes, suggesting a role for *dmrt*1 in gonadal development in both males and females (Guo et al., 2005).

Table 57 Summary of available information on the role played by *Dmy/Dmrt* 1 by as master determination gene and *Dmrt1* gene complex in downstream sex determination in fishes.

Species, Reference	Description
Oryzias curvinotus Matsuda et al. (2003)	In gonadal *DMY*, a Y-chromosome specific DM domain gene, is required for male development and the Y chromosome is homologous to that of some medaka. The Y-specific region in medaka originated from a transposition of an autosomal region
O. latipes Matsuda et al. (2002), Nanda et al. (2002) Zhang (2004)	The 280 bp *DMRT1* gene is a duplicated copy of autosomal *DMRT* gene and is expressed in male embryonic and larval development and in the Sertoli cells of adult testis. The only functional gene *DMY* present in the Y-specific region is the counterpart of *SRY* gene. Only one characterstic amino acid distinguishes all *DMY* sequences from all *DMRT1* sequences, i.e. a single amino acid change is largely responsible for the establishment of *DMY* as male sex determination gene
Oreochromis niloticus Guan et al. (2000)	DM-domain containing (*Double sex/Mab-3* DNA-binding motif) gene is highly conserved across species. Two distinct DM domain DNAs *tDMRT-1* from testis and *tDMO* from ovary are isolated. A male specific motif is absent in *tDMO*. The alternatively spliced male and female types of *double sex tDMRT-1* and *tDMO* cDNAs are encoded by two different genes. The mutually exclusive nature of *tDMRT-1* and *tDMO* expression in testis and ovary indicates that they both play an important role in gonad development and function
Danio rerio Guo et al. (2005)	*dmrt* 1 is expressed in the developing gonads of both sexes
Oncorhynchus mykiss Marchand et al. (2000)	*dmrt* 1 is expressed early in testis development but not in ovary
Paralichthys olivaceus Wen et al. (2009)	*Dmrt 4* is closely related to tilapias *Dmo* (DM domain gene in ovary). It is expressed strongly in testis and very weakly in ovary. It is related to the development of gonads, nervous system and sense organs

Mapping 11 genes including *Amh, Cyp*19, *Dax*1, *Dmrt*2, *Dmrta*2, *Sox*8 and others, and comparing their locations to previously identified QTL for sex determination, Shirak et al. (2006) have proposed that *Amh* and *Dmrta*2 are candidates for master key regulators for sex determination in tilapias. *Amh* is mapped 5 cM from UNH 879 within a QTL region and *Dmrta2*. 4 cM from UNH 848 with in another QTL region for sex determination (Fig. 54). These genes are highly associated with two different quantitative trait loci for sex determination. However, conclusive evidence is yet to be provided in support of their claim that *Amh* and *Dmrta2* either jointly or independently act as master sex determining genes.

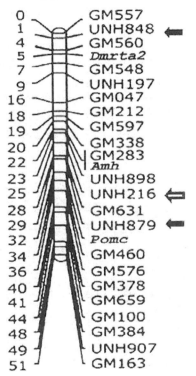

0	GM557
1	UNH848 ⬅
4	GM560
5	*Dmrta2*
7	GM548
9	UNH197
16	GM047
18	GM212
19	GM597
20	GM338
22	GM283
	Amh
23	UNH898
25	UNH216 ⬅
28	GM631
29	UNH879 ⬅
32	*Pomc*
34	GM460
36	GM576
40	GM378
41	GM659
44	GM100
48	GM384
49	UNH907
51	GM163

Fig. 54 The tilapia linkage group 23. Mapping positions of QTL for sex determination and sex specific mortality are denoted by solid and open arrows, respectively (from Shirak et al., 2006).

Spontaneously or hormonally sex reversed XX males and XY females of fishes are generally completely fertile (see Pandian and Sheela, 1995). Apparently, some male fertility genes are also located on the X chromosome or autosomes, and the reverse is true for the Y chromosomes. This experimental evidence clearly indicates that the reproductive male or female specific specializations of the X and Y chromosomes of fishes are not pronounced in fishes, as in mammals. In mammals, YY individuals are inviable and XX males are infertile (Lahn et al., 2001). Consequently, it is to be expected that the expression of one or another sex determination gene is not sexually dimorphic. For instance, *Sox9* and *Sox9a* reveal no male or female specific difference in expression in fishes like *Acipenser sturio* and *Poecilia reticulata*, respectively (Table 58); the same is true of the *dmrt*1 in *Danio rerio* (Table 57). In fact evidences from different species listed in Tables 57 and 58 suggest that the timing and levels of *dmrt1* expression can be associated with sex specific differentiation fates, with high expression levels being correlated with testis development and low levels being related with ovary development. However, there also others like *Oreochromis niloticus* and *Paralichthys olivaceus tDmrt1* and *Dmrt4* are associated with testis development and *tDmo* and *Dmo* are with ovary development, respectively (see also Liu et al., 2009).

Table 58 Summary of available information on the role played by *Sox* gene complex in downstream sex determination/differentiation in fishes.

Species, Reference	Description
Acipenser sturio Hett et al. (2005)	*Sox* 9 consists of 3 exons and 2 introns with completely conserved exon-intron boundaries and shows high levels of homology to *Sox9* sequences of other vertebrates, especially in the N-terminus region containing the HMG box. It reveals no male or female specific difference
Poecilia reticulata Shen et al. (2007)	*Sox9b* coding sequence is a 1,443 bp long fragment, encodes a protein of 480 amino acid residues, and shows high homology to those of *Oryzias latipes* (91.0%) and *Gasterosteus aculeatus* (88.3%). *Sox9* mRNA is detected both in testis and ovary
Cyprinus carpio Du et al. (2007)	*Sox9a* and *Sox9b* differ in having an intron of different lengths (Fig. 55) 704 and 616 bp, respectively in the conserved HMG box region that codes for identical amino acid sequence (Fig. 55). Cc*Soxb* overexpresses in adult brain and testis but is weakly expressed in ovary
Allotetraploid Liu et al. (2007b)	*Sox9a* is highly expressed in testis but weakly in brain and heart and is not expressed in adult female
3n crucian carp Guo et al. (2008)	Isolated and characterized cDNA of *Sox9a*, encodes a protein of 457 amino acids with an HMG box. *Sox9a* is expressed in testis as in zebrafish. But the testicular expression of *Sox9a* gene 3n seldom contributes to the formation of functional spermatozoa, although plays an important role in development of testicular tubules. Thus the contribution of *Sox9* genes differ in different species
Danio rerio von Hofsten and Olsson (2005)	Two *Sox9* genes, *Sox9a* and *Sox9b* have been identified. Both contain HMG box and are able to bind the AACAAAG recognition site. *Sox9a* is expressed in testis and *Sox9b* in ovary. FTZ-F1a (*fushitarazu* factor 1a) genes are involved in regulation of inter-renal development and thereby steroid biosynthesis. Sox9a alone does not determine sex but a combination of *Sox9a*, FF1a, *b* and c, MMH GATA and *Dmrt* seems to perform it

CCAATGAATGCCTTTATGGTCTGGG<u>GCTCAAGCGGCGCGC</u>AGGAAACTGGCGGACCAGTATCCACACCTGCACAACGCCG

P M N A F M V W A Q A A R R K L A D Q Y P H L H N A

AGCTCAGCAAGACCCTCGGCAAACTCTGGAGgtcagagcattcattgtttatgaagtgtaggacaactccagaagccgg

E L S K T L G K L W R

agcaacactgattcatttaactgcccagacaaactcactgtattattaatattaaactgcattgtttcatagcattatt

atataatgtcaggcaatactgataagctgtttgggaaataaagggtaaaactatagcttaggcaaatggtgaattaacc

cttacttcagtgcagaaaaagttgtgtagtcactaaaaactttggtaacactttggaaaagggaacacttactcactat

taactatgacttttccctctataaattcctaatttgctgcttattaatagttagtatggnagcttttaagtttaggtat

gaggtaggattagggatgtagaataaggggcatggaaaaaaaagacattaatatgtgcttaactactactaataaatggc

taatattctagtaatatgcatgctaataagaaactagttaagagaccctaaaataaagtgttacccaaacttttagttt

taagaaatttagttcgatctggattactttattcttttttttcctgaataatattttttcttaagtgaacccaaaaagt

aataatcatagtttccaagaacaatttatacctttttttgtaggtttctttattgttaactaaggtagatcttagtattg

attatttctcatgtgcttttgtagGTTACTGAATGAGGGCGAGAAGCGTCCATTTGTGGAGGAGGCCGAGCGTCTGAGG

L L N E G E K R P F V E E A E R L R

<u>GTCCAGCACAAGAAAGACC</u>ACCCCAACTACAAGTACCAGCC

V Q H K K D H P N Y K Y Q

CCAATGAACGCGTTTATGGTGTGGG<u>GCTCAAGCCGCGCGC</u>AGGAAACTGGCGGATCAGTATCCGCACCTGCACAACGCCG

P M N A F M V W A Q A A R R K L A D Q Y P H L H N A

AGCTCAGCAAGACCCTCGGAAAACTCTGGAGgtgagagagagagacaattcattttactttgcttcgttttttgtttgtt

E L S K T L G K L W R

caggactgctccagaagtcaggaaactttgtgtagtactattaaattgcaatgatttatagcattattatgcaatggtt

ataaaacaatattgataagcttgagaaaatgtaactatttgggtagtaaggtaaaactttagtttagtcaaatagtaaa

ttaaccctaacttcacatgcacttcaacacagaaaaagttgtggagtcggtaaaaacctccagtttttaagaaaagatgc

tgattactttattacttttttgggtttatgtaagaaaagtagtattcacttaaaaaaaaaaaaataattataatagctgnt

tcaagaacaatctataccttttttttaaagnttattttattattaattggacaagataaagaaagtngattacttttctg

aattacttttattgctttgcgggtttacttattaagaaaaaaatagtataaaaaaaaaaaagaatagtcatatttcaaa

aacaatttgtacctttaatgaaaatgtctgtttttgttaattgaagaatgcaaaggaagtaatctgagcattggttaattct

tgtggatatgtgtagGTTACTGAATGAGGGCGAGAAGCGTCCGTTCGTGGAGGAGGCCGAGCGTCTGAGG<u>GTGCAGCAC</u>

L L N E G E K R P F V E E A E R L R V Q H

<u>AAGAAAGAC</u>CACCCCGACTACAAGTACAGACC

K K D H P D Y K Y R

Fig. 55 *Cyprinus carpio*: The DNA sequences and the predicted amino acid sequences of CcSox9a (upper panel) and CcSox9b (lower panel). Unlined sequences were used for primers of PCR reactions to confirm the splicing sites. Lower case letters indicate introns (from Du et al., 2007).

11.4 Sex changers

Among the sequential hermaphrodites, protogynous hermaphrodites, especially groupers have attracted much attention of molecular biologists. The groupers include 159 species belonging to 15 genera. The genus

Epinephelus alone comprises of 98 species. They are fast growing, disease-resistant and amenable for culture (see Zhou and Gui, 2008). They are protogynous hermaphrodites; the age or body size, at which they naturally change sex varies among different species. The sex change occurs at the threshold size of 20 cm (total body length) in the honeycomb grouper *E. merra*, 40–50 cm in the dusky grouper *E. marginatus*, and 70–90 cm between the age of 9 and 16 years in the estuarine grouper *E. tauvina*. The groupers have been considered as a good model for studies on sex determination, as their gonad undergoes transformation from ovary to ovotestis and to testis, and can be subjected to experimentation over a long transition period. Chinese and Japanese ichthyologists have contributed a series of interesting publications, which have been summarized by Zhou and Gui (2008).

A set of SMART (switching mechanism at 5′ end of the RNA transcript) cDNA plasmid libraries from pituitaries, hypothalamus and gonads at different stages have been constructed and molecular studies have been initiated to understand the regulation mechanisms of sex change in the orange grouper *E. coioides*. With FST sequencing and RACE-PCR analysis, over 100 genes have been cloned, and from the analysis of the tissue distribution and expression patterns, three kinds of candidate genes have been categorized. Of them, the third category includes the most interesting three genes that show differential expression between ovary and testis: These genes include (i) thyroid stimulating hormone (*TSHβ*), (ii) *DMRT1* and (iii) *Sox3*. More abundant *DMRT1* and *TSHβ* transcripts have been detected in male testis than in female ovaries (Wang et al., 2004; Xia et al., 2007) but Sox3 transcripts are more abundant in the ovaries than in the testes (Yao HHC et al., 2005; Yao B et al., 2007). Incidentally, *DMART1* is most related to fishes that undergo sex change, as in protogynous hermaphrodites like *Monopterus albus* (Huang et al., 2005), *Halichoeres tenuispinis* (Choi et al., 2004), protandrous hermaphrodites like *Acanthopagrus schlegeli* (He et al., 2003a, b) and temperature sensitive *Odontesthes bonaiensis* and *Oreochromis niloticus* (Miranda et al., 2001). When endocrine sex reversal is in progress, the expressions of *DMRT1* and *TSHβ* are upregulated, while *Sox3* expression is downregulated (Wang et al., 2004, Xia et al., 2007). Hence *Sox3* plays an important role in oogenesis, while *DMRT1* in spermatogenesis. *Sox3* is expressed very early and can be detected in the nuclei of some PGCs along the germinal epithelium. When the *Sox3* positive PGCs develop towards oogonia (cf Kurokawa et al., 2007) and then oocytes, *Sox3* expression continues to increase greatly, but stops, when *Sox3*-positive PGCs develop towards spermatogonia (see Zhou and Gui, 2008). A 252 bp fragment of *Sox9* has been cloned and specific antibody has been produced. The grouper *Sox9* is expressed only in Sertoli cells and is mainly localized in cytoplasm, suggesting that *Sox 9* is a member of the sex determination cascade.

The rice field eel *Monopterus albus,* a commercially important freshwater fish of Southeast Asia undergoes natural sex change from female and to male during its life cycle. Its chromosomes are all telocentric (2n = 24) and its haploid genome size of 600 Mb is among the smallest of the vertebrates. These features make it an ideal vertebrate model to study molecular mechanisms and evolution of sex change. With differential expression of *dmrt*1 isoforms in gonadal epithelium having the bipotential differentiation capacity during gonadal transformation, the eel provides an insight into the role of alternative splicing of *dmrt*1 in sex change (Huang et al., 2005).

Huang et al. (2005) reported that alternative splicing at 3′ region generated four isoforms of *dmrt*1a, b, c and d in the eel's testis, ovotestis and ovary. The 5′ region and DM domain of the *dmrt*1 isoforms are common to all transcripts. But the *dmrt1* a, b and c isoforms were generated from a common splicing site at amino acid 185 (Fig. 56). These four isoform transcripts encode four different sized proteins with 301, 196, 300 and 205 amino acids. Not only *dmrt1* is expressed specifically in gonads, but its multiple isoforms are differentially co-expressed in gonadal epithelium during gonadal transformation. Based on comparisons of mean values of the real-time fluorescent quantitative RT-PFR analysis, the expression level of *Dmrt1*a was found from low to high: ovary < ovotestis I < ovotestis II < ovotestis < testis. Although expression level of *dmrt1b* changed in a pattern similar to that of *dmrt1*a, but the overall expression level of *dmrt1b* was much lower than that of *dmrt1*a. The *dmrt1d* expression remained not only low but it did not also change significantly during sex change. The differential expression of *dmrt1* isoforms may also be regulated by their 3′ untranslated regions (UTRs), although these 3′ UTRs do not contribute to intracellular localization of the *Dmrt1* protein. Thus alternate splicing of *dmrt1* has facilitated the sex change from female to male in the rice field eel.

11.5 Polygenic sex determination

In gonochores characterized by monogenic sex determination system 1♀ : 1♂ ratio is optional in a population with random mating and Mendelian segregation. The occurrence of skewed sex ratio may imply non-random mating and non-Mendelian segregation. Animals, in which sex is determined polygenically, are indeed very few and in fishes it is claimed to occur in the European sea bass *Dicentrarchus labrax* (Vanderputte et al., 2007). However, a series of publications, that described influence of temperature on the delicate sex determination/differentiation processes in 59 freshwater and marine fishes including the European sea bass has been reviewed by Piferrer et al. (2005) and Ospina-Alvarez and Piferrer (2008). Interestingly, even a small change in 1–2°C can significantly alter the sex ratio and this

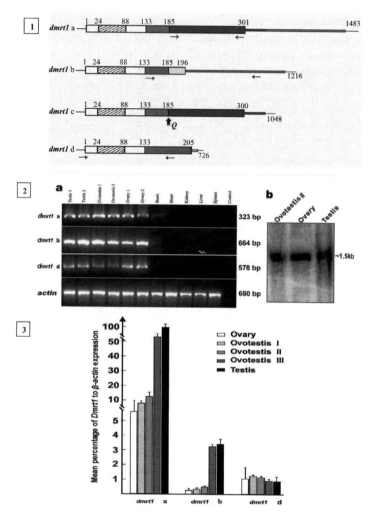

Fig. 56 *Monopterus albus:* ⬚1⬚ Upper figure: Splicing of *dmrt* forms four kinds of mRNAs: *dmrt1a, dmrt1b, dmrt1c* and *dmrt1d,* which encode 4 proteins with 301, 196, 300 and 205 amino acids, respectively. *dmrt* has deleted an amino acid Q compared with that of *dmrt1a* because of alternative splicing. DM domains are indicated by shaded boxes from amino acids 24–88; sequences from amino acids 1–133 are common among the four transcripts. Alternatively spliced region in the 3' are shown by different colours and same colour represents the same DNA sequence. The numbers in the end under the lines indicate nucleotide numbers of these cDNAs. ⬚2⬚ Middle panel: (a) RT-PCR and (b) Northern blot analysis of *dmrt1* of the rice field eel. (a) Expression of *dmrt1* in three kinds of gonads. (b) The expression of *dmrt1* is high on testis, with a slightly lower expression in ovary and ovotestis. ⬚3⬚ Lower panel: The relative expression of *dmrt1* isoforms to *beta-actin* in ovary, ovotestis I (pre-intersex stage), ovotestis II (medium intersex stage), ovotestis III (post intersex stage) and testis measured by Real time PCR (from Huang et al., 2005).

Colour image of figure 56(1) appears in the color plate section at the end of the book.

aspect is considered elsewhere (Pandian, 2011). Incidentally, the polygenic sex determination system is considered to be evolutionarily unstable (Price, 1984) and its maintenance is poorly understood. With reference to fishes, the following are known and accepted facts: (i). the presence of monogenic sex determining master gene, (ii) to date *Dmrt1aY* is the only master sex determining gene, and (iii) there is no universal sex determing master gene. Nevertheless, there are several publications that describe the overriding autosomal genes modifying the sex ratio. However, these autosomal genes may function more along the cascade of genes that facilitate the differentiation and sexualization of an individual fish, whose genetic sex has been determined by one or another master sex determining gene. A mutation in one or another of the cascade of genes may alter phenotypic sex differentiation; for instance, the *fancl* gene is involved in sustenance of developing oocytes through meiosis; a mutation in zebrafish *fancl* causes female to male reversal (Rodriguez-Mari, 2010). As indicated elsewhere, Shirak et al. (2006) have alone proposed the possibility of two genes *Amh* and *Dmrt* a 2 as the master determining genes in tilapias, although conclusive evidence is not yet available.

Vanderputte et al. (2007) designed a mating programme combining 33 males and 23 females in three full factorial sets for a total of 5893 individuals belonging to 253 full sib families of the sea bass. They found that the frequency of females ranged fom 5 to 47% in paternal half sib-families and 1 to 40% in maternal half sib-families, clearly indicating that both sire and dame had a highly significant effect on the progeny sex ratio. Secondly, the heritability of the sex ratio trait was high 0.62. Thirdly, there was strong correlation between sex and size, implying that some of the genes acting on sex determination and growth are at least strongly linked . However, a similar situation does exist in *Oreochromis niloticus* (Barioiller et al., 1996) and *Menidia menidia* (Conover and Heins, 1987). In the absence of molecular studies confirming the presence and operation of two or more master sex determining genes in the sea bass, it may only be suggested that the sea bass and others like *O. niloticus* and *M. menidia* are good models for molecular biologists in search of sex determining genes in fishes.

11.6 Concluding remarks

This comprehensive exposition of sex determination in fishes from cytogenetics through hybrids, gynogenics, androgenics, ploidies, allogenics/xenogenics to sexonomics in gonochores, hermaphrodites and unisexuals has led to the following generalizations:

1. The only sex determining gene thus far identified in fishes is *dmrt1bY* in *Oryzias latipes* and *O. curvinotus*; however, the other species belonging

to *Oryzias* do not have the same the sex determining master gene. All the salmonid species do not employ the same chromosomes as sex chromosomes. The sex determining mechanism is also not common and shared by all fishes, as evidenced by the widely scattered male and female heterogamety across fishes; in many cases, all the species belonging to a genus and in some cases, all the populations belonging to a species do not have the same gamety. As against mammals and birds, fishes do not have a common and universal sex determining master gene, sex chromosomes and sex determining mechanism.

2. Unlike in mammals, the position and function of a host of genes in the cascade that regulates the differentiation process in fishes do not remain unassailable. For instance, *DMRT1* is an autosomal gene involved in the male cascade downstream of *SRY* in mammals. Due to a change in a single amino acid, estimated to have occurred some 10 MYa, this autosomal *Dmy* has been elevated from the downstream of the genetic cascade to the master sex determining gene. Likewise the function of a specific gene in the cascade may also revert to an opposite direction, e.g. *fancl* gene . Not surprisingly, the sex of an individual fish, even after its sex is determined by a master sex determining gene, may revert its sex, when induced by one or another autosomal/environmental factor(s). Whereas mammals have a single pathway in their genetic cascade to regulate the sex differentiation process, the fishes may have as many as three, one each for the primary or differentiated gonochores, the secondary or undifferentiated gonochores and the third for the others. This is further complicated by the fact that within the same genus, like, 'Oncorhynchus' the cascade of genes responsible for gonadal differentiation is diverse; *O. kisutch* is a primary gonochore but *O. masou* is a secondary gonochore .

3. Irrespective of being a gonochore, hermaphrodite, or unisexual, the fishes have retained bisexual potency, as evidenced by the following:

   a. Pioneering surgical experiments performed by Lowe and Larkin have proved that the fighting fish following ovariectomy regenerated functional testes.

   b. SSC transplantation studies led by Yoshizaki have demonstrated the retention of bisexual potency by the rainbow trout even after sexual maturity and spermiation.

   c. Hormonal masculinization of the unisexual female Amazon molly by Schartl has provided irrevocable evidence that the molly has retained not only these structural male genes as functional but also expressed them after the passage of 840,000 generations over a period of 280,000 years.

d. Hormonally sex reversed XX males, XY females, ZW males and ZZ females (Mair et al., 1991b; George and Pandian, 1995; Kobayashi et al., 2003b) have provided evidence for the prevalence of bisexual potency among the gonochores, i.e. unlike in mammals and birds, the emergence of Y chromosome in male heterogametic and W chromosome in female heterogametic fishes has not completely eliminated the genes in their respective counterparts (X or Z) or autosomes, that can regulate sex differentiation in an opposite direction to the genetic sex; for instance, *dmrt1*, known to be involved in spermatogonial differentiation is expressed in hormonally sex reversed XX males but not in XY neofemales.

e. In the self-fertilizing hermaphroditic rivulus *Kryptolebias marmoratus*, hermaphrodites and secondary males occur in nature; primary males and hormonally sex reversed females can be produced in the laboratory. The amenability of this hermaphrodite to become a male or female clearly indicates that the rivulus has retained the bisexual potency for the last 200, 000 years ever since it was phylogenetically separated from its close gonochoric relative *K. ocellatus* . Among the gamete exchanging hermaphrodites too, some like *Serranus baldwini* and *S. fasciatus*, on attaining a larger body size, switch over as pure males and even hold harems each with 1–7 hermaphrodites behaving as pure females. The scarid *Leptoscarus vaigiensis* has become secondarily a gonochore, after shedding its original protogynous hermaphroditic status (see Pandian. 2010). The gonochoric checker board cichlid *Crenicara punctulata* is almost on its 'wings' to become a protogynous hermaphrodite . The other gonochore *Macropodus opercularis*, an almost bidirectionally sex changing serial hermaphrodite, achieves its normal fecundity by the fourth spawning cycle following sex change (Koteeswaran and Pandian, 2011, see Fig. 47). Hence the bisexual potency has been retained by many of these hermaphrodites and gonochores.

References

Abbas, K, Li, MY, Wang, WM, Zhou, XY. 2009. First record of the natural occurrence of hexaploid loach *Misgurnus anguillicaudatus* in Hubei Province, China. J Fish Biol, 75: 435–441.

Aegerter, S, Jalabert, B. 2004. Effects of post-ovulatory oocyte ageing and temperature on egg quality and on the occurrence of triploid fry in rainbow trout, *Oncorhynchus mykiss*. Aquaculture, 231: 59–71.

Aida, S, Arai, K. 1998. Sex ratio in the progeny of gynogenetic diploid marbled sole *Limanda yokohamae* males. Fish Sci, 64: 989–990.

Akhtar, N. 1984. Anesthesia, abdominal surgery, efficacy of ovariectomy and subsequent androgen treatments in inducement of sex reversal in *Tilapia nilotica*. Ph.D. Thesis, Auburn University, Auburn, USA.

Alfaqih, MA, Brunelli, JP, Drew, RE, Thorgaard, GH. 2009. Mapping of five candidate sex determining loci in rainbow trout (*Oncorhynchus mykiss*). BMC Genetics, 10: doi: 10.1186/1471-2156-10-2.

Allsop, DJ, West, SA. 2003. Constant relative age and size at sex change in sequentially hermaphroditic fish. J Evol Biol, 16: 921–929.

Allsop, DJ, West, SA. 2004, Sex ratio evolution in sex changing animals. Evolution, 58: 1019–1027.

Almeida-Toledo, LF. 1978. Contribuicao a citogenetica dos Gymnotoidei (Pisces, Ostariophysi). Ph.D. Thesis, University of Sao Paulo, Sao Paulo, Brazil.

Almeida-Toledo, LF, Foresti, F, Daniel, MFZ, Toledo-Filho, SA. 2000. Sex chromosome evolution in fish: The formation of the neo-Y chromosome in *Eigenmannia* (Gymnotiformes). Chromosoma, 109: 197–200.

Alves, AL, Oliveira, C, Nirchio, M, Grando, A, Foresti, F. 2006. Karyotypic relationships among the tribes of *Hypostominae* (Siluriformes: Loricariidae) with description of XO sex chromosomes system in a Neotropical fish species. Genetica, 128: 1–9.

Alves, MJ, Coelho, MM, Collares-Pereira, MJ. 1996. Evidence for non-clonal reproduction in triploid *Rutilus alburnoides*. Isozyme Bull, 29: 23.

Alves, MJ, Coelho, MM, Collares-Pereira, MJ. 1997. The *Rutilus alburnoides* complex (Cyprinidae): evidence for a hybrid origin. J Zool Syst Evol Res, 35: 1–10.

Alves, MJ, Coelho, MM, Collares-Pereira, MJ. 1998. Diversity in the reproductive modes of females of the *Rutilus alburnoides* complex (Teleostei: Cyprinidae): A way to avoid the genetic constraints of unisexualism. Mol Biol Evol, 15: 1233–1242.

Alves, MJ, Coelho, MM, Prospero, MI, Collares-Pereira, MJ. 1999. Production of fertile unreduced sperm by hybrid males of the *Rutilus alburnoides* complex (Teleostei: Cyprinidae): An alternative route to tetraploidization in unisexuals. Genetics, 151: 277–283.

Alves, MJ, Coelho, MM, Collares-Pereira, MJ. 2001. Evolution in action through hybridization and polyploidy in an Iberian freshwater fish: a genetic review. Genetica, 111: 375–385.

Alves, MJ, Collares-Pereira, MJ, Dowling, TE, Coelho, MM. 2002. The genetics of maintenance of an all-male lineage in the *Squalius alburnoides* complex. J Fish Biol, 60: 649–662.

Alves, MJ, Gromicho, M, Collares-Pereira, MJ, Crespo-Lopez, E, Coelho, MM. 2004. Simultaneous production of triploid and haploid eggs by triploid *Squalius alburnoides* (Teleostei: cyprinidae). J Exp Zool, 301A: 552–558.

Amer, MA, Miura, T, Miura, C, Yamaguchi, K. 2001. Involvement of sex steroid hormones in the early stages of spermatogenesis in Japanese huchen (*Hucho perryi*). Biol Reprod, 65: 1057–1066.

Andreata, AA, Almeida-Toledo, LF, Oliveira, MC, Toledo-Filho, SA. 1993. Chromosome studies in Hypoptopomatidae (Pisces, Siluriformes, Laricariidae). 2. ZZ/ZW Sex chromosome system, B chromosomes and constitutive heterochromatin differentiation in *Microlepidogaster leucofrenatus*. Cytogenet Cell Genet, 63: 215–220.

Anon. 1980. Nuclear transplantation in teleosts. 1. Hybrid from the nucleus of carp and the cytoplasm of crucian. Scientist Sinica, 23: 517–523.

Aparicio, SJ, Chapman, E, Stupka, N, Putnam, J-M, Chia, P, Brenner, S, et al. 2002. Whole genome shotgun assembly and analysis of the genome of *Fugu rubripes*. Science, 297: 1301–1310.

Arai, K. 1986. Effect of allotriploidization on development of the hybrids between female chum salmon and male brooktrout. Bull Jap Soc Sci Fish, 52: 823–829.

Arai, K. 2000. Chromosome manipulation in aquaculture: Recent progress and perspective. Suisan Zoshoku, 48: 295–303.

Arai, K. 2001. Genetic improvement of aquaculture finfish species by chromosome manipulation techniques in Japan. Aquaculture, 197: 205–208.

Arai, K, Ishimoto, M. 1996. A preliminary note on DNA fingerprints of the first and second generation of gynogenetic diploids in the loach. The Second Japan-Korea Joint Meeting and Symposium on Aquaculture. National Fisheries University, Shimonoseki. p 48.

Arai, K, Mukaino, M. 1997. Clonal nature of gynogenetically induced progeny of triploid (diploid x tetraploid) loach *Misgurnus anguillicaudatus* (Pisces: Cobitidae). J Exp Zool, 278: 412–421.

Arai, K, Mukaino, M. 1998. Electrophoretic analyses of the diploid progenies from triploid-diploid crosses in the loach *Misgurnus anguillicaudatus* (Pisces: Cobitidae). J Exp Zool, 280: 368–374.

Arai, K, Inamori, Y. 1999. Viable hyperdiploid progeny between diploid female and induced triploid male in the loach *Misgurnus anguillicaudatus*. Suisan Zoshoku, 47: 489–495.

Arai, K, Matsubara, K, Suzuki, R. 1991a. Karyotype and erythrocyte size of spontaneous tetraploidy and triploidy in the loach *Misgurnus anguillicaudatus*. Nippon Suisan Gakkaishi, 57: 2167–2172.

Arai, K, Matsubara, K, Suzuki, R. 1991b. Chromosomes and development potential of progeny of spontaneous tetraploidy loach *Misgurnus anguillicaudatus*. Nippon Suisan Gakkaishi, 57: 2173–2178.

Arai, K, Matsubara, K, Suzuki, R. 1993. Production of polyploids and viable gynogens using spontaneously occurring tetraploid loach *Misgurnus anguillicaudatus*. Aquaculture, 117: 227–235.

Arai, K, Ikeno, M, Suzuki, R. 1995. Production of androgenetic diploid loach *Misgurnus anguillicaudatus* using sperm of natural tetraploids. Aquaculture, 137: 131–138.

Arai, K, Taniura, K, Zhang, Q. 1999. Production of second generation progeny of hexaploid loach. Fish Sci, 65: 186–192.

Arai, K, Morishima, K, Momotani, S-I, Kudu, N, Zhang, Q. 2000. Clonal nature of gynogens induced from spontaneous diploid eggs in the loach *Misgurnus anguillicaudatus*. Folia Zool, 49: 31–36.

Araki, K, Shinma, H, Nagoya, H, Nakayama, I, Onozato, H. 1995. Androgenetic diploids of rainbow trout (*Oncorhynchus mykiss*) produced by fused sperm. Can J Fish Aquat Sci, 52: 892–896.

Argue, BJ. 1996. Performance of channel catfish *Ictalurus punctatus*, blue catfish *I. furcatus* and their F_1, F_2, F_3 and backcrossed hybrids. Ph. D. Thesis, Auburn University, Auburn, USA.

Argue, BJ, Dunham, RA. 1999. Hybrid fertility, introgression and backcrossing in fish. Rev Fish Sci, 7: 137–195.

Arkhipchuk, VV. 1995. Role of chromosomal and genome mutations in the evolution of bony fishes. Hydrobiol J, 31: 55–65.

Arnold, ML. 2006. Evolution through Genetic Exchange. Oxford University Press, Oxford, UK.

Artieri, CG, Mitchell, LA, Ng, SSH, Parisotto, SE, Danzmann, RG, Hoyheim, B, Phillips, RB, Morasch, M, Koop, BF, Davidson, WS. 2006. Identification of the sex determining locus of Atlantic salmon (*Salmo salar*) on chromosome 2. Cytogenet Gen Res, 112: 152–159.

Artoni, RF, Bertollo, LAC. 2002. Evolutionary aspects of the ZZ/ZW sex chromosome system in the Characidae fish genus *Triportheus*. A monophyletic state and NOR. Heredity, 89: 15–19.

Artoni, RF, Falco, JN, Moreira-Filho, O, Bertollo, LAC. 2001. An uncommon condition for a sex chromosome system in Characidae fish. Distribution and differentiation of ZZ/ZW system in *Triportheus*. Chromosome Res, 9: 449–456.

Asoh, K, Shapiro, DY. 1997. Bisexual juvenile gonad and gonochorism in the fairy basslet *Gramma loreto*. Copeia, 1997: 22–31.

Aspinwall, N, McPhail, JD. 1995. Reproductive isolating mechanisms between the peamouth *Mylocheilus caurinus* and redside shiner *Richardsonius balteatus* at Stave Lake, British Columbia, Canada. Can J Zool, 73: 330–338.

Avise, JC. 2001. Cytonuclear genetic signature of hybridization phenomena: Rationale, utility and empirical examples from fishes and other aquatic animals. Rev Fish Biol Fish, 10: 253–263.

Avise, JC, Saunders, NC. 1984. Hybridization and introgression among species of sunfish (*Lepomis*) analysis by mitochondrial DNA and allozyme markers. Genetics, 108: 237–255.

Avise, JC, Nelson, WS, Arnold, J, Koehn, RK, Williams, GC, Thorsteinsson, V. 1990. The evolutionary status of Islandic eels. Evolution, 44: 1254–1262.

Avise, JC, Quattro, JM, Vrijenhoek, RC. 1992. Molecular clones within organismal clones. Evol Biol, 26: 225–246.

Azevedo, MFC, Oliveira, C, Pardo, BG, Martinez, P, Foresh, F. 2007. Cytogenetic characterization of six species of flatfishes with comments to karyotype differentiation patterns in Pleuronectiformes (Teleostei). J Fish Biol, 70 A: 1–15.

Balsano, JS, Randle, EJ, Rasch, EM, Monaco, PJ. 1985. Reproductive behavior and maintenance of all female *Poecilia*. Env Biol Fish, 12: 251–263..

Balsano, JS, Rasch, EM, Monaco, PJ. 1989. The evolutionary ecology of *Poecilia formosa* and its triploid associate. In: Ecology and Evolution of Live-bearing Fishes (Poeciliidae) GK Meffe, FF Jr Snelson (eds) Prentice Hall Englewood Cliffs, New Jersey, pp 277–299.

Baroiller, JF, Nakayama, I, Foresti, F, Chourrout, D. 1996. Sex determination studies in two species of teleost fish *Oreochromis niloticus* and *Loporinus elongatus*. Zool Stud, 35: 279–285.

Baroiller, JF, Guiguen, Y, Fostier, A. 1999. Endocrine and environmental aspects of sex differentiation in fish. Cell Mol Life Sci, 55: 910–931.

Baroiller, JF, D'Cotta, H, Bezault, E, Wessels, S, Horstgen-Schwark, G. 2009a. Tilapia sex determination: Where temperature and genetics meet. Comp Biochem Physiol, 153A: 30–38.

Baroiller, JF, D'Cotta, H, Saillant, E. 2009b. Environmental effects of fish sex determination and differentiation. Sex Dev, 3: 118–135.

Barton, NH, Hewitt, GM. 1985. Analysis of hybrid zones. Annu Rev Ecol Syst, 16: 113–148.

Basolo, AL. 2001. The effect of intrasexual fitness differences on genotype frequency stability in Fisherian ratio equilibrium. Ann Zool Fennici, 38: 297–304.

Beardmore, JA, Mair, GC, Lewis, RI. 2001. Monosex male population in finfish as exemplified by tilapia: applications, problems and prospects. Aquaculture, 197: 283–301.

Bekkevold, D, Hansen, MM, Loeschoke, V. 2002. Male reproductive competition in spawning aggregations of cod (*Gadus morhua*). Mol Ecol, 11: 91–102.

Bellafronte, E, Vicari, MR, Artoni, RF, Margarido, VP, Moreira-Filho, O. 2009. Differentiated ZZ/ZW sex chromosomes in *Apareiodon ibitiensis* (Teleostei, Parodontidae): Cytotaxonomy and biogeography. J Fish Biol, 75: 2313–2325.

Bellafronte, E, Moreira-Filho, O, Vicari, MR, Artoni, RF, Bertollo, LAC, Margarido, VP. 2010. Cytogenetic identification of invasive fish species following connections between hydrographic basins. Hydrobiologia, 649: 347–354.

Bematchez, LH, Glemer, H, Wilson, CC, Danzmann, RG. 1995. Introgression and fixation of Arctic charr (*Salvelinus alpinus*) mitochondrial genome in an allopatric population of brook trout (*Salvelinus fontinalis*). Can J Fish Aquat Sci, 52: 179–185.

Benfey, TJ. 1989. A bibliography of tetraploid fish 1943 to 1988. Can Tech Rep Fish Aquat Sci 1682, 33p.

Benfey, TJ. 1999. The physiology and behaviour of triploid fishes. Rev Fish Sci, 7: 39–67.

Bercsenyi, M, Magyary, I, Urbani, B, Orban, L, Horvath, L. 1998. Hatching out goldfish from common carp eggs: interspecific androgenesis between two cyprinid species. Genome, 41: 573–579.

Berrebi, P. 1995. Speciation of the genus *Barbus* in the north Mediterranean basin. Recent advances from biochemical genetics. Biol Conserv, 71: 237–249.

Berrebi, P, Rab, P. 1998. The '*Barbus intermedius*' species flock in Lake Tana (Ethiopia): 3. Cytogenetic and molecular genetic data. Ital J Zool, 65: 15–20.

Berrebi, P, Cataneo-Berrebi, G, LeBrun, N. 1993. Natural hybridization of two species of tetraploid barbells: *Barbus meridionalis* and *Barbus barbus* (Osteichthyes, Cyprinidae) in southern France. Biol J Linn Soc, 48: 319–333.

Bertollo, LAC, Cavallaro, ZI. 1992. A highly differentiated ZZ/ZW sex chromosome system in Characidae. Cytogenet Cell Genet, 60: 60–63.

Bertollo, LAC, Fontes, MS, Fenocchio, AS, Cano, J. 1997. The X_1X_2Y sex chromosomes system in the fish *Hoplias malabaricus* I.G, C and chromosome replication banding. Chromosome Res, 5: 493–499.

Bertollo, LAC, Oliveiro, C, Molina, WF et al. 2004. Chromosome evolution in the erythrinid fish *Erythrinus erythrinus* (Teleostei, Characiformes). Heredity, 93: 228–233.

Beukeboom, LM. 2007. Sex to some degree. Heredity, 98: 123–124.

Beulbens, K, Eding, EH, Gilson, P, Olivier, F, Komen, J, Richter, CJJ. (1997). Gonadal differentiation, intersexuality and sex ratios of European eel (*Anguilla anguilla L.*) maintained in captivity. Aquaculture, 153: 135–150.

Bhowmick, RM, Kowtal, GV, Gupta, SD, Jana, RK. 1987. Some observations on the catla x calbasu hybrid produced by hypophysation. In: Proc World Symp on Selection, Hybridization and Genetic Engineering in Aquaculture, K Tiews (ed) Bordeaux Berlin 2: 101–107.

Birstein, VJ, Hanner, R, DeSalle, R. 1997. Phylogeny of the Acipenseriformes; Cytogenetic and molecular approaches. Env Biol Fish, 48: 127–155.

Bishop, RD. 1967. Evaluation of the striped bass (*Roccus sextalis*) and white bass (*R. chrysops*) hybrids after two years. Proc Annu Con SE Asso Fish Wildlife Agencies, 21: 245–254.

Black, DA, Howell, WM. 1979. The North American mosquitofish *Gambusia affinis*: a unique case in sex evolution. Copeia, 1979: 509–513.

Blanc, JM, Poisson, H, Escaffre, AM, Aquirre, P, Vallee, F. 1993. Inheritance of fertilizing ability in male tetraploid rainbow trout (*Oncorhynchus mykiss*). Aquaculture, 110: 61–70.

Bobyrev, A, Burmensky, V, Vasil'ev, V, Krisunov, E, Lebedeva, E. 2002. Coexistence of triploid and diploid forms of spined loach *Cobitis taenia*: a model based approach. Proc Sec Internatl Con Loaches Genus *Cobitis* and Related Genera, Olsztyn, Poland, Abract.10.

Bogart, JP, Bi, K, Fu, J, Noble, DWA, Niemietz, A. 2007. Unisexual salamanders (Genus *Amblystoma*) present a new reproductive mode for eukaryotes. Genome, 50 : 119–136.

Bongers, ABJ, Abarca, BJ, Doulabi, BZ, Eding, EH, Komen, J, Richter, CJJ. 1995. Maternal influence on the development of androgenetic clones of common carp *Cyprinus carpio L.* Aquaculture, 137: 139–147.

Bongers, ABJ, Doulabi, BZ, Voorthuis, PK, Bovenhuis, H, Komen, J, Richter, CJJ. 1997a. Genetic analysis of testis development in all male F_1 hybrid strains of common carp *Cyprinus carpio*. Aquaculture, 158: 33–41.

Bongers, ABJ, Ben-Ayed, MZ, Doulabi, BZ, Komen, J, Richter, CJJ. 1997b. Origin of variation in isogenic gynogenetic and androgenetic strains of common carp *Cyprinus carpio*. J Exp Zool, 277: 72–79.

Bongers, ABJ, Doulabi, BZ, Richter, CJJ, Komen, J. 1999. Viable androgenetic YY genotypes of common carp (*Cyprinus carpio* L.). J Hered, 90: 195–198.

Borin, LA, Martins-Santos, IC, Oliveria, C. 2002. A natural triploid in *Trichomycterus davisi* (Siluriformes, Trichomycteridae): Mitotic and meiotic characterization by chromosome banding and synaptonemal complex analyses. Genetica, 115: 253–258.

Braat, AK, Speksnijder, JE, Zivkonic, D. 1999. Germ line development in fishes. Int J Dev Biol, 43: 745–760.

Brantley, RK, Wingfield, JC, Bass, AH. 1993. Sex steroid levels in *Porichthys notatus*, a fish with alternative reproductive tactics and a review of the hormonal basis of male dimorphisms among teleost fishes. Hormones and Behavior, 27: 332–347.

Brinster, RL, 2002. Germline stem cell transplantation and trangenesis. Science, 296: 2174–2175.

Bromage, NR, Jones, J, Randall, C, Thrush, M, DeVries, B, Springate, J, Duston, J, Barker, G. 1992. Broodstock management fecundity, egg quality and the timing of egg production in the rainbow trout (*Oncorhynchus mykiss*). Aquaculture, 100: 141–166.

Brooks, S, Tyler, CR, Sumpter, JP. 1997. Egg quality in fish: What makes a good egg? Rev Fish Biol Fish, 7: 384–416.

Brown, KH. 2008. Fish mitochondrial genomics: sequence inheritance and functional variation. J Fish Biol, 72: 355–374.

Brum, MJI. 1996. Cytogenetic studies of Brazilian marine fish. Braz J Genet, 19: 421–427.

Brusle, S. 1988. Sex differentiation in teleosts: primordial germ cells as stem cells. Reproduction in fish: basic and applied aspects in endocrinology and genetics. Colloq Inst Natl Rech, Agron, 44: 21–24.

Cal, RM, Vidal, S, Gomez, C, Alvarez-Blazquez, B, Martinez, P, Piferrer, F. 2006. Growth and gonadal development in diploid and triploid turbot (*Scophthalmus maximus*). Aquaculture, 251: 99–108.

Campos-Ramos, R, Harvay, SC, Masabanda, JS, Carrasco, LAP, Griffin, DK, McAndrew, BJ, Bromage, NR, Penman, DJ. 2001. Identification of putative sex chromosomes in the blue tilapia *Oreochromis aureus* through synaptonemal complex and FISH analysis. Genetica, 111: 143–153.

Carmona, JA, Sanjur, OI, Dadrio, I, Machordom, A, Vrijenhoek, RC. 1997. Hybridogenic reproduction and maternal ancestry of polyploidy Iberian fish: The *Tropidophoxinellus alburnoides* complex. Genetics, 146: 983–993.

Carillo, M, Bromage, NR, Zunuy, S, Serrano, R, Prat, F. 1989. The effects of modifications in photoperiod on spawning time, ovarian development and egg quality in the sea bass (*Dicentrarchus labrax* L.). Aquaculture, 81: 351–365.

Carrasco, LAP, Penman, DJ, Bromage, NR. 1999. Evidence for the presence of sex chromosomes in the Nile tilapia (*Oreochromis niloticus*) from synaptonemal complex analysis of XX, XY and YY genotypes. Aquaculture, 173: 207–218.

Carruth, LL. 2000. Freshwater cichlid *Crenicara punctulata* is a protogynous sequential hermaphrodite. Copeia, 2000: 71–82.

Cassani, JR. 1990. A new method for early evaluation of grass carp larvae. Prog Fish Cult, 52: 207–210.

Cassani, JR, Caton, WE, Clark, B. 1984. Morphological comparisons of diploid and triploid hybrid grass carp *Ctenopharyngodon idella* female x *Hypophthalmichthys nobilis* male. J Fish Biol, 25: 269–278.

Castelli, M. 1994. Study on sex determination in the common barbell (*Barbus barbus* L.) (Pisces, Cyprinidae) using gynogenesis. In: Genetics and Evolution of Aquatic Organisms, AR Beaumont, (ed) Chapman and Hall, London, pp 509–519.

Carvalho, RA, Martins-Santos, IC, Dias, AL. 2008. B chromosomes: an update about their occurrence in freshwater Neotropical fishes (Teleostei). J Fish Biol, 72: 1907–1932.

Centofante, L, Bertollo, LAC, Moreira-Filho, O. 2001. Comparative cytogenetics among sympatric species of *Characidium* (Pisces, Characiformes). Diversity analysis with the description of ZW sex chromosome system and natural triploidy. Caryologia, 54: 139–150.

Centofante, L, Bertollo, LAC, Buckup, PA, Moreira-Filho, O. 2002. ZZ/ZW sex chromosome system in new species of the genus *Paradon* (Pisces, Paradontidae). Caryologia, 55: 139–150.

Centofante, L, Bertollo, LAC, Moreira-Filho, O. 2003. Chromosomal divergence and maintenance of sympatric characidiin fish species (Crenuchidae, Characidiinae). Hereditas, 138: 213–218.

Chao, N, Chao, W, Liu, K, Liao, I. 1987. The properties of tilapia sperm and its crypreservation. J Fish Biol, 30: 107–118.

Charlesworth, D, Charlesworth, B, Marias, G, 2005. Steps in the evolution of heteromorphic sex chromosomes. Heredity, 95: 118–128.

Charnov, EL. 1982. The Theory of Sex Allocation. Princeton University Press, Princeton, NJ, USA.

Charnov, EL, Bull, JJ. 1989. Non-Fisherian sex ratios with sex change and environmental sex determination. Nature, 338: 148–150.

Chaudhury, RC, Prasad, R, Dass, CC. 1982. Karyological studies in five Tetradontiform fishes from Indian Ocean. Copeia, 3: 728.

Chen, J, Fu, Y, Xiang, D, Zhao, G, Long, H, Liu, J, Yu, Q. 2008. XX/XY heteromorphic sex chromosome system in two bullhead catfish species *Liobagrus marginatus* and *L. styani* (Amblycipitidae; Siluriformes). Cytogenet Genome Res, 122: 169–174.

Chen, SL, Li, J, Deng, SP, Tian, YS, Wang, QY, Zhuang, ZM, Sha, ZX, Xu, JY. 2007. Isolation of female specific AFLP markers and molecular identification of genetic sex in half smooth tongue sole (*Cynoglossus semilaevis*). Mar Biotechnol, 9: 273–280.

Chen, SL, Tian, YS, Yang, JF, Shao, C-W, Ji, XS, Zhai, JM, Liao, XL, Zhuang, ZM, Su, PZ, Xu, JY, Sha, ZX, Wu, PF, Wang, N. 2009. Artificial gynogenesis and sex determination in half smooth tongue sole (*Cynoglossus semilaevis*). Mar Biotechnol, 11: 243–251.

Chenuil, A, Crespin, L, Ponyaud, L, Berrebi, P. 2004. Autosomal differences between males and females in hybrid zones: a first report from *Barbus barbus* and *Barbus meridionalis* (Cyprinidae). Heredity, 93: 128–134.

Cherfas, NB. 1966. Natural triploidy in females of the unisexual variety of the silver crucian carp (*Carassius auratus gibelio*). Genetika, 2: 16–24.

Cherfas, NB. 1981. Gynogenesis in fishes. In: Genetic Basis of Fish Selection, VS Kirpichnikov (ed) Springer Verlag, Berlin, pp 255–273.

Cherfas, NB, Rothbard, S, Hulata, G, Kozinsky, O. 1991. Spontaneous diploidization of maternal chromosome set in ornamental (koi) carp *Cyprinus carpio* L. J Appl Ichthyol, 7: 72–77.

Cherfas, NB, Gomelsky, B, Peretz, Y, Ben-Dom, N, Hulata, G, Moaz, B. 1993. Induced gynogenesis and polyploidy in the Israeli common carp line Dor-70. Isr J Aquacult, 45: 59–72.

Cherfas, NB, Gomelsky, B, Emelyanova, OV, Recoubratsky, AV. 1994a. Induced diploid gynogenesis and polyploidy in crucian carp *Carassius auratus gibelio* (Bloch) x common carp *Cyprinus carpio* L. hybrids. Aquacult Fish Mgmt, 25: 943–954.

Cherfas, NB, Gomelsky, B, Ben-Dom, N, Peretz, Y, Hulata, G. 1994b. Assessment of triploid common carp (*Cyprinus carpio* L.) for culture. Aquaculture, 127: 11–18.

Cherfas, NB, Gomelsky, B, Ben-Dom, N, Hulta, G. 1995. Evidence for the heritable nature of spontaneous diploidization in common carp *Cyprinus carpio* L. eggs. Aquacult Res, 289–292.

Chevassus, B. 1983. Hybridization in fish. Aquaculture, 17: 113–128.

Childers, WF. 1967. Hybridization of four species of sunfishes (Centrarchidae). Hist Surv Bull, 29: 159–214.

Childers, WF. 1971. Hybridization of fishes of North America (Family Centrarchidae). Rep FAO/UNDP (TA) Rome, 2926: 113–142.

Choi, I, Oh, J, Cho, BN, Jung, YK, Hau Kim, D, Cho, C. 2004. Characterization and comparative genomic analysis of intronless Adams with testicular gene expression. Genomics, 83: 636–646.

Chopelet, J, Waples, RS, Mariani, S. 2009. Sex change and the genetic structure of marine fish. Fish Fish 10: 329–343.

Chourrout, D. 1984. Pressure induced retention of second polar body and suppression of first cleavage: production of all triploids and tetraploids, heterozygous, homozygous diploid gynogenesis. Aquaculture, 36: 111–126.

Chourrout, D. 1989. Revue sur le determinisme genetique du sexe des poissons teleosteens. Bull Soc Zool France, 113: 123–143.

Chourrout, D, Quillet, E. 1982. Induced gynogenesis in the rainbow trout: sex ratio and survival of progenies production of all–triploid populations. Theor Appl Genet, 63: 201–205.

Chourrout, D, Nakayama, I. 1987. Chromosome studies of progenies of tetraploid female rainbow trout. Theor Appl Genet, 74: 687–692.

Chourrout, D, Chevassus, B, Krieg, F, Happe, A, Burger, G, Renard, P. 1986. Production of second generation of triploid and tetraploid rainbow trout by mating tetraploid males and diploid females-Potential tetraploid fish. Theor Appl Genet, 72: 193–206.

Cimino, MC. 1972. Meiosis in triploid all-female fish (*Poecilopsis*, Poeciliidae). Science, 175: 1484–1486.

Cole, KS. 2002. Gonad morphology, sexual development and colony composition in the obligate coral-dwelling damselfish *Dascyllus aruanas*. Mar Biol, 140: 151–163.

Cole, KS, Hoese, DF. 2001. Gonad morphology, colony demography and evidence for hermaphroditism in *Gobiodon okinawae* (Teleostei, Gobiidae). Env Biol Fish, 61: 161–173.

Cole, KS, Robertson, DR. 1988. Protogyny in the Caribbean reef goby *Coryphopterus personatus*: Gonad ontogeny and social influences on sex change. Bull Mar Sci, 42: 317–333.

Collares-Pereira, MJ. 1985. The *"Rutilus alburnoides"* (Steindachner, 1866) complex (Pisces: Cyprinidae). II First data on the karyology of a well-established diploid-triploid group. Arq Mus Bocage Ser, 3A: 69–90.

Collares-Pereira, MJ, Madeira, JM, Rab, P. 1995. Spontaneous triploidy in the stone loach *Noemacheilus barbatulus* (Balitoridae). Copeia, 2: 483–484.

Comai, L. 2005. The advantages and disadvantages of being polyploidy. Nat Rev Genet, 6: 836–846.

Conover, DO, Heins, SW. 1987. Adaptive variation in environment and genetic sex determination in fishes. Nature, 326: 496–498.

Coreley-Smith, GE, Lion, CJ, Brandhorst, BP. 1996. Production of androgenetic zebrafish (*Danio rario*). Genetics, 142: 1265–1276.

Cotter, D, O'Donovan, V, O'Maoileidigh, Rogan, G, Roche, N, Wilkins, NP. 2000. An evaluation of the use of triploid Atlantic salmon (*Salmo salar L.*) in minimizing the impact of escaped farmed salmon on wild populations. Aquaculture, 186: 61–75.

Coughlan, T, Schartl, M, Hornung, U, Hope, I, Stewart, A. 1999. PCR-based sex test for *Xiphophorus maculatus*. J Fish Biol, 54: 218–222.

Crabtree, RE, Bullock, LH. 1998. Age, growth and reproduction of black grouper *Mycteroperca bonaci* in Florida Waters. Fish Bull, 96: 735–753.

Cui, J-Z, Shen, X-Y, Gong, Q-L, Yang, GP, Gu, Q-Q. 2006. Identification of sex markers by cDNA-AFLP in *Takifugu rubripes*. Aquaculture, 257: 30–36.

Cunado, N, Terrones, J, Sanchez, L, Martinez, P, Sants, JL. 2002. Sex dependent synaptic behaviour in triploid turbot *Scophthalmus maximus* (Pisces : Scophthalmidae). Heredity, 89: 460–469.

Cunha, C, Doadrio, I, Coelho, MM. 2008. Speciation towards tetraploidization after intermediate processes of non-sexual reproduction. Phil Trans R Soc, 363B: 2921–2929.

Dabrowski, K, Blom, JH. 1994. Ascorbic acid deposition in rainbow trout (*Oncorhynchus mykiss*) eggs and survival of embryos. Comp Biochem Physiol, 108A: 129–135.

Dabrowski, K, Rinchard, J, Lin, F, Garcia-Abiado, MA, Schmidt, D. 2000. Induction of gynogenesis in muskellunge with irradiated sperm of yellow perch proves diploid muskellunge male homogamety. J Exp Zool, 287: 96–105.

David, CJ. 2004. Experimental sperm preservation and genetic studies in selected fish. Ph.D. Thesis, Madurai Kamaraj University, Madurai, India.

David, CJ, Pandian, TJ. 2006a. *GFP* reporter gene confirms paternity in the androgenote Buenos Aires tetra *Hemigrammus caudovittatus*. J Exp Zool, 305A: 83–95.

David, CJ, Pandian, TJ. 2006b. Cadaveric sperm induces intergeneric androgenesis in the fish *Hemigrammus caudavittatus*. Theriogenology, 65: 1048–1070.

David, CJ, Pandian, TJ. 2006c. Maternal and paternal hybrid triploids of tetras. J Fish Biol, 69: 1–18.

David, CJ, Pandian, TJ. 2008. Dispermic induction of interspecific androgenesis in the fish Buenos Aeres tetra using surrogate eggs of widow tetra. Curr Sci, 95: 63–74.

Davies, M, Yaekel, A, Dawley, RM. 1990. All female hybrids of the Killifishes *Fundulus heteroelitus* and *F diaphanus* 2. Ploidy Karyotype and hybrid origin. J Penn Acad Sci, 635: 209.

Dawley, RM. 1987. Hybridization and polyploidy and a community of three sunfish species (Pisces) Centrarchidae). Copeia, 1987: 326–335.

Dawley, RM. 1992. Clonal hybrids of the common laboratory fish *Fundulus heteroclitus*. Proc Natl Acad Sci, USA, 89: 2485–2488.

Dawley, RM, Goddard, KA. 1988. Diploid triploid mosaics *Phoxinus eos* and *Phoxinus neogaeus*. Evolution, 42: 649–659.

DeCordier, I, Cundari, E, Kirsch-Volders, M. 2008. Mitotic check points and the maintenance of chromosome karyotype. Mutation Res, 651: 3–13.

deGirolamo, M, Scaggiante, M, Rasotto, MB. 1999. Social organization and sexual pattern in the Mediterranean parrotfish *Sparisoma cretense* (Teleostei: Scaridae). Mar Biol, 135: 353–360.

Deng, Y, Oshiro, T, Higaki, S, Takashima, F. 1992. Survival, growth and morphometric characteristics in diploid and triploid hybrids of rainbow trout (*Oncorhynchus mykiss*). Suisan Zoshoku, 114: 121–129.

deOliveira, RR, Feldberg, E, DosAnjos, MB, Zuanon, J. 2009. Mechanisms of chromosomal evolution and its possible relation to natural history characteristics in *Ancistrus* catfishes (Siluriformes: Loricariidae). J Fish Biol, 75: 2209–2225.

Desai, VR, Rao, KJ. 1970. On the occurrence of natural hybrid catla-rohu in Madhya Pradesh. J Zool Soc India, 22: 35–40.

Desprez, D, Briand, C, Hoareau, MC, Melard, C, Bosc, P, Baroiller, JF. 2006. Study of sex ratio in progeny of a complex *Oreochromis* hybrid the Florida red tilapia. Aquaculture, 25: 231–237.

Devlin, RH, Nagahama, Y. 2002. Sex determination and sex differentiation in fish: an overview of genetic, physiological and environmental influences. Aquaculture, 208: 191–364.

Devlin, RH, Biagi, CA, Smailus, DE. 2001. Genetic mapping of Y-chromosomal DNA markers in Pacific salmon. Genetica, 111: 43–58.

DeWoody, JA, Avise, JC. 2001. Genetic perspectives on the natural history of fish mating systems. J Hered, 92: 167–172.

Diniz, D, Laudicina, A, Cioffi, MB, Bertollo, LAC. 2009. Microdissection and whole chromosome painting: improving sex chromosome analysis in *Triportheus* (Teleostei, Characiformes). Cytogenet Genome Res, 122: 163–168.

Doitsidou, M, Reichman-Fried, M, Stebler, J, Koprunner, M, Dorries, J, Meyer, D, Esqueera, CV, Leung, T, Raz, 2002. Guidance of primordial germ cell migration by the chemokine SDF-1. Cell, 111: 647–659.

Dowling, TE, Smith, GR, Brown, WM. 1989. Reproductive isolation and introgression between *Notropis cornutus* and *Notropis chrysocephalus* (Family Cyprinidae): comparison of morphology, allozymes and mitochondrial DNA. Evolution, 43: 620–634.

Du, Q-Y, Wang, F-Y, Hua, H-Y, Chang, Z-J. 2007. Cloning and study of adult tissue specific expression of *Sox9* in *Cyprinus carpio*. J Genet, 86: 85–91.

Du, SJ, Devlin, RH, Hew, CL. 1993. Genomic structure of growth hormone genes in Chinook salmon (*Oncorhynchus tshawytscha*): presence of two functional genes, GH-I and GH-II and a male specific pseudogene GH-ψ. DNA Cell Biol, 12: 739–751.

Duan, W, Qin, Q-B, Chen, S, Liu, SJ, Wang, J, Zhang, C, Sun, YD. 2007. The formation of improved tetraploid population of red cruciancarp x common carp hybrids by androgenesis. Sci China, Life Sci 50C: 753–763.

Dunham, RA, Smitherman, RO. 1987. Genetics and breeding of catfish. South Coop Ser Bull, 325, Auburn University, Auburn, USA.

Ebeling, AW, Chen, TR 1970. Heterogamety in teleostean fishes increases in some deep-sea fishes. Am Nat, 105: 549–561.

Echelle, AA, Echelle, AP, Decault, LE, Durham, DW. 1988. Ploidy levels in silverride fishes (Atherinidae : *Menidia*) on the Texas coast: flow cytometric analysis of occurrence of allotriploidy. J Fish Biol, 32: 835–344.

Echelle, AA, Echelle, AF, Middaugh, DP. 1989. Evolutionary biology of the *Menidia clarkhubbsi* complex of unisexual fishes (Athernidae): Origins, clonal diversity and mode of reproduction. In : Evolution and Ecology of Unisexual Vertebrates, RM Dawley, JP Bogart (eds) New York State Museum, Albany, Bulletin 466: 144–152.

Elder, JF, Schlosser, IJ. 1995. Extreme clonal uniformity of *Phoxinus eos neogaeus* gynogens (Pisces, Cyprinidae) among variable habitats in northern Minnesota beaver ponds. Proc Natl Acad Sci USA, 92: 5001–5005.

Erisman, BE, Rosale-Casian, JA, Hastings, PA. 2008. Evidence for gonochorism in a grouper *Mycteroperca rosacea* from Gulf of California, Mexico. Env Biol Fish, 82: 23–33.

Ezaz, MT, Myers, JM, Powell, SF, McAndrew, BJ, Penman, DJ. 2004a. Sex ratios in progeny of an androgenetic and gynogenetic YY male in Nile tilapia *Oreochromis niloticus* L. Aquaculture, 232: 205–214.

Ezaz, MT, McAndrew, BJ, Penman, DJ. 2004b. Spontaneous diploidization of the maternal chromosome set in Nile tilapia (*Oreochromis niloticus* L.) eggs. Aquacult Res, 35: 271–277.

Falco, JN. 1988. Characeracao cariotipica em peixes do genero *Triportheus* (Teleostei, Characiformes, Characidae) Ph.D. Thesis, University of Sao Paulo, Sao Paulo, Brazil.

Felip, A, Martinez-Rodringuez, G, Piferrer, F, Carillo, M, Zanuy, S. 2000. AFLP analysis confirms exclusive maternal genomic contribution of meiogynogenetic sea bass *Dicentrachus labrax* L. Mar Biotechnol, 2: 301–306.

Ferguson-Smith, M. 2007. The evolution of sex chromosomes and sex determination in vertebrates and key role of DMRT1. Sex Dev 1: 2–11.

Fernandes-Matioli, FMC, Almeida-Toledo, LF, Toledo-Filho, SA. 1998. Natural triploidy in the Neotropical species *Gymnotus carapo* (Pisces, Gymnotiformes). Caryologia, 51: 319–322.

Ferreira, IA, Bertollo, LAC, Martins, C. 2007. Comparative chromosome mapping of 5s rDNA and 5S Hind III repetitive sequences in Erythrinidae fishes (Characiformes) with emphasis on the *Hoplias malabaricus* "species complex". Cytogenet Genome Res, 118: 78–83.

Fineman, R, Hamilton, J, Chase, G. 1975. Reproductive performance of male and female phenotypes in three sex chromosome genotypes (XX, XY, YY) in the killifish *Oryzias latipes*. J Exp Zool, 192: 349–354.

Fischer, EA, Petersen, CW. 1987. The evolution of sexual patterns in the sea basses. Bioscience, 37: 482–488.

Fishelson, L. 1970. Protogynous sex reversal in the fish *Anthias squamipinnis* (Teleostei: Anthidae) regulated by the presence or absence of male fish. Nature, 227: 90–91.

Fishelson, L, Hilzerman, F. 2002. Flexibility in reproductive styles of male St. Peter's tilapia *Sarotherodon galilaeus* (Cichlidae). Env Biol Fish, 63: 173–182.

Flajshans, M. 1997. Reproduction sterility caused by spontaneous triploidy in tench (*Tinca tinca*). Pol Arch Hydrobiol, 44: 39–45.

Flajshans, M, Linhart, O, Kvasnicka, P. 1993. Genetic studies of tench (*Tinca tinca L.*): induced triploidy and tetraploidy and first performance data. Aquaculture, 113: 301–312.

Flajshans, M, Kohlmann, K, Rab, P. 2007. Autotriploid tench *Tinca tinca* (*L.*) larvae obtained by fertilization of eggs previously subjected to post-ovulatory ageing *in vitro* and/or *in vivo*. J Fish Biol, 71: 868–876.

Flajshans, M, Rodina, M, Halacka, K, Kvetesnik, L, Gela, D, Luskova, V, Lusk, S. 2008. Characteristic of sperm of polyploidy Prussian carp *Carassius gibelio* (Bloch). J Fish Biol, 73: 323–328.

Flores, JA, Burns, JR. 1993. Ultrastructural study of embryonic and early adult germ cells and their support cells in both sexes of *Xiphophorus*, Teleostei, Poeciliidae. Cell Tissue Res, 271: 263–270.

Flynn, SR, Matsuoka, M, Reith, M, Martin-Robichaud, DJ. 2006. Gynogenesis and sex determination in short nose sturgeon (*Acipenser brevirostrum*). Aquaculture, 253: 721–727.

Fopp-Bayat, D. 2008. Inheritance of microsatellite loci in polyploid Siberian sturgeon (*Acipenser baeri* Brandt) based on uniparental haploids. Aquacult Res, 39: 1787–1792.

Fopp-Bayat, D, Kolman, R, Woznicki, P. 2007. Induction of meiotic gynogenesis in starlet (*Acipenser ruthenus*) using UV-irradiated bester sperm. Aquaculture, 264: 54–58.

Foresti, F, Oliveira, C, Galetti, PM, Almeida-Toledo, LF. 1993. Silver stained NOR and synaptonemal complex analysis during male meiosis of *Tilapia rendalli*. J Hered, 74: 127–128.

Foster, JW, Graves, JAM. 1994. An Sry-related sequence on the marsupial X-chromosome implications for the evolution of mammalian testis determining gene. Proc Natl Acad Sci USA, 91: 1927–1931.

Francis, RC. 1984. The effect of bidirectional selection for social dominance on agnostic behaviour and sex ratios in the Paradise fish *Macropodus opercularis*. Behaviour, 90: 25–45.

Francis, RC, Barlow, GW. 1993. Social control of primary sex differentiation in the Midas cichlid. Proc Natl Acad Sci. USA, 90: 10673–10675.

Fricke, HW, Fricke, S. 1977. Monogamy and sex change by aggressive dominance in coral reef. Anim Behav, 28: 561–569.

Fridolfsson, AK, Cheng, H, Copeland, NG, et al. 1998. Evolution of the avian sex chromosome from an ancestral pair of autosomes. Proc Natl Acad Sci USA, 95: 8147–8152.

Frisch, A. 2004. Sex change and gonadal steroids in sexually hermaphroditic fish. Rev Fish Biol Fish, 14: 481–499.

Frolov, SV. 1990. Differentiation of sex chromosomes in Salmonidae. 3. Multiple sex chromosomes in *Coregonus sardinella*. Tsitologiya, 32:1391–1394.

Fu, P, Neff, BD, Gross, MR. 2001. Tactic specific success in sperm competition. Proc R Soc London, 268B: 1105–1112.

Fujiwara, A, Abe, S, Yamaha, E, Yamasaki, F, Yoshida, MC. 1997. Uniparental chromosome elimination in the early embryogenesis of the inviable salmonid hybrids between masu salmon female and rainbw trout male. Chromosoma, 106: 44–52.

Galbreath, PF, Thorgaard, GH. 1995. Sexual maturation and fertility of diploid and triploid Atlantic salmon x brown trout hybrids. Aquaculture, 137: 299–311.

Galbursera, P, Volckaert, FAM, Olevier, F. 2000 . Gynogenesis in the African catfish *Clarias gariepinus* (Burcell, 1822) 3. Induction of endomitosis and the presence of residual genetic variation. Aquaculture, 185: 25–42.

Galetti, PM Jr. 1998. Chromosome diversity in Neotropical fishes: NOR studies. Ital J Zool, 65: 53–56.

Galetti, PM Jr, Foresti, F. 1986. Evolution of the ZZ/ZW system in *Leporinus* (Pisces, Anostomidae). Role of constitutive heterochromatin. Cytolgenet Cell Genet, 43: 43–46.

Galetti, PM Jr, Foresti, F. 1987. Two new cases of ZZ/ZW heterogamety in *Leporinus* (Anostomidae, Characiformes) and their relationships in the phylogeny of the group. Rev Bras Genet, 10: 135–140.

Galetti, PM Jr, Aguilar, CT, Molina, WF. 2000. An overview of marine fish cytogenetics. Hydrobiologia, 420: 55–62.

Gante, HF, Collares-Pereira, MJ, Coelho, MM. 2004. Introgressive hybridization between two Iberian *Chondrosoma* species (Teleostei, Cyprinidae) revisited: new evidence from morphology, mitochondrial DNA, allozyme and NOR-phenotypes. Folia Zool, 53: 423–432.

Gao, ZX, Wang, HP, You, H, Tiu, L, Wang, WM. 2010. No sex specific markers detected in bluegill sunfish. *Lepomis macrochirus* by AFLP. J Fish Biol, 76: 408–414.

Garcia-Vazquez, E, Moran, R, Martinez, JL, et al. 2001. Alternative mating strategies in Atlantic salmon and brown trout. J Hered, 92: 146–149.

Gehring, WJ. 1998. Master Control Genes in Development and Evolution: The Homebox story. Yale University Press, New Haven,USA.

George, T, Pandian, TJ. 1995. Production of ZZ females in the female heterogametic black molly *Poecilia sphenops* by endocrine sex reversal and progeny testing. Aquaculture, 136: 81–90.

George, T, Pandian, TJ. 1997. Interspecific hybridization in poeciliids. Indian J Exp Biol, 35: 628–637.

George, T, Pandian, TJ, Kavumpurath, S. 1994. Inviability of the YY zygote of the fighting fish *Betta splendens*. Isr J Aquacult, 46: 3–8.

Gervai, J, Csanyi,V. 1984. Artificial gynogenesis and mapping of gene contromere distance in the Paradise fish *Macropodus opercularis*. Theor Appl Genet, 68: 481–485.

Gillet, C. 1994. Egg production in Arctic charr (*Salvelinus alpinus* L.) broodstock. Effect of photoperiod on the timing of ovulation and egg quality. Can J Zool, 72: 334–338.

Glamuzina, B, Kozul, Tutman, P, Skardmuca, B. 1999. Hybridization of Mediterranean groupers *Epinephelus marginatus* ♀ x *E. aeneus* ♂ and early development. Aquacult Res, 30: 625–628.

Goddard, KA, Dowling, TE. 1989. Origin and genetic relatives of diploid, triploid, and diploid-triploid mosaic biotypes in the *Phoxinus eos neogaeus* complex. In: Evolution and Ecology of Unisexual Vertebrates, RM Dawley, JP Bogard (eds) New York State Museum, Albany, Bulletin 446: 268–280.

Goddard, KA, Dawley, RM. 1990. Clonal inheritance of a diploid nuclear genome by a hybrid freshwater minnow (*Phoxinus eos neogaeus*, Pisces, Cyprinidae). Evolution, 44: 1052–1065.

Gold, JR. 1979. Cytogenetics. In: Fish Physiology, WS Hoar, DJ Randall, JR Brett (eds) Academic Press, New York, 8: 353–405.

Golubstov, AS, Krysanov, EY. 1993. Karyological study of some cyprinid species from Ethiopia. The ploidy differences between large and small *Barbus*. J Fish Biol 42: 445–455.

Gomelsky, B. 2003. Chromosome set manipulation and sex control in common carp: a review. Aquat Living Resour, 16: 408–415.

Gomelsky, B, Emelyanova, OV, Recoutratsky, AV. 1992. Application of scale cover gene (N) to identification of type of gynogenesis and determination of ploidy in common carp. Aquaculture, 106: 233–237.

Gordon, M. 1952. Sex determination in *Xiphophorus* (*Platypoecilus*) *maculatus*. 3. Differentiation of gonads in platyfish from broods having a sex ratio of the three female to one male. Zoologica, 37: 91–100.

Goudie, CA, Simco, BA, Davis, KB, Liu, Q. 1995. Production of gynogenetic and polyploid catfish by pressure-induced chromosome manipulation. Aquaculture, 133: 185–198.

Grant, BR, Grant, PR. 2008. Fission and fusion of Darwins finches populations. Phil Trans R Soc, 363B, 2821–2829.

Graves, JA. 2006. Sex chromosome specialization and degeneration in mammals. Cell, 124: 901–914.

Gray, AK, Evans, MA, Thorgaard, GH. 1993. Viability and development of diploid and triploid salmond hybrids. Aquaculture, 112: 125–142.

Green, DM. 1990. Muller's ratchet and the evolution of supernumerary chromosomes. Genome, 3: 818–824.

Gregory, TR, Mable, BK. 2005. Polyploidy in animals. In: The Evolution of the Genome, TR Gregory (ed) Elsevier, San Diego, pp 427–451.

Gromicho, M, Collares-Pereira, MJ. 2004. Polymorphism of major ribosomal gene chromosomal site (NOR- phenotypes) in the hybridogenetic fish *Squalius alburnoides* complex (Cyprinidae) assessed through crossing experiments. Genetica, 122: 291–302.

Gromicho, M, Coutanceau, J-P, Ozouf-Costaz, C, Collares-Pereira, MJ. 2006. Contrast between extensive variation of 28s rDNA and stability of 5s rDNA and telometric repeats in the diploid-triploid *Squalius alburnoides* complex and in its maternal ancestor *Squalius pyrenaicus* (Teleostei, Cyprinidae). Chromosome Res, 14: 297–306.

Guan, G, Kobayashi, T, Nagahama, Y. 2000. Sexually dimorphic expression of two types of DM (Doublesex/Mdb-3)- domain genes in a teleost fish the tilapia *Oreochromis niloticus*. Biochem Biophy Res Commun, 272: 662–666.

Guegan, JF, Morand, S. 1996. Polyploid hosts: strange attractors for parasites? Oikas, 77: 366–370.

Guegan, JF, Rab, P, Machordom, A, Doadrio, I. 1995. New evidence of hexaploidy in large. African *Barbus* with some considerations on the origin of hexaploidy. J Fish Biol, 47: 192–198.

Guest, WC. 1984. Trihybrid sunfishes: their growth, catchability and reproductive success compared to parentals and hybrids. Proc Annu Conf Southeast Ass Fish Wildlife Agen, 38: 421–435.

Gui, JF, Jia, SC, Liang, YG. 1992. Meiotic chromosome behavior in male triploid transparent coloured crucian carp *Carassius auratus* L. J Fish Biol, 41: 317–326.

Guo, XH, Liu, SJ, Liu, Y. 2006. Evidence for recombination of mitochondrial DNA in triploid crucian carp. Genetics, 172: 1745–1749.

Guo, X, Yan, J, Liu, S, Xiang, B, Liu, Y. 2008. Isolation and expression analysis of the *Sox9a* gene in triploid crucian carp. Fish Physiol Biochem, DO1. 10. 1007/s 10695-008-92092.

Guo, Y, Cheng, H, Huang, X, Gao, S, Yu, H, Zhou, R. 2005. Gene structure, multiple alternative splicing and expression in gonads of zebrafish (Dmrt1). Biochem Biophys Res Commun, 330: 950–957.

Haaf, T, Schmid, M. 1984. An early stage of ZZ/ZW sex chromosome differentiation in *Poecilia sphenops* var *melanosticta* (Poeciliidae, Cyprinodontiformes). Chromosoma, 89: 37–41.

Haas, R. 1979. Intergeneric hybridization in a sympatric pair of Mexican Cyprinodontid fishes. Copeia, 1979: 149–152.

Habicht, C, Seeb, JE, Gates, RB, Brock, IR, Ofito, CA. 1994. Triploid coho salmon outperform diploid and triploid hybrids between coho salmon and chinook salmon during their first year. Can J Fish Aquat Sci, 51: 31–37.

Haffray, P, Braunt, J-S, Facqueur, J-M, Fostier, A. 2005. Gonad development, growth, survival and quality tracts in triploids of the protandrous hermaphrodite gilthead seabream *Sparus aurata* (L.). Aquaculture, 247: 107–117.

Hamaguchi, S, Sakaizumi, M. 1991. Autosomal genes involved in primary sex determination in the teleost *Oryzias latipes*. Zool Sci, 8: 1126.

Hamaguchi, S, Sakaizumi, M. 1992. Sexually differentiated mechanisms of sterility in interspecific hybrids between *Oryzias latipes* and *O. curvinotus*. J Exp Zool, 263: 323–329.

Han, Y, Liu, M, Zhang, LL, Simpson, B, Zhang, GX. 2010. Comparison of reproductive development in triploid and diploid female rainbow trout *Oncorhynchus mykiss*. J Fish Biol, 76: 1742–1750.

Hardie, DC, Hebert, PDN. 2004. Genome size evolution in fishes. Can J Fish Aquat Sci, 61: 1636–1646.

Harrington, RW. 1967. Environmentally controlled induction of primary male gonochorists from eggs of the self fertilizing hermaphroditic fish *Rivulus marmoratus* Poey. Physiol Zool, 41: 47–460.

Harrington, RW. 1971. How ecological and genetic factors interact to determine, when self-fertilizing hermaphrodites of *Rivulus marmoratus* change into functional males with a reappraisal of the modes of intersexuality among fishes. Copeia, 1971: 389–342.

Harrison, RG. 1990. Hybrid zones: windows of evolutionary process. In: Oxford Survey in Evolutionary Biology, DJ Futuyma, J Antonovics, (eds) Oxford University Press, Oxford, UK, 7: 69–128.

Harvey, SC, Masabanda, J, Carassco, LAP, Bromage, NR, Penman, DJ, Griffin, DK. 2002. Molecular-cytogenetic analysis reveals sequence difference between sex chromosomes of *Oreochromis niloticus;* evidence for an early stage of sex chromosome differentiation. Cytogenet Genome Res, 97: 76–80.

Harvey, SC, Boonphakdee, C, Campos-Ramos, R, Ezaz, MT, Griffin, DK, Bromage, NR, Penman, DJ. 2003. Analysis of repetitive DNA sequences in the sex chromosomes of *Oreochromis niloticus*. Cytogenet Genome Res, 101: 314–319.

Hashimoto, Y, Maegawa, S, Nagai, T, Yamaha, E, Suzuki, H, Yasuda, K, Inoue, K. 2004. Requirement of localized maternal factors of zebrafish germ cell formation. Dev Biol, 268: 152–161.

Hawkins, MB, Thorton, JW, Crews, D, Skipper, JK, Dotte, A, Thomas, P. 2000. Identification of a third distinct estrogen receptor and reclassification of estrogen receptors in teleosts. Proc Natl Acad Sci USA, 97: 10751–10756.

Hayes, T. 1998. Sex determination and primary sex differentiation in amphibians: Genetic and developmental mechanism. J Exp Zool, 281: 373–399.

He, CL, Du, JL, Wu, GC, Lee, YH, Sun, IT, Chang, CF. 2003a. Differential *Dmrt1* transcripts in gonads of the protandrous black porgy *Acanthopagrus schlegeli*. Cytogenet Genome Res, 101: 309–313.

He, C, Chen, L, Simmons, M, Li, P, Kim, S, Liu, ZJ. 2003b. Putative SNP discovery in interspecific hybrids of catfish by comparative EST analysis. Anim Genet, 34: 445–448.

Heath, DD, Rankin, L, Bryden, CA, Heath, JW, Shrimpton, JM. 2002. Heritability and Y chromosome influence in the jack male life history of chinook salmon (*Oncorhynchus tshawytscha*). Heredity, 89: 311–317.

Hett, AK, Pitra, C, Jenneckens, I, Ludwig, A 2005. Characterization of *Sox9* in European Atlantic sturgeon (*Acipenser sturio*). J Hered, 96: 150–154.

Hewitt, GM. 1988. Hybrid zones-natural laboratories for evolutionary studies. Trends Ecol Evol, 3: 158–167.

Hickling, CF. 1968. Fish hybridization. FAO Fish Rep. 44: 1–11.

Hinegardner, R, Rosen, DE. 1972. Cellular DNA content and the evolution of teleostean fishes. Am Nat, 106: 621–644.

Hobbs, J-PA, Munday, PL. 2004. Intraspecific controls spatial distribution and social organization of the coral-dwelling goby *Gobiodon histrio*. Mar Ecol Prog Ser, 278: 253–259.

Hobbs J-PA, Munday, PL, Jones, GP. 2004. Social induction of maturation and sex determination in a coral reef fish. Proc R Soc London, 2713: 2109–2114.

Hoffman, HA, Benson, ME, Fernald, RD. 1999. Social status regulates growth rate: Consequences for life history strategies. Proc Natl Acad Sci USA, 96: 14171–14176.

Horstgen-Schwark, G. 1993. Initiation of tetraploidy breeding line developments in rainbow trout *Oncorhynchus mykiss* (Walbaum). Aquacult Fish Mgmt, 24: 642–652.

Horvath, L, Orban, L. 1995. Genome and gene manipulation in the common carp. Aquaculture, 129: 157–181.

Hrbek, T, Seckinger, J, Meyer, A. 2007. A phylogenetic and biographic perspectives on the evolution of poeciliid fishes. Mol Phylogenet Evol, 43: 986–988.

Huang, X, Guo, Y, Shui, Y, Gao, S, Yu, H, Cheung, H, Zhou, R. 2005. Multiple alternative splicing and differential expression of *dmrt1* during gonadal transformation of the rice field eel. Biol Reprod, 73: 1017–1024.

Hubbs, CL, Hubbs, LC. 1932. Apparent parthenogenesis in nature in a form of fish of hybrid origin. Science, 76: 628–630.

Hubbs, Drewry, GE, Warburton, B. 1959. Occurrence and morphology of a phenotypic male of a gynogenetic fish. Science, 129: 1227–1229.

Hulak, M, Kaspar, V, Psenicka, M, Gela, D, Li, P, Linhart, O. 2009. Does triploidization produce functional sterility of triploid males of tench *Tinca tinca*? Rev Fish Biol Fish, Doi 10. 1007/s 11160-009-9139-9.

Hussain, MG, Chatterji, A MC, Andrew, BJ, Johnstone, R. 1991. Triploidy induction in the Nile tilapia *Oreochromis niloticus L.* using pressure, heat and cold shocks. Theor Appl Genet, 81: 6–12.

Hussain, MG, Penman, DJ, McAndrew, BJ, Johnstone, R. 1993. Suppression of first cleavage in Nile tilapia *Oreochromis niloticus L.* – a comparison of the relative effectiveness of pressure and heat shocks. Aquaculture, 111: 263–270.

Hutchings, JA, Bishop, TD, McGregor-Shaw, CR. 1999. Spawning behaviour of Atlantic cod *Gadus morhua*: evidence of mate competition and male choice in a broadcast spawner. Can J Fish Aquat Sci, 56: 97–104.

Imai, H, Kashiwagi, F, Cheng, J-H, Chen, T-I, Tachihara, K, Yoshino, T. 2009. Genetic and morphological evidence of hybridization between *Nematolosa japonica* and *N. come* (Clupeiformes: Clupeidae) off Okinawa island, Ryuku Archipelago, Japan. Fish Sci, 75: 343–350.

Immler, S, Mazzodi, C, Rasotto, MA. 2004. From sneaker to parental male: change of reproductive traits in black goby *Gobius niger* (Teleostei, Gobiidae). J Exp Zool, 301A: 177–185.

Itono, M, Morishma, K, Kujimoto, T, Bando, E, Yamaha, E, Arai, K. 2006. Premeiotic endomitosis produces diploid eggs in the natural clone loach *Misgurnus anguillicaudatus* (Teleostei, Cobitidae). J Exp Zool, 305A: 513–523.

Itono, M, Okayabashi, N, Morishma, K, Fujimoto, T, Yoshikawa, H, Yamaha, E, Arai, K. 2007. Cytological mechanisms of gynogenesis and sperm incorporation in unreduced diploid eggs of the clonal loach *Misgurnus anguillicaudatus* (Teleostei: Cobitidae). J Exp Zool, 307A: 35–50.

Jaillon, O, Aury, JM, Brunet, F, Petit, N, et al. 2004. Gome duplication in the teleost fish *Tetraodon nigroviridis* reveals the early vertebrate proto-karyotype. Nature, 431: 946–957.

Jamieson, BGM. 1991. Fish Evolution and Systematic Evidence from Spermatozoa. Cambridge University Press, Cambridge, UK.

Jensen, GL, Shelton, WL, Yang, S-L, Wilken, LU. 1978. Sex reversal in gynogenetic grass carp by implantation of methytestosterone. Trans Am Fish Soc, 112: 79–85.

Jesus, CM, Moreira-Filho, O. 2000. Cytogenetic studies in some *Apareiodon* species (Pisceas, Paradontidae). Caryologia, 65: 397–402.

Ji, X-S, Chen, S-L, Liao, X-L, Yang, J-F, Xu, TJ, Ma, HY, Tian, Y-S, Jiang, Y-L, Wu, P-F. 2009. Microsatellite-centromere mapping in *Cynoglossus semilaevis* using gynogenetic diploid families produced by the use of homologous and non-homologous sperm. J Fish Biol, 75: 422–434.

Ji, X-S, Tian, Y-S, Yang, J-F, Wu, P-F, Jiang, Y-L, Chan, S-L. 2010. Artificial gynogenesis in *Cynoglossus semilaevis* with homologous sperm and its verification using microsatellite markers. Aquacult Res, 41: 913–920.

Jia, ZY, Shi, LY, Sun, XW, Lei, QQ. 2008. Inheritance of microsatellite DNA in bisexual gynogenesis of Fengzheng silver crucian carp *Carassius auratus gibelio* (Bloch). J Fish Biol, 73: 1161–1169.

Jianxun, C, Xiuhai, R, Qixing, Y. 1991. Nuclear DNA content variation in fishes. Cytologia, 56: 425–429.

Johnson, KR, Wright, JE. 1986. Female brown trout x Atlantic salmon hybrids produce gynogens and triploids, when backcrossed to male Atlantic salmon. Aquaculture, 57: 345–358.

Jones, AG, Ostlund-Nilsson, S, Avise, JC. 1998. A microsatellite assessment of sneaked fertilization and egg thievery in the fifteen spine stickleblack. Evolution, 52: 848–858.

Jones, AG, Walker, DeE, Kvarnemo, C, Lindstrom, K, Avise, JC. 2001. How cuckoldry can decrease the opportunity for sexual selection. Data and theory from a genetic parentage analysis of the sand goby *Pomatoschistus minutus*. Proc Natl Acad Sci USA, 98: 9151–9156.

Juchno, D, Lackowska, B, Boron, A, Kilarski, W. 2009. DNA content of hepatocyte and erythrocyte nuclei of the spined loach (*Cobitis biwae* L.) and its polyploid forms. Fish Physiol Biochem, Doi 10, 1007/s/10/95-009-9322.x

Kallman, KD. 1984. The new look at sex determination in poeciliid fishes. In: Evolutionary Genetics of Fishes, BJ Turner (ed) Plenum Publishing Corporation, New York, pp 95–171.

Kamler, E. 2005. Parent-egg-progeny relationships in teleost fishes: an energetic perspective. Rev Fish Biol Fish, 15: 399–421.

Kanoh, Y. 2000. Reproductive success associated with territoriality, sneaking and grouping in male rose bitterling *Rhodeus ocellatus* (Pisces: Cyprinidae). Env Biol Fish, 57: 143–154.

Karayucel, S, Karayucel, I, Penman, DJ, McAndrew, BJ. 2004. Evidence for two unlinked "sex reversal" loci in the Nile tilapia *Oreochromis niloticus* and for linkage of one of these to red body colour gene. Aquaculture, 234: 51–63.

Kato, K, Hayashi, R, Yuasa, D, Yamomoto, S, Miyashita, S, Kumai, H. 2002. Production of cloned red sea bream *Pagrus major*, by chromosome manipulation. Aquaculture, 207: 19–27.

Kavumpurath, S. 1992. Ploidy induction in ornamental fish. Ph.D Thesis, Madurai Kamaraj University, Madurai, India.

Kavumpurath, S, Pandian, TJ. 1990. Induction of triploidy in the zebrafish *Brachydanio rerio*. Hamilton. Aquacult Fish Mgmt, 21: 299–306.

Kavumpurath, S, Pandian, TJ. 1992a. Production of YY male in the guppy *Poecilia reticulata* by endocrine sex reversal and progeny testing. Asian Fish Sci, 5: 265–276.

Kavumpurath, S, Pandian, TJ. 1992b. The development of all-male sterile triploid fighting fish (*Betta splendens* Regan) by integrating hormonal sex reversal of broodstook and chromosome set manipulation. Isr J Aquacult, 44: 111–119.

Kavumpurath, S, Pandian, TJ. 1992c. Hybridization and gynogenesis in two species of the genus *Brachydanio*. Aquaculture, 105: 107–116.

Kavumpurath, S, Pandian, TJ. 1994. Induction of heterozygous and homozygous diploid gynogenesis in *Betta splendens* (Regan) using hydrostatic pressure. Aquacult Fish Mgmt, 25: 133–142.

Kawamura, K. 1998. Sex determination system of the rosy barb *Rhodeus ocellatus ocellatus*. Env Biol Fish, 52: 251–260.

Kawamura, K, Hosoya, K. 2000. Masculinizing mechanisms of hybrids in bitterlings (Teleostei: Cyprinae). J Hered, 91: 464–473.

Kawamura, K, Ueda, T, Aoki, K, Hosoya, K. 1999. Spermatozoa in triploids of the rosy bitterling *Rhodeus ocellatus ocellatus*. J Fish Biol, 55: 420–432.

Kazianis, S, Nairn, RS, Walter, RB, Johnston, DA, Kumar, J, et al. 2004. The genetic map of *Xiphophorus* fishes represented by 24 multipoint linkage groups. Zebrafish, 1: 287–304.

Kersten, CA, Krisfalusi, M, Parsons, JE, Cloud, JG. 2001. Gonadal regeneration in masculinized female or steroid treated rainbow trout (*Oncorhynchus mykiss*). J Exp Zool, 290: 396–401.

Keyvanshokooh, S, Pourkazermic, M, Kalbassi, MR. 2007. The RAPD technique failed to identify sex specific sequence in beluga (*Huso huso*). J Appl Ichthyol, 23: 1–2.

Khoo, L, Lim, TM, Chan, WK, Phang, VPE. 1999a. Genetic basis of variegated tail pattern in the guppy *Poecilia reticulata*. Zool Sci, 16: 431.

Khoo, L, Lim, TM, Chan, WK, Phang, VPE. 1999b. Linkage analysis of three sex-linked colour pattern genes in the guppy *Poecilia reticulata*. Zool Sci, 16: 893–903.

Khoo, L, Lim, TM, Chan, WK, Phang, VPE. 1999c. Sex linkage on the black tail candle-peduncle and red tail genes in the tuxedo strain of the guppy *Poecilia reticulata*. Mar Biotechnol, 5: 273–293.

Khuda-Bukhsh, AR, Datta, AR. 1997. Sex specific difference in NOR-location on metaphase chromosome of mosquito fish *Aplocheilus panchax* (Cyprinodontidae). Ind J Exp Biol, 35: 1111–1114.

Kidwell, MG. 2002. Transposable elements and the evolution of genome size in eurykaryotes. Genetica, 115: 49–63.

Kiflawi, M, Mareroll, AI, Goulet, D. 1998. Does mass spawning enhance fertilization in coral reef fish? A case study of the brown surgeon fish. Mar Ecol Prog Ser, 172: 107–114.

Kim, DS, Jo, JY, Lee, TY. 1994. Induction of triploidy in mud loach (*Misgurnus mizolepis*) and its effect on gonad development and growth. Aquaculture, 120: 203–210.

King, J, Withler, RE. 2005. Male nest site fidelity and female serial polyandry in lingcod (*Ophiodon elongatus*, Hexagrammidae). Mol Ecol, 14: 653–660.

Kirankumar, S. 2003. Induction of intra specific and inter specific androgenesis in fish. Ph.D Thesis, Madurai Kamaraj University, Madurai, India.

Kirankumar, S, Pandian, TJ. 2003. Production of androgenetic tiger barb *Puntius tetrazona*. Aquaculture, 228: 37–51.

Kirankumar, S, Pandian, TJ. 2004a. Interspecific androgenetic restoration of rosy barb using cadaveric sperm. Genome, 47: 66–73.

Kirankumar, S, Pandian, TJ. 2004b. Production and progeny testing of androgenetic rosy barb *Puntius conchonius*. J Exp Zool, 301A: 477–490.

Kirankumar, S, Pandian, TJ. 2004c. Use of heterologous sperm for the dispermic induction of androgenesis in barbs. J Fish Biol, 64: 1485–1497.

Kirankumar, S, Anathy, V, Pandian, TJ. 2003. Hormonal induction of supermale golden rosy barb and isolation of Y chromosome specific markers. Gen Comp Endocrinol, 134: 62–71.

Kirasznai, Z, Marian. T, 1986. Shock-induced triploidy and its effect on growth and gonadal development of the European catfish *Silurus glanis*. J Fish Biol, 29: 519–528.

Kirpichnikov, VS. 1981. Genetic Basis of Fish Selection. Springer Verlag, Berlin.

Kitamura, S, Ogata, H, Onozato, H. 1991a. Triploid male masu salmon *Oncorhynchus masou* shows normal courtship behavior. Nippon Suisan Gakkaishi, 57: 2157.

Kitamura, H, Teong, OY, Arakawa, T. 1991b. Gonadal development of artificially induced triploid red sea bream *Pagrus major*. Nippon Suisan Gakkaishi, 57: 1657–1660.

Kitamura, H, Nakao, S, Okado, E, Arakawa, T. 1992. Gonadal development of triploid Black Sea bream *Acanthopagrus schlegeli*. Suisan Zoshoku, 40: 411–415.

Klinkhardt, MB. 1998. Some aspects of karyoevolution in fishes. Ani Res Dev, 48: 7–36.

Knapp, R. 2004. Endocrine mediation of vertebrate male alternate reproductive tactics: the next generation of studies. Integ Comp Biol, 43: 658–668.

Kobayashi, H. 1976. A cytological study on the maturational division in the oogenic process of the triploid ginbuna (*Carassius auratus langsdorfi*). Jap J Ichthyol, 22: 234–240.

Kobayashi, K, Suzuki, K. 1990. Gonadogenesis and sex succession in the protogynous wrasse *Cirrhilabrus temmincki* in Suruga Bay, Central Japan. Jap J Ichthyol, 37: 256–264.

Kobayashi, H, Kawashima, Y, Takeuchi, N. 1976. Comparative chromosome studies in the genus *Carassius*, especially with a finding of polyploidy in the ginbuna (*Carassius auratus langsdorfi*). Jap J Ichthyol, 17: 153–160.

Kobayashi, H, Nakano, K, Nakamura, M. 1977. On the hybrids 4n ginbuna (*Carassius auratus langsdorfi* x kinbuna (*C. auratus* sub sp) and their chromosomes. Bull Jap Soc Sci Fish, 43: 31–37.

Kobayashi, M, Aida, K, Stacey, NE. 1991. Induction of testis development by implantation of 11-ketotestosterone in female goldfish. Zool Sci, 8: 389–393.

Kobayashi, S, Yamada, M, Asaoka, M, Kitamura, T. 1996. Essential role of the posterior morphogen *nanos* for germline development in *Drosophila*. Nature, 380: 708–711.

Kobayashi, T, Sakai, N, Fushiki, S, Nagahama, Y, Amano, M, Aida, K. 1993. Testicular development and changes in the levels of reproductive hormones in triploid male rainbow trout. Nippon Suisan Gakkaishi, 59: 9 81–989.

Kobayashi, T, Ide, A, Hiasa, T, Fushiki, S, Ueno, K. 1994. Production of cloned amago salmon *Oncorhynchus rhodurus*. Fish Sci, 60: 275–281.

Kobayashi, T, Fushiki, Sakai, N, Hara, A, Amano, M, Aida, K, Nakamura, M, Nagahama, Y. 1998. Oogenesis and changes in the levels of reproductive hormones in triploid female rainbow trout. Fish Sci, 64: 206–215.

Kobayashi, T, Takeuchi, Y, Yoshizaki, G, Takeuchi, T. 2003a. Cryopreservation of trout primordial germ cells. Fish Physiol Biochem, 28: 479–480.

Kobayashi, T, Kajura-Kobayashi, H, Nagahama, Y. 2003b. Induction of XX sex reversal by estrogen involves altered gene expression in a teleost tilapia. Cytogenet Genome Res, 101: 289–294.

Kobayashi, T, Yoshizaki, G, Takeuchi, Y, Takeuchi, T. 2004a. Isolation of highly pure and viable primordial germ cells from rainbow trout by *GFP* dependent flow cytometry. Mol Reprod Dev, 67: 91–100.

Kobayashi, T, Matsuda, M, Kajiura,-Kobayashi, H, Suzuki, A, Saito, N, et al. 2004b. Two DM domain genes *DMY* and *DMRT*1 involved in testicular differentiation and development in the medaka *Oryzias latipes*. Dev Dyn, 231: 518–526.

Kobayashi, T, Takeuchi, Y, Takeuchi, T, Yoshizaki, G. 2007. Generation of viable fish from cryopreserved primordial germ cells. Mol Reprod Dev, 74: 207–213.

Kojima, K, Matsubara, K, Kawashima, M, Kajishima, T. 1984. Studies on the gametogenesis in polyploidy ginbuna *Carassius auratus langsdorfi*. J Fac Sci Shinshu Uni, 19: 37–52.

Kokokris, L, Brusle, S, Kentouri, M, Fostier, A. 1999. Sexual maturity and hermaphroditism of the red porgy *Pagrus pagrus* (Teleostei: Sparidae). Mar Biol, 134: 621–629.

Komen, J. 1990. Clones of common carp. Ph.D. Thesis, Agricultural University, Wageninger, The Netherlands.

Komen, J, Richter, CJJ. 1990. Sex control in carp (*Cyprinus carpio L.*). Rec Adv Aquacult, 4: 78–86.

Komen, J, Richter, CJJ. 1993. Sex control in carp. In: Recent Advances in Aquaculture, JF Thur, RJ Roberts (eds) Blackwell Scientific Publications, Oxford, pp 78–86.

Komen, J, Thorgaard, GH. 2007. Androgenesis and gynogenesis and the production of clones in fishes: A review. Aquaculture, 269: 150–173.

Komen, J, de Boer, P, Richter, CJJ. 1992. Male sex reversed in gynogenetic XX females of common carp *Cyprinus carpio L.* by a recessive mutation in a sex-determining gene. J Hered, 83: 431–434.

Komen, J, Duynhouwer, J, Richter, CJJ, Huisman, EA. 1988. Gynogenesis in common carp (*Cyprinus carpio L.*) 1. Effects of genetic manipulation of sexual products and incubation condition of eggs. Aquaculture, 69: 227–239.

Kondo, M, Nanda, I, Hornung, U, Asakawa, S, Shimizu, N, et al. 2003. Absence of the candidate male sex-determining gene *dmrt1*(Y) of medaka from other fish species. Curr Biol, 14: 1664–1669.

Kondo, M, Hornung, U, Nanda, I, Imai, S, Sasaki, T et al. 2006. Genomic organization of the sex determining and adjacent regions of the sex chromosomes of medaka. Genome Res, 16: 815–826.

Kondo, M, Nanda, I, Schmid, M, Schartl, M. 2009. Sex determination and sex chromosome evolution: Insights from medaka. Sex Dev 3: 88–98.

Koteeswaran, R, Pandian, TJ. 2002. Live sperm from post-mortem preserved Indian catfish. Curr Sci, 82: 447–450.

Koteeswaran, R, Pandian, TJ. 2011, Density dependent sex change in the Paradise fish *Macropodus operculatus*. (in prep.).

Kovacs, B, Egedi, S, Bartfai, R, Orban, L. 2000. Male specific DNA markers from African catfish. (*Clarias gariepinus*). Genetica, 110: 267–276.

Koya, Y, Fujita, A, Niki, F, Ishihara, E. 2003. Sex differentiation and pubertal development of gonads in the viviparous mosquitofish *Gambusia affinis*. Zool Sci, 20: 1231–1242.

Kraak, SBM, de Looze, EMA. 1993. A new hypothesis on evolution of sex determination in vertebrates : big females ZW, big males XY. Neth J Zool, 43: 260–273.

226   *Sex Determination in Fish*

Krisfalusi, M, Cloud, JG. 1999. Gonadal sex reversal in triploid rainbow trout *Oncorhynchus mykiss*. J Exp Zool, 284: 466–472.
Krisfalusi, M, Wheeler, PA, Thorgaard, GH, Cloud, JC. 2000. Gonad morphology of female diploid gynogenetic and triploid rainbow trout. J Exp Zool, 286: 505–512.
Kurita, J, Sakaizumi, M, Takashima, F. 1992. Production of trigenomic allotriploid medaka *Oryzias latipes sinensis-curvinotus celebensis*. Nippon Suisan Gakkaishi, 58: 2311–2314.
Kurita, J, Oshiro, T, Takashima, F, Sakaizumi, M. 1995.Cytogenetic studies on diploid and triploid oogenesis in interspecific hybrid fish between *Oryzias latipes* and *O. curvinotus*. J Exp Zool, 273: 234–241.
Kurokawa, H, Saito, D, Nakamura, S, Katoh-Fukui, Y, Ohta, K, Baba, T, Morohashi, K-I, Tanaka, M. 2007. Germ cells are essential for sexual dimorphism in the medaka gonad. Proc Natl Acad Sci USA, 104: 16958–16963.
Kuwamura, T. 1997. The evolution of parental care and mating systems among Tankanykan cichliods. In: Parental care and Mating Systems. H Kawanabe, M Hori, M Nagoshi (eds). Chukyo University Press Yagota, Nagoya. pp 59–86
Kuwamura, T, Nakashima, Y, Yogo, Y. 1994. Sex change in either direction by growth rate advantage in the monogamous goby *Paragobiodon echinocephalus*. Behav Ecol, 5: 434–438.
Kuwamura, T, Suzuki, S, Tanaka, N, Ouchi, E, Karino, K, Nagashima, Y. 2007. Sex change of primary males in a diandric labrid *Helichoeres trimaculatus*: coexistence of protandry and protogyny within a species. J Fish Biol, 70: 1898–1906.
Kusunoki, T, Arai, K, Suzuki, R. 1994a. Production of viable gynogens without chromosome duplication in the spinous loach *Cobitis biwae*. Aquaculture, 119: 11–23.
Kusunoki, T, Arai, K, Suzuki, R. 1994b. Viability and karyotypes of interacial and intergeneric hybrids in loach species. Fish Sci, 60: 415–422.
Laarman, PW. 1979. Reproduction of F$_1$ hybrid sunfishes in small ponds. Prog Fish Cult, 41: 145–147.
Lacerda, SMSN, Batlouni, SR, Homem, CSP, Franca, LR. 2006. Germ cells transplantation in fish: the Nile tilapia model. Anim Reprod, 3: 146–159.
Lacerda, SMSN, Batlouni, SR, Assis, LH, REsende, FM, Campos-Silva, SM, Campos-Silva, R, Sagetelli, TM, Franca, LR. 2008. Germ cell transplantation in tilapia (*Oreochromis niloticus*). Cybium, 32: 115–118.
Lahn, BT, Pearson, NM, Jegalian, K. 2001. The human Y chromosome in the light of evolution. Nature Rev Genet, 2: 207–216.
Lajbner, Z, Slechtova, V, Sletcha, M, Svatora, M, Berrebi, P, Kotlik, P. 2009. Rare and asymmetrical hybridization of the endemic *Barbus carpathicus* with its wide spread congener *Barbus barbus*. J Fish Biol, 74: 418–436.
Lam, T. 1994. Hormones and egg/larval quality in fish. J Wld Aquacult Soc, 25: 2–12.
Lamatsch, DK, Nanda, I, Epplen, JT, Schmid, M, Schertl, M. 2000. Unusual triploid males in a microchromosome-carrying clone of the Amazon molly *Poecilia formosa*. Cytogenet Cell Genet, 9: 148–159.
Lamatsch, DK, Nanda, I, Schlupp, I, Epplen, JT, Schmid, M, Schartl, M 2004. Distribution and stability of supernumerary microchromosomes in natural populations of the Amazon molly *Poecilia formosa*. Cytogenet Genome Res, 106: 189–194.
Lamatsch, DR, Lampert, KP, Fischer, P, Geiger, M, Schlupp, I, Schartl, M. 2009. Diploid Amazon mollies (*Poecilia formosa*) show a higher fitness than triploids in clonal competitive experiments. Evol Ecol, 23: 687–697.
Lampert, KP, Schartl, M. 2008. The origin and evolution of a unisexual hybrid: *Poecilia formosa*. Philos Trans R Soc. B doi: 10. 1098/rstli. 2008.0040.
Lampert, KP, Lamatsch, DK, Fischer, P, Schartl, M. 2008. A tetraploid Amazon molly *Poecilia formosa*. J Hered, doi: 10.1093/hered/esm 102.
Lawrence, C, Ebersole, JP, Kesseli, RV. 2008. Rapid growth and out crossing promote female development in zebra fish (*Danio rerio*). Env Fish Biol, 81: 239–246.

Leary, RF, Allendorf, FW, Forbes, SH. 1993. Conservation genetics of bull trout in the Columbia and Klamath river drainages. Conserv Biol, 7: 856–865.

Le Comber, SC, Smith, C. 2004. Polyploidy in fishes: patterns and processes. Biol J Linn Soc, 82: 431–434.

LeGac, F, Loir, M. 1999. Male reproductive system in fish.In: Encyclopedia of Reproduction, E Knotil, JD Neill (eds) Academic Press, San Diego, 3: 20–30.

Legendre, MG, Teugels, G, Canty, C, Jalabert, B. 1992. A comparative study on morphology growth rate and reproduction of *Clarias gariepinus* (Burchell, 1822) *Heterobranchus longifilis* Valenciennes 1840 and reciprocal hybrids (Pisces, Clariidae). J Fish Biol, 40: 59–79.

Leggatt, RA, Iwama, GK. 2003. Occurrence of polyploidy in fishes. Rev Fish Biol Fish, 13: 237–246.

Lejeune, P. 1987. The effect of local stock density on social behaviour and sex change in the Mediterranean labrid *Coris julis*. Env Biol Fish, 18: 135–141.

Li, S-F, Zou, S-M, Cai, W-Q, Yang, H-Y. 2006. Production of interploid triploid by 4n x 2n blunt snout bream (*Megalobrama amblycephala* Yih) and their first performance data. Aquacult Res, 37: 374–379.

Libertini, A, Sola, L, Rampin, M, Rossi, AR, Iijima, K, Ueda, T. 2008. Classical and molecular cytogenetic characterization of allochthonous European bitterling *Rhodeus amarus* (Cyprinidae, Acheilognathinae) from Northern Italy. Genes Genet Syst, 83: 417–422.

Lincoln, RF. 1981. The growth female diploid and triploid plaice (*Pleuronectes platessa*) x flounder (*Platichthys flesus*) hybrids over one spawning season. Aquacultuure, 25: 259–268.

Lincoln, RF, Scott, AP. 1984. Sexual maturation in triploid rainbow trout (*Salmo gaidneri*, Richardson). J Fish Biol, 25: 385–392.

Linhart, O, Billard, R. 1995. Survival of ovulated oocytes of the European catfish (*Silurus glanis*) after *in vivo* and *in vitro* storage or exposure to saline solutions and urine. Aquat Living Resour, 8: 317–322..

Linhart, O, Kivasnicka, P, Flajshans, M, Kasal, A, Rab, P, Palecek, J, Sletcha, V, Hemaokova, J, Prokes, M. 1995. Genetic studies with tench *Tinca tinca* L.: induced meiotic gynogenesis and sex reversal. Aquaculture, 132: 239–251.

Linhart, O, Radina, M, Flajshans, M, Mavrodiev, N, Nebesarova, J, Gela, D, Kocour, M. 2006. Studies on sperm of diploid and triploid tench *Tinca tinca* L. Aquacult Int, 14: 9–25.

Lindholm, A, Breden, F. 2002. Sex chromosomes and sexual selection in poeciliid fishes. Am Nat, 160: S214–S224.

Lindholm, AK, Brooks, R, Breden, F. 2004. Extreme polymorphism in a Y-linked sexually selected trait. Heredity, 92: 156–162.

Liu, D, Liu, S, You, C, Chen, L, Lin, Z, Lin, L, Wang, J, Liu, Y. 2009. Identification and expression analysis of genes involved in early ovary development in diploid gynogenetic hybrid red crucian carp x common carp. Mar Biotechnol, Do 110, 1007a/s/10126-009-92133.

Liu, M, Sadovy, Y. 2004a. The influence of social factors on adult sex change and juvenile sexual differentiation in a diandric protogynous epinepheline *Cephalophlis boenak* (Pisces, Serranidae). J Zool Lond, 264: 239–248.

Liu, M, Sadovy, Y. 2004b. Early gonadal development and primary males in the protogynous epinepheline *Cephalopholis boenak*. J Fish Biol, 65: 987–1002.

Liu, Q, Goudie, CA, Simco, BA, Davis, KB, Morizot, DC. 1992. Gene-centromere mapping of six enzymes loci in gynogenetic channel catfish. J Hered, 83: 245–248.

Liu, Q, Goudie, CA, Simco, BA, Davis, KB. 1996. Sex linkage of glucose phosphate isomerase B and mapping of the sex determining gene in channel catfish. Cytogenet Cell Genet, 73: 182–285.

Liu, SJ. 2010. Distant hybridization leads to different ploidy fishes. Sci China Life Sci, 53C: 416–425.

Liu, SJ, Liu, Y, Zhou, GJ, Zhang, XJ, Luo, et al. 2001. The formation of tetraploid stocks of red crucian carp x common carp hybrid s, as an effect of intersepcific hybridization. Aquaculture, 192: 172–186.

Liu, SJ, Sun, YD, Zhou, GJ, Zhang, XJ, Liu, Y. 2003. The ultra-microstructure of the mature gonads and erythrocytes in allotetraploids. Prog Nat Sci, 13: 194–197.

Liu, SJ, Sun, YD, Zhang, C, Luo, KK, Liu, Y. 2004c. Production of gynogenetic progeny from allotetraploid hybrids red crucian carp x common carp. Aquaculture, 236: 193–200.

Liu, SJ, Qin, Q, Xiao, J, Lu, W, Shan, J, Li, W, Liu, J, Duan, W, Zhang, C, Tau, M, Zhao, R, Yan, J, Liu, Y. 2007a. The formation of the polyploid hybrids from different sub family fish crossings and evolutionary significance. Genetics, 176: 1023–1034.

Liu, SJ, Liu, S, Tao, M, Li, W, Liu, Y. 2007b. Isolation and expression analysis of testicular type *Sox9b* in allotetraploid fish. Mar Biotechnol, 9: 329–334.

Liu, Y, Zhou, G. 1986. Cytological study on the gonadal development of F₁ hybrids produced by crossing *Carassius auratus* (♀) with *Cyprinus carpio* (♂). Acta Hydrobiol Sinica, 10: 101–108.

Liu, ZH, Zhang, YG, Wang, DS. 2008. Studies on feminization sex determination and differentiation of the southern catfish *Silurus meridionalis*- a review. Fish Physiol Biochem, Doi 10. 1007/s 10965-008-92817.

Lodi, E. 1979. Instances of sex inversion in the domesticated swordtail *Xiphophorus helleri*. Heckel (Pisces, Osteichthyes). Experientia, 35: 1440–1441.

Lodi, E. 1980. Hermaphroditic and genochoric populations of *Cobitis taenia bilineata* Canestrini (Cobitidae, Osteichthyes). Monit Zool Ital, 14: 235–243.

Loewe, L, Lamatsch, DK. 2008. Quantifying the threat of extinction from Mullers ratchet in the diploid Amazon molly (*Poecilia formosa*). BMC Evol Biol, 8: 88 doi: 10.1186/1471-2148/8/88.

Lo Nostro, F, Guerrero, GA. 1996. Presence of primary and secondary males in a population of the protogynous *Synbranchus marmoratus*. J Fish Biol, 49: 788–800.

Lo Nostro, F, Grier, H, Andreone, L, Guerrero, GA. 2003. Involvement of the gonadal germinal epithelium during sex reversal and seasonal testicular cycling in the protogynous swamp eel *Synbranchus marmoratus* Bloch 1975 (Teleostei, Symbranchidae). J Morphol, 257: 107–126.

Love, MS, Yoklarich, M, Thornstein, L. 2002. The rockfishes of the northeast Pacific. University of California Press, Berkeley CA, USA.

Lovshin, LL. 1982. Tilapia hybridization. In: The Biology and Culture of Tilapias, RSV Pullin RH Lowe-McConnel (eds) ICLARM Conf Proc, 7 Manila.

Lowe, TP, Larkin, JR. 1975. Sex reversal in *Betta splendens* Regan with emphasis on the problem of sex determination. J Exp Zool, 191: 25–32.

Lu, R, Chen, H. 1993. Advances in fish cell engineering in China. Aquaculture 111: 41–50.

Lucinda, PHF. 2003. Poeciliidae. In: Check list of the Freshwater Fishes of south and central America, RE Reis, SO Kullander, CJ Ferraris (eds) EDIPUCRS Porto Alegre, Brazil, pp 555–581.

Luckenbach, JA, Godwin, J, Daniels, HV, Beasley, JM, Sullivan, CV, Borski, RJ. 2004. Induction of diploid gynogenesis in southern flounder (*Paralichthys lethostigma*) with homologus and heterologous sperm. Aquaculture, 237: 499–516.

Lutz, CG. 1997. What do you get when you cross.....Aquaculture Magazine March/April 84–90.

Lynch, M, Burger, R, Butcher, D, Gabriel, W. 1993. The mutatioanal meltdown in asexual populations. J Hered, 84: 339–344.

Mable, BK. 2004. Why polyploidy is rarer in animals than in plants: myths and mechanisms. Boil J Linn Soc, 82: 453–466.

MacDonald, RI. 1985. Membrane fusion due to dehydration by polyethylene glycol, dextran or sucrose. Biochemistry, 24: 4058–4066.

Madhu, K, Madhu, R. 2007.Influence of lunar rhythm on spawning of clown anemonefish *Amphiprion percula* under captive conditions in Andaman and Nicobar Islands. J Mar Biol Ass India, 49: 58–64.

Mair, GC, Scott, AG, Penman, DJ, Beadmore, JA, Skibinski, DOF. 1991a. Sex determination in the genus *Oreochromis*. 1. Sex reversal, gynogenesis and triploidy in *O. niloticus* (*L.*). Theor Appl Genet, 82: 144–152.

Mair, GC, Scott, AG, Penman, DJ, Skibinski, DOF, Beardmore, JA. 1991b. Sex determination in the genus *Oreochromis* 2. Sex reversal gynogenesis and triploidy in *O. aureus* Steindachner. Theor Appl Genet, 82: 153–160.

Mair, GC, Abucay, JS, Skibinski, DOF, Abella, TA, Beadmore, JA. 1997. Genetic manipulation of sex ratio for the large scale production of all male tilapia *Oreochromis niloticus*. Can J Fish Aquat Sci, 54: 396–404.

Maistro, EL, Dias, AL, Foresti, F, Oliveira, C, Moreira-Filho, O. 1995. Natural triploidy in *Astyanax scabripinnis* (Pisces, Characidae) and simultaneous occurrence of macro B chromosomes. Caryologia, 47 : 233–239.

Mamcarz, A, Kucharczyk, D, Kujawa, R. 2006. Reciprocal hybrids of tench *Tinca tinca* (*L.*) x bream *Abramis brama* (*L.*), and tench x carp *Cyprinus carpio* L. and some characteristics of their early development. Aquacult Int, 14: 27–33.

Mani, I, Kumar, R, Singh, M, Kushwaha, B, Nagpure, NS, Srivastava, PK, Murmu, K, Rao, DSR, Lakra, WS. 2009. Karyotypic diversity and evolution of seven masher species (Cyprinidae) from India. J Fish Biol, 75: 1079–1091.

Mank, JE, Avise, JC. 2006a. The evolution of reproductive and genomic diversity in ray-finned fishes : insights from phylogeny and comparative analysis. J Fish Biol, 69: 1–27.

Mank, JE, Avise, JC. 2006b. Comparative phylogenetic analysis of male alternative reproductive tactics in ray-finned fishes. Evolution, 60: 1311–1316.

Manna, GK. 1984. Progress in fish cytogenetics. The Nucleus 27: 203–231.

Manning, AJ, Burton, MPM. 2003. Cytological abnormalities in the ovaries of triploid yellowtail flounder *Limanda ferruginea*. (Storer). Fish Physiol Biochem, 29: 269–273.

Manning, MJ, Nakanishi, T. 1996. The specific immune system: cellular defences. In: The Fish Immune System, G Iwana, T Nakanishi (eds) Academic Press, New York, pp 159–205.

Mantelman, II. 1969. On the possibility of polyspermy in bony fishes. Dokl Acad Nauk SSR Biol Sci, 189: 820–823.

Marchand, U, Govoroun, M, D'cotta, H, McMeel, Lareyre, J, Bernot, A, Laudet, V, Guiguen, Y. 2000. DMRT1 expression during gonadal differentiation and spermiogenesis in rainbow trout *Oncorhynchus mykiss*. Biochem Biophy Acta, 1493: 180–187.

Marconato, A, Shapiro, DY, Petersen, CW, Warner, RR, Yoshikawa, T. 1997. Methodological analysis of fertilization rate in the blue head wrasse *Thalassoma bifaciatum*: pair versus group spawns. Mar Ecol Prog Ser, 161: 61–70.

Margarido, VP, Bellafronte, E, Moreira-Filho, O. 2007. Cytogenetic analysis of three sympatric *Gymnotus* (Gymnotiformes, Gymnotidae) species verifies invasive species in the Upper Prana River basin, Brazil. J Fish Biol, 70B: 155–164.

Marian, IA, Pandian, TJ. 1992. Ploidy inductuion as a method for monosex production of fish. Biol Edu, Madurai Kamaraj University, Jan-Mar, 5–11.

Marie, AD, van Herdwerden, L, Choat, JH, Hobbs, JPA. 2007. Hybridization of reef fishes at the Indo Pacific biogeographic barrier: a case study. Coral Reefs, 26: 841–856.

Mariotto, T, Myiazawa, CS. 2006. *Ancistrus dubius* (Siluriformes, Ancistrinae) a complex of species 1. Chromosomic characterization of four populations and occurrence of sexual chromosomes of type XX/XY in the Patanas basin of Mato Grosso, Brazil. Caryologia, 59: 299–304.

Martin, SB. 2007. Association behaviour of self fertilizing *Kryptolebias marmoratus* (Poecy): the influence of microhabitat use on potential for a complex mating system. J Fish Biol, 71: 1383–1392.

Martins, MJ. 1998. Diploids vs triploids of *Rutilus alburnoides*: spatial segregation and morphological differences. J Fish Biol, 52: 817–828.

Martinez, JL, Moran, P, Perez, J, DeGaudemar, B, Beall, E, Gargia-Vazquez, E. 2000. Multiple paternity increases effective population size of southern Atlantic: salmon population. Mol Ecol, 9: 976–782.

Mateos, M, Sanjur, OL, Vrijenhoek, RC. 2002. Historical biogeography of the fish genus *Poeciliopsis* (Cyprinodontiformes). Evolution, 56: 972–984.

Matonda, N, Nlemvo, AB, Ovidio, M, Poncin, P, Phillipart, JL. 2008. Fertility in first generation hybrids of roach *Rutilus rutilus* (L.) and silver bream, *Blicca bjoerkna* (L.). J Appl Ichthyol, 24: 63–67.

Matsubara, K, Arai, K, Suzuki, R. 1995. Survival potential and chromosomes of progeny of triploid and pentaploid females in the loach *Misgurnus anguillicaudatus*. Aquaculture, 131: 37–48.

Matsuda, M, Matsuda, C, Hamaguchi, S, Sakaizumi, M. 1998. Identification of the sex chromosomes of the medaka *Oryzias latipes* by fluorescence *in situ* hybridization. Cytogenet Cell Genet, 82: 257–262.

Matsuda, M, Nagahama, Y, Shinomiya, A, Sato, T, Matsuda, C, Kobayashi, T, et al. 2002. *DMY* is a Y specific DM-domain gene required for male development in the medaka fish. Nature, 417: 555–563.

Matsuda, M, Sato, T, Toyazaki, Y, Nagahama, Y, Hamaguchi, S, Sakaizumo, M. 2003. *Oryzias curvinotus* has *DMY*, a gene that is required for male development in the medaka *O. latipes*. Zool Sci, 20: 159–161.

McCombie, H, Lapegue, S, Cornette, F, Ledu, C, Boudry, P. 2005. Chromosome loss in bi-parental progenies of tetraploid Pacific oyster *Crassostrea gigas*. Aquaculture, 247: 97–105.

McElroy, DM, Kornfield, I. 1993. Novel jaw morphology in hybrids between *Pseudotropheus zebra* and *Labeotropheus fuellebarni* (Teleostei: Cichlidae) from Lake Malawi, Africa. Copeia, 1993: 933–945.

McKay, SJ, Devlin, RH, Smith, MJ. 1996. Phylogeny of Pacific salmon and trout based on growth hormone type-2 and mitochondrial NADH dehydrogenase subunit 3 DNA sequences. Can J Fish Aquat Sci, 53: 1165–1176.

McMillan, M, Wilcove, D. 1994. Gone but not forgotten: Why have species protected by the Endangered species Act become extinct? Endangered Species Update, 11: 5–6.

Mei, W, Cao, Y, Li, R, Shi, B, Hu, A. 1993. A preliminary observation on sex reversal of the mud eel *Monopterus albus*. J Zhenjiang Cole/Zhenjuang Schuichan Xueyuan Xuebao, 12: 53–58.

Mestriner, A, Bertollo, LAC, Galetti, PM Jr. 1995. Chromosome banding and synaptonemal complexes in *Loporinus lacustris* (Pisces, Anostomidae): analysis of a sex system. Chromosome Res, 3: 440–443.

Meyer, A, Schartl, M. 1999. Gene and genome duplications in vertebrates: the one to four (to-eight in fish) rule and the evolution of novel gene functions. Curr Opin Cell Biol, 11: 699–704.

Meyers, LA, Levin, DA. 2006. On the abundance of polyploids in the flowering plants. Evol Int J Org Evol, 60: 1198–1206.

Mia, MY, Taggart, JB, Gilnour, AE, et al. 2005. Detection of hybridization between Chinese carp species (*Hypophthalmichthys molitrix Arishichthys nobilis*) in hatchery broodstock of Bangaldesh using DNA microsatellite loci. Aquaculture, 247: 267–273.

Mims, SD, Shelton, WL, Linhart, O, Wang, C. 1997. Induced meiotic gynogenesis of paddlefish (*Polyodon spathula*). J World Aquacult Soc, 28: 334–343.

Miranda, LA, Strussmann, CA, Somoza, GM. 2001. Immuno-cytochemical identification of GtH1 and GtH2 cells during the temperature-sensitive period for sex determination in pejerrey *Odontesthes bonariensis*. Gen Comp Endocrinol, 124: 45–52.

Mirza, JA, Shelton, WL. 1998. Induction of gynogenesis and sex reversal in silver carp. Aquaculture, 68: 1–14.

Miura, C, Miura, T, Yamashita, M, Yamaguchi, K, Nagahama, Y. 1996. Hormonal induction of all stages of spermatogenesis in germ-somatic cell culture from immature. Japanese eel testis. Dev Growth Differ, 38: 257–262.

Miura, C, Kuwahara, R, Miura, T. 2007. Transfer of spermatogenesis-related cDNAs into eel testis germ somatic cell coculture pellets by electroporation: Methods for analysis of gene function. Mol Reprod Dev, 74: 420–427.

Miura, T. 1999. Spermatogenic cycle in fish. In : Encyclopedia in reproduction. E Knobil, JD Neill (eds) Academic Press, San Diego, 4: 571–578.

Miura, T, Miura, C. 2003. Molecular control mechanisms of fish spermatogenesis. Fish Physiol Biochem, 28: 181–186.

Miura, T, Miura, C, Ohta, T, Nader, MR, Todo, T, Yamaguchi, K. 1999. Estradial 17β stimulates the renewal of spermatogonial stem cells in male. Biochem Biophy Res Com, 264: 230–234.

Mobley, KB, Amundsen, Forsgren, E, Svensson, PA, Jones, AG. 2009. Multiple mating and a low incidence of cuckoldry for nest-holding males in the two-spotted goby *Gobiusculus flavescens*. BMC Evol Biol, doi: 10.1186/1471-21489-6.

Moenkhaus, WJ. 1904. The development of hybrids between *Fundulus heteroclitus* and *Menidia notata* with special reference to the behaviour of the maternal chromatin. Am J Anat, 3: 29–66.

Molloy, PP, Goodwin, NB, Cote, IM, Reynolds, ID, Gage, MJG. 2007. Sperm competition and sex change: A comparative analysis across fishes. Evolution, 61: 640–652.

Momotani, S, Morishma, K, Zhang, Q, Arai, K. 2002. Genetic analysis of the progeny of triploid gynogens induced from unreduced eggs of triploid (diploid female x tetraploid male) loach. Aquaculture, 204: 311–322.

Moran, P, Martinize, JL, Garcia-Vazquez, E, Pendas, AM. 1996. Sex chromosome linkage of 5s rDNA in rainbow trout (*Oncorhynchus mykiss*). Cytogenet Cell Genet, 75: 145–150.

Moreira-Filho, O, Bertollo, LAC, Galetti, PM Jr. 1980. Evidence for a multiple sex chromosome system with female heterogamety in *Apareiodon affinis* (Pisces, Parodontidae). Caryologia, 33: 83–91.

Moreira-Filho, O, Bertollo, LAC, Galetti, PM Jr. 1993. Distribution of sex chromosome mechanisms in Neotropical fish and description of ZZ/ZW system in *Parodon hilarii*. Caryologia, 46: 115–125.

Morescalchi, A, Hureau, JC, Olmo, L, Ozouf-Costaz, E, Pisano, E, Stanyon, R. 1992. A multiple sex chromosome system in Antartic ice fishes. Polar Biol, 11: 655–661.

Mori, T, Saito, S, Koshioka, C, Arai, K. 2006. Aquaculture performance of triploid barfin flounder *Verasper moseri*. Fish Sci, 72: 270–277.

Morishma, K, Nakayama, I, Arai, K. 2001. Microsatellite centromere mapping in the loach *Misgunus anguillicaudatus*. Genetica, 111: 59–69.

Morishma, K, Horie, S, Yamaha, E, Arai, K. 2002. A cyptic clonal line of the loach *Misgurnus anguillicaudatus* (Teleostei, Gobitidae) evidenced by induced gynogenesis, interspecific hybridization microsatellite genotyping and multilocous DNA fingerprinting. Zool Sci, 19: 565–575.

Morishma, K, Yoshikawa, H, Arai, K. 2008. Meiotic hybridogenesis in triploid *Misgurnus anguillicaudatus* loach derived from a clonal lineage. Heredity, 100: 581–586.

Muller-Belecke, A, Horstgen-Schwark, G. 1995. Sex determination in tilapia (*Oreochromis niloticus*) sex ratios in homozygous gynogenetic progeny and their offspring. Aquaculture, 137: 57–65.

Muller-Belecke, A, Horstgen-Schwark, G. 2007. A YY male *Oreochromis niloticus* strain developed from an exceptional mitotic gynogenetic male and growth performance of genetically all-male progenies. Aquacult Res, 38: 773–775.

Munday, PL. 2001. Changing sex. Nature, Australia, September 2001: 51–59.

Munday, PL, Caley, MJ, Jones, GP. 1998. Bidirectional sex change in a coral-dwelling goby. Behav Ecol Sociobiol, 43: 371–377.

Munday, PL, Burston, PM, Warner, RR. 2006a. Diversity and flexibility of sex change strategies in animals. Trends Ecol Evol, 21: 89–95.

Munday, PL, Wilson, W, Warner, RR. 2006b. A social basis for the development of primary males in a sex changing fish. Proc R Soc London, 273B: 2845–2851.

Munehara, H, Takano, K, Koya, Y. 1989. Internal gametic association and external fertilization in the Elkhorn sculpin *Alcichthys alcicornis*. Copeia, 1989: 675–678.

Muniayandi, N, Haniffa, MA, Gopalakrishnan, A, Basheer, VS, Muner, A. 2006. Genetic variability of *Channa punctatus* populations using randomly amplified polymorphic DNA. Aquacult Res, 37: 1151–1155.

Munoz, RC, Warner, RR. 2003. Alternative context of sex change with social control in the buck tooth parrotfish *Sparisoma radians*. Env Biol Fish, 61: 877–887.

Myers, JM, Penman, DJ, Rana, JK, Bromage, N, Powell, SF, McAndrew, BJ. 1995. Applications of induced androgenesis with tilapia. Aquaculture 137: 150.

Naddafi, R, Abdoli, A, Kiabi, HB, Amiri, MB, Karami, M. 2005. Age, growth and reproduction of the Caspian roach (*Rutilus rutilus caspicus*) in the Anzali and Gomishan wetlands North Iran. J Appl Ichthyol, 21: 492–497.

Nagai, T, Yamaha, E, Arai, K. 2001. Histological differentiation of primordial germ cells in zebrafish. Zool Sci, 18: 215–223.

Nagasawa, K, Takeuchi, Y, Miwa, M, Higuchi, K, Morita, T, Mitsuboshi, T, Miyaki, K, Kadomura, K, Yoshizaki, G. 2009. cDNA cloning and expression analysis of a *vasa*-like gene in Pacific bluefin tuna *Thunnus orientalis*. Fish Sci, 75: 71–79.

Nagler, JJ, Cavileer, T, Steinhorst, K, Devlin, RH. 2004. Determination of genetic sex in chinook salmon (*Oncorhynchus tshawytscha*) using the male-linked growth hormone psuedogene by Real-Time PCR. Mar Biotechnol, 6: 186–191.

Nagoya, H, Kawamura, K, Ohta, H. 2010. Production of androgenetic amago salmon *Oncorhynchus masou* ishikawae with dispermy fertilization. Fish Sci, 76: 305–313.

Nagy, A, Rajki, K, Horvath, Csanyi, V. 1978. Investigation on carp *Cyprinus carpio* L. gynogenesis. J Fish Biol, 13: 215–224.

Nakamura, M. 1978. Morphological and experimental studies on sex differentiation of the gonad in several teleost fishes. Ph. D. Thesis. Hukkaido University, Hukkaido, Japan.

Nakamura, M, Tsuchiya, F, Ishahashi, M, Nagahama, Y. 1993. Reproductive characteristics of precociously mature triploid male masu *Oncorhynchus masou*. Zool Sci, 10: 117–125.

Nakayama, I, Biagi, CA, Koide, N, Devlin, RH. 1998. Identification of a sex linked *GH* pseudogene in two species of Japanese salmon (*Oncorhynchus masou* and *O. rhodurus*) Aquaculture, 173: 65–72.

Nakazona, A, Kuwamura, T. 1987. Sex Change in Fishes. Tokai University Press, Tokyo.

Nam, YK, Kim, DS. 2004. Ploidy status of progeny from the crosses between tetraploid males and diploid females in mud loach (*Misgurnus mizolepis*). Aquaculture, 236: 575–582.

Nam, YK, Cho, YS, Kim, DS. 2000. Isogenic transgenic homozygous fish induced by artificial parthenogenesis. Transgenic Res, 9: 463–469.

Nam, YK, Cho, HJ, Im, JH, Park, I-S, Choi, GC. 2001a. Production of all female diploid and triploid far eastern catfish *Silurus asotus* (Linnaeus): survival and growth performance. Aquacult Res, 32: 991–997.

Nam, YK, Choi, GC, Park, DJ, Kim, DS. 2001b. Survival and growth of induced tetraploid mud loach (*Misgurnus mizolepis*). Aquacult Int, 9: 61–71.

NaNakorn, U, Sidthikraiwong, P, Tanchala-Nukit, W, Roberts, TR. 1993a. Chromosome study of hybrid and gynogenetic offspring of artificial crosses between members of the catfish families Clariidae and Pangasiidae. Env Biol Fish, 37: 317–322.

NaNakorn, U, Rangsin, W, Witchasunscul, S. 1993b. Suitable conditions for induction of gynogenesis in the catfish *Clarias macrocephalus* using sperm of *Pangasius sutchi*. Aquaculture, 118: 53–68.

Nanda, I, Feichtinger, W, Schmid, M, Schroder, JH, Zischier, H, Epplen, JT. 1990. Simple repetitive sequences are associated with differentiation of the sex chromosomes in the guppy fish. J Mol Evol, 30: 456–462.

Nanda, I, Schartl, M, Feichtinger, W, Epplen, JT, Schmid, M. 1992. Early stages of sex chromosome differentiation in fish as analysed by simple repetitive DNA sequence. Chromosoma, 101: 301–310.

Nanda, I, Volff, JN, Weis, S, Korting, C, Froschauer, A, et al. 2000. Amplification of a long terminal repeat-like element on the Y chromosome of the platyfish *Xiphophorus maculatus*. Chromosoma, 109: 173–180.

Nanda, I, Kondo, M, Hornung, U, Asakawa, S, Winkler, C, Shimizu, A, Shanz, Z, Haaf, T, Shimizu, N, Shima, A, Schmid, M, Schartl, M. 2002. A duplicated copy of *DMRT1* in the sex determination region of the Y chromosome of the medaka *Oryzias latipes*. Proc Natl Acad Sci, USA, 99: 11778–11783.

Nanda, I, Schlupp, I, Lamatsch, DK, Lampert, K, Schmid, M, Schartl, M. 2007. Stable inheritance of host species derived microchromosomes in the gynogenetic fish *Poecilia formosa*. Genetics, 177: 917–926.

Naruse, K. Ijiri, K, Shima, A, Egami, N. 1985. The production of cloned fish in the medaka (*Oryzias latipes*). J Exp Zool, 236: 335–341.

Neff, BD. 2004. Increased performance of offspring sired by parasitic males in bluegill sunfish. Behav Ecol, 15: 327–331.

Nikoljukin, NI. 1972. Distant hybridization in Acipenseridae and teleostei, Theory and Practice. Ordalennay gilbridizashiy osetrovyh i kostishyh rhy, toriy i praktika, Moscva, Pischevay Promyshlennosty.

Nilsson, EE, Cloud, JC. 1993. Extent of mosaicism in experimentally produced diploid/triploid chimeric trout. J Exp Zool, 266: 47–50.

Noble, GK, Kumpf, K. 1936. The sexual behaviour and secondary sex characters of gonadectomized fish. Anat Supl Rec, 67: 113–116.

Noleto, RB, Amorim, AP, Vicari, MR, Artoni, RF, Cestard, MM. 2009. An unusual ZZ/ZW sex chromosome system in *Characidium* fish (Crenuchidae, Characiformes) with the presence of rDNA sites. J Fish Biol, 75: 448–453.

Nomura, T, Arai, K, Hayashi, T, Suzuki, R. 1998. Effect of temperature on sex ratio of normal gynogenetic diploid loach. Fish Sci, 64: 753–758.

Norberg, B, Valkner, V, Huse, J, Karlson, I, Grung, GL. 1991. Ovulatory rhythms and egg viability in the Atlantic halibut *Hippoglossus hippoglossus*. Aquaculture, 97: 365–371.

Norbrega, RH, Baltouni, SR, Franca, LR. 2009. An overview of functional and steroidological evaluation of spermatogenesis and germ cell transplantation in fish. Fish Physiol Biochem, 35: 197–206.

Nwadukwe, FO. 1995. Hatchery propagation of five hybrid groups by artificial hybridization of *Clarias gariepinus* (B.) and *Heterobranchus longifilis* (Val.) (Clariidae) using dry powdered carp pituitary hormone. J Aquacult Tropics, 10: 1–11.

Ocalewicz, K, Dobosz, S, Kuziminski, H, Goryczko, K. 2009. Formation of chromosome aberrations in androgenetic rainbow trout *Oncorhynchus mykiss*. J Fish Biol, 75: 2372–2379.

Ocalewicz. K, Dobosz, S, Kuziminski, H, Goryczko, K. 2010. Chromosome rearrangements and survival of androgenetic rainbow trout (*Oncorhynchus mykiss*). J Appl Genet, 51: 309–317.

Ohno, S. 1967. Sex chromosomes and sex linked genes. In: Monographs on Endocrinology, vol 1. A Labhart, LT Mann, LT Samuels, J Zander (eds) Springer Verlag, Berlin.

Ohno, S, Wolf, U, Atkin, NB. 1968. Evolution from fish to mammals by gene duplication. Heriditas, 59: 169–187.

Ohno, S. 1970. The enormous diversity in genome sizes of fish as a reflection of nature's extensive experiments with gene duplication. Trans Am Fish Sci, 99: 120–130.

Ohno, S. 1974. Animal cytogenetics, vol 4. Chordata 1: Prochordata, Cyclostoma and Pisces Gebruder Borntraeger, Berlin.

Ohta, K, Sundaray, JSK, Okada, T, Sakai, M, Kitano, T, Yamaguchi, A, Takeda, T, Matsuyama, M. 2003. Bidirectional sex change and its steroidogenesis in the wrasse *Psuedolabrus sieboldi*. Fish Physiol Biochem, 28: 173–174.

Ojima, Y, Uedo, H. 1982. A karyotypic study of the conger eel (*Conger myriaster*) in *in vitro* cells with special regards to the identification of the sex chromosome. Proc Jap Acad, 58B: 56–59.

Okada, YK, Yamashita, H. 1994. Experimental investigation of the manifestation of secondary sexual characters in fish using the medaka *Oryzias latipes* as material. J Fac Sci Tokyo Univ, Tokyo Sect IV, 6: 383–347.

Okutsu, T, Yano, A, Nagasawa, K, Shikina, S, Kobayashi, T, Takeuchi, Y, Yoshizaki, G. 2006a. Manipulation of fish germ cell. Visualization cryopreservation and transplantation. J Reprod Dev, 52: 685–693.

Okutsu, T, Suzuki, K, Takeuchi, Y, Takeuchi, T, Yoshizaki, G. 2006b. Testicular germ cells can colonize sexually undifferentiated embryonic gonad and produce functional egg in fish. Proc Natl Acad Sci USA, 103: 2725–2729.

Okutsu, T, Shikina, S, Kanno, M, Takeuchi, Y, Yoshizaki, G. 2007. Production of trout offspring from triploid salmon parents. Science, 317: 1517.

Oldfield, RG. 2005. Genetic abiotic and social influences on sex differentiation in cichlid fishes and the evolutionary sequential hermaphroditism. Fish Fish, 6: 93–110.

Oliveira, AV, Prioli, AJ, Prioli, SMAP, Bignotto, TS, Julio, HF Jr, Carrer, H, Agostinho, CS, Prioli, LM. 2006. Genetic diversity of invasive and native (Pisces: Perciformes) populations in Brazil with evidence of interspecific hybridizations. J Fish Biol, 69B: 260–277.

Oliveira, C, Almeida-Toledo, LF, Foresti, F. 1988. Chromosome formulae of Neotropical freshwater fishes. Rev Bras Genet, 11: 577–624.

Oliveira, C, Foresti, F, Hilsdorf, HWS. 2009. Genetics of Neotropical fish: from chromosomes to populations. Fish Physiol Biochem, 35: 81–100..

Olsen, LC, Aasland, R, Fjose, AA. 1997. *vasa*-like gene in zebrafish identifies putative primordial germ cells. Mech Dev, 66: 95–105.

Onozato, H. 1993. Diet production of the supermale (YY) by androgenesis in amago salmon. Biol Int, 28: 69–71.

Onozato, H, Torisawa, M, Kusama, M. 1983. Distribution of the gynogenetic polyploidy crucian carp *Carassius auratus* in Hokkaido, Japan. Jap J Ichthyol, 30: 184–190.

Orlando, EF, Katsu, Y, Miyagawa, S, Iguchi, T. 2006. Cloning and differential expression of estrogen receptor and aromatase genes in the self fertilizing hermaphrodite and male mangrove rivulus *Karyptolebias marmoratus*. J Mol Endocrinol, 37: 353–365.

Oshima, K, Morishma, K, Yamaha, E, Arai, K. 2005. Reproductive capacity of triploid loaches obtained from Hokkaido Island, Japan. Ichthyol Res, 52: 1–8.

Oshiro, T. 1987. Cytological studies on diploid gynogenesis induced in the loach *Misgurnus amguillicaudatus*. Bull J Soc Sci Fish, 53: 933–942.

Oshiro, T, Deng, Y, Higaki, Takashima, F. 1991. Growth and survival of diploid and triploid hybrid masu salmon *Oncorhynchus masou*. Nippon Suisan Gakkaishi, 57: 1851–1857.

Ospina-Alvarez, N, Piferrer, F. 2008. Temperature-dependent sex determination in fish revisited: Prevalence, a single sex ratio response pattern and possible effects of climate change. PLoSONE, 3: e2837.doi: 10.1371/journal pone.0002837.

Otani, S, Maegawa, S, Inoue, K, Arai, K, Yamaha, E. 2002. The germ cell lineage identified by *vas*-mRNA during the embryogenesis in goldfish. Zool Sci, 19: 519–526.

Otani, S, Iwai, T, Nakahata, S, Sakai, C, Yamashita, M. 2009. Artificial fertilization by intracytoplasmic sperm injection in a teleost fish, the medaka (*Oryzias latipes*). Biol Reprod, 80: 175–183.

Otto, SP. 2007. The evolutionary consequences of polyploidy. Cell, 131: 452–462.

Otto, SP, Whiton, J. 2000. Polyploidy incidence and evolution. Annu Rev Genet, 34: 401–437.

Owusu-Frimpong, M, Hargreaves, JA. 2000. Incidence of conjoined twins in tilapia after shock induction in polyploidy. Aquacult Res, 37: 374–379.

Padhi, BK, Mandal, RF. 2001. Applied Fish Genetics. Fishing Chimes, Visakapatnam, India.

Pala, I, Schartl, M, Thorsteinsdóttir, S, Coelho, MM. 2009. Sex Determination in the *Squalius alburnoides* complex: An initial characterization of sex cascade elements in the context of a hybrid polyploid genome. PLoS ONE 4: e6401. doi:10.1371/journal.pone. 00006401.

Pajuelo, JC, Lorenzo, JM, Bilbao, OA, Ramos, AG. 2006. Reproductive characteristic of the benthic coastal fish *Diplodus vulgaris* (Teleostei: Sparida) in the Canarian Archipelago, Northwest Africa. J Appl Ichthyol, 22: 414–418.

Pandey, N, Lakra, WS. 1997. Evidence of female heterogamety, B chromosome and natural tetraploidy in the Asian catfish *Clarias batrachus* used in aquaculture, Aquaculture, 149: 31–38.

Pandian, TJ. 2001. Guidelines for research and utilization of genetically modified fish. Curr Sci, 81: 1172–1178.

Pandian, TJ. 2003. Transplanting the fish. Curr Sci, 85: 101.

Pandian, TJ. 2010. Sexuality in Fishes. Science Publishers, Enfield, NH, USA.

Pandian, TJ. 2011. Sex Differentiation in Fishes (in prep.).

Pandian, TJ, Varadaraj, K. 1987. Techniques to regulate sex ratio and breeding in tilapia. Curr Sci, 56: 337–343.

Pandian, TJ, Varadaraj, K. 1988a. Sterile female triploidy in *Oreochromis mossambicus*. Bull Aquacult Ass Canada, 88: 134–135.

Pandian, TJ, Varadaraj, K. 1988b. Techniques for producing all male and allotriploid *Oreochromis mossambicus*. In: Tilapia in Aquaculture, RS Pullin, T Bhukuswan, K Thonguthai, JL MacClean (eds) ICLARM Cont Proc, Bangkok, pp 243–249.

Pandian, TJ, Varadaraj, K. 1990a. The development of monosex female *Oreochromis mossambicus* broodstock by integrating gynogenetic techniques with endocrine sex reversal. J Exp Zool, 255: 88–96.

Pandian, TJ, Varadaraj, K. 1990b. Techniques to produce 100 % male tilapia. Naga (ICLARM), 13: 3–5.

Pandian, TJ, Sheela, SG. 1995. Hormonal induction of sex reversal in fish. Aquaculture, 138: 1–22.

Pandian, TJ, Koteeswaran, R. 1998. Ploidy induction and sex control in fish. Hydrobiologia, 384: 167–243.

Pandian, TJ, Koteeswaran, R. 1999. Natural occurrence of monoploid and polyploids in the Indian catfish *Heteropneustes fossilis*. Curr Sci, 76: 1134–1137.

Pandian, TJ, Kirankumar, S. 2003. Androgenesis and conservation of fishes. Curr Sci, 85: 917–931.

Pandian, TJ. 1993. Endocrine and chromosome manipulation techniques for the production of all male and all female population in food and ornamental fishes. Proc Ind Natl Sci Acad, 59B: 549–566.

Parenti, LR, Grier, H. 2004. Evolution and phylogeny of gonad morphology in bony fishes. Integr Comp Biol, 44: 333–348.

Parise-Maltempi, PP, Martins C, Oliveira C, et al. 2007. Identification of a new repetitive element in the sex chromosomes of *Leporinus elongatus* (Teleostei: Characiformes: Anostomidae): new insights into the sex chromosomes of *Leporinus*. Cytogenet Genome Res, 116: 218–223.

Parker, A, Kornfield, I. 1996. Polygynandry in *Pseudotropheus zebra*, a cichlid from Lake Malawi. Env Biol Fish, 47: 345–352.

Parsons, JE, Busch, RA, Thorgaard GH, Scheerer, PD. 1986. Increased resistance of triploid rainbow trout x coho salmon hybrids to infectious hematopoietic necrosis virus. Aquaculture, 57: 337–343.

Pavlidis, M, Kokokris, L, Paspatis, M, Somarakis, S, Kentouri, M, Divinach, P. 2006. Gonadal development in hybrids of Mediterranean sparids: *Sparus aurata* (female) x *Pagrus pagrus* (male). Aquacult Res, 37: 302–303.

Paull, GC, Fibby, AL, Tyler, CR. 2009. Growth rate during early life affects sexual differentiation in roach (*Rutilus rutilus*). Env Biol Fish, 85: 277–284.

Pedersen, RA. 1971. DNA content, ribosomal gene multiplicity, and cell size in fish. J Exp Zool, 177: 65–79.

Peichel, CL. 2005. Fishing for the secrets of vertebrate evolution in threespine sticklebacks. Dev Dyn, 234: 815–823.

Pellegrini, E, Menuet, A, Lethimonier, C, Adrio, F, Gueguen, MM, Tascon, C, Anglade, I, Pakdel, F, Kah, O. 2005. Relationships between aromatase and estrogen receptor in the basin of teleost fish. Gen Comp Endocrinol, 142: 60–66.

Penman DJ, Shah MS, Beardmore JA, Skibinski DOF. 1987. Sex ratio of gynogenetic and triploid tilapia. In: Proc World Symp Selection, Hybridisation Genetic Eng in Aquacult, K Tiews (ed) Bordeaux, 2: 267–276.

236 *Sex Determination in Fish*

Pendas, AM, Moran, P, Frije, JP, Garcia-Vazquez, E. 1994. Chromosomal mapping and nucleotide sequence of two tandem repeats of Atlantic salmon 5S rDNA. Cytogenet Cell Genet, 67: 31–36.

Perruzi, S. Scott, AG, Domaniewski, CJ, Warner, GF. 1993. Initiation of gynogenesis in *Oreochromis niloticus* following heterologous fertilization. J Fish Biol, 43: 585–591.

Perruzi, S, Rudolfsen, G, Primicerio, R, Frangten, M, Kauric, G. 2009. Milt characteristics of diploid and triploid Atlantic cod (*Gadus morhua* L.). Aquacult Res, 40: 1160–1169.

Peters, HM. 1975. Hermaphroditism in cichlid fishes. In: Intersexuality in the Animal Kingdom. R Reinboth (ed) Springer Verlag, New York, pp 228–235.

Petersen, CW. 1987. Reproductive behaviour and gender allocation in *Serranus fasciatus*, a hermaphroditic reef fish. Anim Behav, 35: 1601–1614.

Petersen, CW, Warner, RR, Cohen, S, Hess, HC, Hewell, AT. 1992. Variable pelagic fertilization success: implications for mate and spatial patterns of mating. Ecology, 73: 391–401.

Pfeifer, P, Goedecke, W, Kuhfitting-Kulle, S, Obe, G. 2004. Pathways of DNA double-strand break repair and their impact on the prevention and formation of chromosomal aberrations. Cytogenet Genome Res, 104: 7–13.

Phillipart, JC, Berrebi, P. 1990. Experimental hybridization of *Barbus barbus* and *Barbus meridionalis* physiological, morphological and genetic aspects. Aquatic Living Resour, 3: 325–332.

Phillips, RB, Ihssen, PE. 1985. Identification of sex chromosomes in lake trout (*Salvelinus namaycush*). Cytogenet Cell Genet, 39: 14–18.

Phillips, R, Rab, P. 2001. Chromosome evolution in the Salmonidae (Pisces): an update. Biol Rev Camb Phil Soc, 76: 1–25.

Piferrer, F, Donaldson, EE. 1989. Gonadal differentiation in coho salmon *Oncorhynchus kisutch* after a single treatment with androgen or estrogen at different stages during oogenesis. Aquaculture, 77: 2–3.

Piferrer, F, Benfey, TJ, Donaldson, EM. 1994. Gonadal morphology of normal and sex reversed triploid and gynogenetic diploid coho salmon (*Oncorhynchus kisutch*). J Fish Biol, 45: 541–553.

Piferrer, F, Blazquez, M, Navarro, L, Gonzalez, A. 2005. Genetic endocrine and environmental components of sex determination and differentiation in the European sea bass (*Dicentrarchus labrax* L.). Gen Comp Endocrinol, 142: 102–110.

Piferrer, F, Beaumont, A, Falguiere, J-C, Flajshans, M, Haffray, P, Colombo, L. 2009. Polyploid fish and shellfish: Production, biology and applications to aquaculture for performance improvement and genetic containment. Aquaculture, 293: 125–156.

Poleo, GA, Dennison, RS, Reggio, BC, Godke, RA, Tiersch, TR. 2001. Fertilization of eggs of zebrafish *Danio rerio* by intracytoplasmic sperm injection. Biol Reprod, 65: 961–966.

Poleo, GA, Godke, RA, Tiersch, TR. 2005. Intracytoplasmic sperm injection using cryopreserved, fixed and freeze dried sperm in eggs of Nile tilapia. Mar Biotechnol, 7: 104–111.

Pongthana, N, Penman, DJ, Karnasuta, JB, McAndrew, BJ. 1995. Induced gynogenesis in the silver barb (*Puntius gonionotus* Bleeker). Aquaculture, 135: 267–276.

Pongthana, N, Penman, DJ, Baoprasertkul, P, Hussain, M, Islam, MS, Powell, SF, McAndrew, BJ. 1999. Monosex female production in the silver barb (*Puntius gonionotus* Bleeker). Aquaculture, 173: 247–256.

Price, DJ. 1984. Genetics of determination in fishes a brief review. In: Fish Reproduction: Strategies and Tactics, GW Potts, RJ Wooton (eds) Academic Press, New York, pp 77–89.

Pruginin, Y, Rothbard, S, Wohlfarth, GW, Halevy, A, Mao, R, Hulata, G. 1975. All-male broods of *Tilapia nilotica* x *T. aurea* hybrids. Aquaculture, 6: 11–21.

Purdom, CE. 1983. Genetic engineering by the manipulation of chromosomes. Aquaculture, 33: 287–300.

Quillet, E. 1994. Survival growth and reproductive traits of mitotic gynogenetic rainbow trout females. Aquaculture, 123: 223–236.

Quillet, E, Gaignon, JL. 1990. Thermal induction of gynogenesis and triploidy in Atlantic salmon (*Salmo salar*) and their potential interest for aquaculture. Aquaculture, 89: 351–364.

Quillet, E, Chevassus, Blanc, J-M, Krieg, F, Chourrout, D. 1988. Performance of auto and allotriploids in salmonids. 1. Survival and growth in freshwater farming. Aquat Living Resour, 1: 29–43.

Quillet, E, Aubard, G, Quean, I. 2002. Mutation in sex determining gene in rainbow trout: detection and genetic analysis. J Hered, 93: 91–99.

Quattro, JM, Avise, JC, Vrijenhoek, RC. 1991. Molecular evidence for multiple origins of hybridogenetic fish clones (Poeciliidae, *Poeciliopsis*). Genetics, 127: 391–398.

Quattro, JM, Avise, JC, Vrijenhoek, RC. 1992. An ancient clonal lineage in the fish genus *Poeciliopsis* (Atheriniformes: Poeciliidae). Proc Natl Acad Sci USA, 89: 348–352.

Rab, P. 1981. Karyotype of European catfish *Silurus glanis* (Siluridae, Pisces) with remarks on cytogenetics of silurid fishes. Folia Zool, 30: 271–286.

Rab, P, Rabova, M, Pereira, CS, Collares-Pereira, MJ, Pelikanova, S. 2008. Chromosome studies of European cyprinid fishes; interspecific homology of leuciscine cytotaxonomic marker-the largest subtelocentric chromosome pair as revealed by cross species painting. Chromosome Res, 16: 863–873.

Raicu, P, Taisescu, E, Banarescu, P. 1981. *Carassius carassius* and *C. auratus* a pair of diploid and tetraploid representative species (Pisces, Cyprinidae). Caryologia, 46: 233–240.

Rasch, EM, Balsano, JS. 1989. Trihybrids related to unisexual molly fish *Poecilia formosa.* In: Evolution and Ecology of Unisexual Vertebrates, RM Dawley, J Bogart (eds) New York State Museum Bulletin, 466: 252–267.

Raymond, CS, Shamu, CE, Shen, MM, Seifert, KJ, Hirsch, B, et al. 1998. Evidence for evolutionary conservation of sex determining genes. Nature, 391: 691–995.

Reed, KM, Phillips, RB. 1997. Polymorphism of the nucleolus organizer region (NOR) on the putative sex chromosomes of Arctic charr (*Salvelinus alpinus*) is not sex related. Chromosome Res, 3: 221–226.

Reis, RE, Kullander, SO, Ferraris, CJ Jr. 2003. Checklist of the freshwater fishes of South America. Edipucrs, Porto Alegre, Brazil.

Reinboth, R. 1970. Intersexuality in fishes. Mem Soc Endocrinol, 18: 515–543.

Ribeiro, F, Cowx, IG, Tago, P, Filipe, AF, Costa, LM, Collares-Pereira, MJ. 2003. Growth and reproductive traits of diploid and triploid forms of the *Squalius alburnoides*. Cyprinid complex in the tributary of the Guadiana River, Portugal. Arch Hydrobiol, 156: 471–484.

Rico, C, Kuhnlein, U, Fitzgerald, GJ. 1992. Male reproductive tactics in the three. spined stickleback-an evaluation by DNA fingerprinting. Mol Ecol, 1: 79–87.

Rinchard, J, Garcia-Abiad, MA, Dabrowski, K, Ottobre, J, Schmidt, D. 2002. Induction of gynogenesis and gonad development in the muskellunge. J Fish Biol, 60: 427–441.

Rishi, KK. 1989. Current status of fish cytogenetics. In: Fish Genetics in India, PR Das, AG Jhingran (eds)Today and Tomorrow Publishers, New Delhi, pp 1–18.

Robalo, JI, Sousa-Santos, C, Levy, A, Almada, VC. 2006. Molecular insights on the taxonomic position of the paternal ancestor of the *Squalius alburnoides* hybridogenetic complex. Mol Phylogenet Evol, 39: 276–281.

Robertson, DR. 1981. The social and mating systems of two labroid fishes *Halichoeres maculipinna* and *H. garnoti* off the Caribbean coast of Panama. Mar Biol, 64: 327–340.

Robertson, OH. 1961. Prolongation of the life span of kokanee salmon (*Oncorhynchus nerka kennerlyi*) by castration before beginning of gonad development. Proc Natl Acad Sci USA, 47: 609–611.

Rock, J, Eldrige, M, Champion, A, Johnson, P, Joss, J. 1996. Karyotype and nuclear DNA content of the Australian lungfish *Neoceratodus forsteri* (Ceratodidae: Dipnoi). Cytogenet Cell Genet, 73: 187–189.

Rodriquez-Mari, A, Canestro, C, BreMiller, RA, Nguyen-Johnson, A, Asakawa, K, Kawakami, K, Postlethwait, JH. 2010. Sex reversal in zebrafish *fancl* mutants is caused by Tp-53 mediated germ cell apoptosis. PLoS Genet, 6: e1001034. doi: 10.1371/journal. pgen.1001034.

Romashov, DD, Beleyaeva, VN. 1964. Cytology of radiation gynogenesis and androgenesis in the loach (*Misgurnus fossilis L.*). Doklady Akad. Nauk. USSR, 157: 964–967.

Rosa, R, Bellafronte, E, Moreira-Filho, O, Margarido, VP. 2006. Constitutive heterochromatin 5S and 18S rDNA genes in *Apareiodon* sp (Characiformes, Parodontidae) with a ZZ/ZW sex chromosome system. Genetika, 128: 159–166.

Rosa, R, Caetano-Filho, M, Shibatta, OA, Giuliano-Caetano, L. 2009. Cytotaxonomy in distinct populations of *Hoplias* aff. *malabaricus* (Characiformes, Erythrinidae) from lower Paranapanema River basin. J Fish Biol, 75: 2682–2694.

Ross, RA, Urton, JR, Boland, J, Shapiro, MD, Peichel, CL. 2009. Turnover of sex chromosomes in the stickleback fishes (Gasterosteidae). PLoS Genet, 5: e1000391.

Ross, RM, Losey, GS, Diamond, M 1983. Sex change in a coral reef fish: Dependence of stimulation and inhibition on relative size. Science, 221: 574–575.

Ruiguang, Z, Zhang, S, Wanguo, L. 1986. Studies on karyotype and nuclear DNA contents of some cyprinid fishes with notes on fish polyploids in China. In: Proc Second Int Conf Indo-Pacific Fish, Ichthyological Society of Japan, Tokyo, pp 877–885.

Ruis-Carus, R. 2002. Chromosome analysis of the sexual phases of the protogynous hermaphrodites *Epinephelus guttatus* and *Thalassoma bifasciatum* (Serranidae, Labridae, Teleostei). Carib J Sci, 38: 44–51.

Saber, MH, Noveiri, SB, Pourkazemi, M, Yarmohammadi, M. 2008. Induction of gynogenesis in stellate sturgeon (*Acipenser stellatus*, Pallas, 1771) and its verification using microsatellite markers. Aquacult Res, 39: 1483–1487.

Sadovy, Y, Liu, M. 2008. Functional hermaphroditism in teleosts. Fish Fish, 9: 1–43.

Saito, T, Goto-Kazeto, R, Arai, K, Yamaha, E. 2008. Xenogenesis in teleost fish through generation of germ line chimeras by single primordial germ cell transplantation. Biol Reprod, 78: 159–166.

Sakai, Y, Kohda, M, Kuwamura, T. 2001. Effect of changing harem on timing of sex change in female cleaner fish *Labroids dimidiatus*. Behaviour, 62: 251–257.

Sakaizumi, M, Shimizu, Y, Hamaguchi, S. 1992. Electrophoretic studies of meiotic segregation in inter and intraspecific hybrids among East Asian species of the genus *Oryzias* (Pisces: Oryziatidae). J Exp Zool, 264: 85–92.

Sakaizumi, M, Shimizu, Y, Matsuzaki, T, Hamaguchi, S. 1993. Unreduced diploid eggs produced by interspecific hybrids between *Oryzias latipes* and *O. curvinotus*. J Exp Zool, 266: 312–318.

Sandra, G-E, Norma, MM. 2009. Sexual determination and differentiation in teleost fish. Rev Fish Biol Fish, DOI 10. 1007/s 11160-009-9123-4.

Sarder, MR, Penman, DJ, Myers, JM, McAndrew, BJ. 1999. Production and propogation of fully inbred clonal lines in the Nile tilapia (*Oreochromis niloticus*). J Exp Zool, 284: 675–685.

Schaefer, SA. 1998. Conflict and resolution impact of new taxa in phylogenetic studies of the Neotropical cascudinhos (Siluroidei, Labricariidae). In: Phylogeny and Classification of Neotropical Fishes, LR Malabarba, RE Reis, RP Vari (eds) Edupucrs, Porto Alegre, Brazil.

Schafhauser-Smith, D, Benfey, T. 2001. The reproductive physiology of three age classes of adult female diploid and triploid brook trout (*Salvelinus fontinalis*). Fish Physiol Biochem, 25: 319–333.

Schartl, M. 2004. A comparative view on sex determination in medaka. Mech Dev, 121: 639–645.

Schartl, M, Schlupp, I, Schartl, A, Meyer, MK, Nanda, I, Schmid, M, Epplen, JT, Parzefall, J. 1991. On the stability of dispensable constituents of the eukaryotic genome: Stability of coding sequences versus truly hypervariable sequences in a clonal vertebrate the Amazon molly *Poecilia formosa*. Proc Natl Acad Sci USA, 88: 8759–8763.

Schartl, M, Nanda, Schlupp, I, Wilde, B, Epplen, JT, Schmid, M, Parzefall, J. 1995a. Incorporation of subgenomic amounts of DNA as compensation for mutational load in a gynogenetic fish. Nature, 373: 68–71.

Schartl, M, Wilde, B, Schlupp, I, Parzefall, J. 1995b. Evolutionary origin of a parthenoform the Amazon molly *Poecilia formosa* on the basis of molecular geneology. Evolution, 49: 827–835.

Schaschl H, Tobler M, Plath M, Penn DJ, Schlupp I. 2008. Polymorphic MHC loci in an asexual fish, the amazon molly *(Poecilia formosa*; Poeciliidae). Mol Ecol, 17: 5220–5230.

Scheerer, PD, Thorgaard, GH, Allendorf, F. 1991. Genetic analysis of androgenetic rainbow trout. J Exp Zool, 260: 382–390.

Scheerer, PD, Thorgaard, GH, Allendorf, F, Knudsen, K. 1986. Androgenetic rainbow trout produced from inbred and outbred sperm sources show similar survival. Aquaculture, 57: 289–298.

Schulz, RW, Miura, T. 2002. Spermatogenesis and its endocrine regulation. Fish Physiol Biochem, 26: 43–56.

Schultheis, C, Zhou, Q, Froschauer, A, Nanda, I, Selz, Y, et al.. 2006. Molecular analysis of the sex determining region of the platyfish *Xiphophorus maculatus*. Zebrafish, 3; 295–305.

Schultheis, C, Bohne, A, Schartl, M, Volff, JN, Galiana-Arnoux, D. 2009. Sex determination diversity and sex chromosome evolution in poeciliid fish. Sex Dev, 3: 68–77.

Schultz, RJ. 1977. Evolution and ecology of unisexual fishes. Env Biol Fish, 10: 277.

Schultz, RJ. 1989. Origins and relationships of unisexual poeciliids. In: Ecology and Evolution of Live-bearing Fishes, GK Meffe, FF Jr Snelson (eds) Prentice Hall, Cliffs, NJ, New York, pp 69–87.

Schwartz, FJ. 1981. World literature to fish hybrids with an analysis by family, species, and hybrid. Suppl. NOAA Tech Rep NMFS- SSRF-750, US Dept Comm p 507.

Schwenk, K, Breder, N, Streit, B. 2008. Introduction, extent, processes and evolutionary impact of interspecific hybridization in animals. Phil Trans R Soc, 363B: 2805–2811.

Scott, AG, Penman, DJ, Beardmore, JA, Skinbinski, DOF. 1989. The 'YY' supermale in *Oreochromis niloticus* (L.) and its potential in aquaculture. Aquaculture, 78: 237–251.

Scribner, KT, Avise, JC. 1993. Molecular evidence for phylogeographic structuring and introgressive hybridization in mosquitofish. Mol Ecol, 2: 139–149.

Scribner, KT, Avise, JC. 1994a. "Population cage" experiments with a vertebrate: the temporal demography and cytonuclear genetics of hybridization in *Gambusia* fishes. Evolution, 48: 155–171.

Scribner, KT, Avise, JC. 1994b. Cytonuclear genetics of experimental fish hybird zones inside Biosphere II. Proc Natl Acad Sci USA 91: 5066–5069.

Scribner, KT, Page, KS, Bartron, ML. 2001. Hybridization in freshwater fishes: a review of case studies and cytonuclear methods of biological inference. Rev Fish Biol Fish, 10: 293–323.

Seeb, JE, Thorgaard, GH, Tynan, T. 1993. Triploid hybrids between chum salmon female X chinook salmon male have early sea-water tolerance. Aquaculture, 1 17: 37–45.

Shah, MS. 1988. Female homogamety in tilapia *(Oreochromis niloticus)* revealed by gynogenesis. Asian Fish Sci, 1: 215–219.

Shapiro, DY. 1981. Size, maturation and the social control of sex reversal in the coral-reef fish *Anthias squamipinnis*. J Zool Lond, 193: 105–128.

Shapiro, DY, Rasotto, MB. 1993. Sex differentiation and gonadal development in the diandric protogynous wrasse *Thalassoma bifasciatum* (Pisces Labridae). J Zool Lond, 230: 231–245.

Shapiro, DY, Marconato, A, Yoshikawa, T. 1994. Sperm economy of a coral reef fish. Ecology, 75: 1334–1344.

Shelton, WL. 1986. Control of sex in cyprinids for aquaculture. In: Aquaculture of Cyprinids, R Billard, J Marcel (eds), INRA, Paris, pp 1372–1374.

Shen, J, Fan, Z, Wang, E. 1983. Karyotype studies of male triploid crucian carp (Fengzhenk crucian carp) in Heilongjian. Acta Genet Sci, 10: 133–136.

Shen, XY, Cui, JZ, Gong, QL, Nagahama, Y. 2007. Cloning of the full length coding sequence and expression analysis of *Sox9b* in guppy *(Poecilia reticulata)*. Fish Physiol Biochem, 33: 195–202.

Shimizu, Y, Oshiro, T, Sakaizumi, M. 1993. Electrophoretic studies of diploid, triploid and tetraploid forms of the Japanese silver crucian carp *Carassius auratus langsdorfi*. Jap J Ichthyol, 40: 65–75.

Shimizu, YH, Shibata, N, Yamashita, M. 1997. Spermiogenesis without preceding meiosis in the hybrid medaka *Oryzias latipes* and *O. curvinotus*. J Exp Zool, 279: 102–112.

Shinomiya, AI, Shibata, N, Sakaizumi, M, Hamaguchi, S. 2002. Sex reversal of genetic female (XX) induced by the transplantation of XY somatic cells in the medaka *Oryzias latipes*. Int J Dev Biol, 46: 711–717.

Shirak, A, Serrousi, E, Gnaani, A, Howe, AE, Domokhovsky, R, Zilberman, N, Kocher, TD, Hulata, G, Ron, M. 2006. *Amh* and *Dmrta2* genes map to tilapia (*Oreochromis* spp) linkage group 23 within Quantitative Trait Locus regions for sex determination. Genetics, 174: 1573–1581.

Shikina, S, Ihara, S, Yoshizaki, G. 2008. Culture conditions for the maintaining survival and mitotic activity of rainbow trout transplantable Type A spermatogonia. Mol Reprod Dev, 75: 529–537.

Siegfried, KR. 2010. In search of determinants: gene expression during gonadal sex differentiation. J Fish Biol, 76: 1879–1902.

Slanchev, K, Stebler, J, Cueva-Mendez, de la, G, Raz, E. 2005. Development without germ cells : The role of the germ line in zebrafish sex differentiation. Proc Natl Acad Sci USA, 102: 4074–4079.

Smith, EM, Gregory, TR. 2009. Patterns of genome size diversity in the ray-finned fishes. Hydrobiologia, 625: 1–25.

Smitherman, RO, Hester, FE. 1962. Artificial propagation of sunfishes, with meristic comparisons of three species of *Lepomis* and five of their hybrids. Trans Am Fish Soc, 91: 333–341.

Sofikitis, N, Kaponis, A, Mio, Y, Makredimas, D, et al. 2003. Germ cell transplantation: a review and progress report on ICSI from spermatozoa generated in xenogenic testes. Human Reprod Update, 9: 291–307.

Sogard, PS, Gilbert-Hovarth, E, Anderson, EC, Fischer, R, Berkeley, SA, Garza, JC. 2007. Multiple paternity in viviparous kelp reef fish *Sebastes altrovirens*. Env Biol Fish, 81: 7–13.

Sola, L, Iselli, V, Rossi, AR, Rasch, EM, Monaco, PJ. 1992. Cytogenetics of bisexual/unisexual species of *Poecilia*. 3. Karyotype of *Pocilia formosa*, a gynogenetic species of hybrid origin. Cytogenet Cell Genet, 60: 236–240.

Sola, L, Cipelli, O, Gornung, E, Rossi, AR, Andaloro, F, Crosetti, D. 1997a. Cytogenetic characterization of the greater amberjack *Seriola dumerili* (Pisces: Carangidae) by different staining techniques and fluorescence *in situ* hybridization. Mar Biol, 128: 573–577.

Sola, L, Marzovillo, m, Rossi, A, Gornung, E, Bressanello, S, Turner, B. 1997b. Cytogenetic analysis of self-fertilizing fish *Rivulus marmoratus*; remarkable chromosomal constancy over a vast geographic range. Genome, 40: 945–949.

Som, C, Bagheri, HC, Reyer, H-U. 2007. Mutation accumulation and fitness effects in hybridogenetic populations: a comparison of sexual to asexual system. BMC Evol Biol, 7: 80, doi: 10.1186/1471-2148-7-80.

Sousa-Santos, C, Collares-Pereira, MJ, Almada, V. 2007a. Reading the history of a hybrid fish complex from its molecular record. Mol Phytogenet Evol, 45: 981–996.

Sousa-Santos, C, Collares-Pereira, MJ, Almada, V. 2007b. Fertile triploid males-An uncommon case among hybrid vertebrates. J Exp Zool, 307A: 220–225.

Sridhar, S, Haniffa, MA. 1999. Interspecific hybridization between the catfishes *Heteropneustes fossilis* (Bloch) and *Heteropneustes microps* (Gunther) (Siluridae: Heteropneustidae). Curr Sci, 76: 871–873.

Stanley, JG. 1983. Gene expression in haploid embryos of Atlantic salmon. J Hered, 74: 19–22.

Stanley, JG, Biggers, CJ, Schultz, DE. 1976. Isozymes in androgenetic and gynogenetic white amur gynogenetic carp and carp-amur hybrids. J Hered, 67: 129–134.

Stein, J, Reed, KM, Wilson, CC, Phillips, RB. 2002. A sex-linked microsatellite locus isolated from the Y chromosome of lake charr *Salvelinus namaycush*. Env Biol Fish, 64: 211–216.

Steinemann, S, Steinemann, M. 2005. Y chromosomes: born to be destroyed. Bioessays, 27: 1076–1083.

Stenike, D, Salzburger, W, Meyer, A. 2006. Novel relationship among ten fish model species revealed based on a phylogenomic analysis using ESTs. J Mol Evol, 62: 772–784.

St Mary, CM. 1998. Characteristic gonad structure in the gobiid genus *Lythrypnus* and comparison with other hermaphroditic gobies. Copeia, 1998: 720–724.

Storchova, Z, Breneman, A, Candle, J, Dunn, J, Burbank, K, O'Toole, E, Pellman, D. 2006. Genome-wide genetic analysis of polyploidy in yeast. Nature, 443: 541–547.

Sun, YD, Liu, SJ, Zhang, C, Li, JZ, et al. 2003. Chromosome number and gonadal structure of F_0-F_{11} allotetraploid crucian carp. China J Genet, 30: 37–41.

Sun, YD, Zhang, C, Liu, SJ, et al. 2007. Induced interspecific androgenesis using diploid sperm from allotriploid common carp x red crucian carp hybrids. Aquaculture, 264: 47–53.

Sutterlin, AM, Holder, J, Benfey, TJ. 1987. Early survival rate and subsequent morphological abnormalities in land locked anadromous and hybrid (land locked x anadromous) diploid and triploid Atlantic salmon. Aquaculture, 64: 154–164.

Suwa, M, Arai, K, Suzuki, R. 1994. Suppression of the first cleavage and cytogenic studies on the gynogenetic loach. Fish Sci, 60: 673–681.

Suzuki, A, Taki, Y. 1981. Karyotype of tetraploid Asian cyprinid *Acrossocheilus sumatransis*. Jap J Ichthyol, 28: 173–176.

Suzuki, Y, Nijhout, F. 2006. Evolution of a polyphenism by genetic accommodation. Science, 311: 650–652.

Suzuki, R, Oshiro, T, Nakanishi, T 1985. Survival growth and fertility of gynogenetic diploids induced in the cyprinid loach *Misgurnus anguillicaudatus*. Aquaculture, 48: 45–65.

Swarup, H. 1959. Production of triploidy in *Gasterosteus oculeatus L.* J Genet, 56: 129–142.

Tabata, K. 1991. Induction of gynogenetic diploid males and presumption of sex determination mechanisms in the hirame *Paralichthys olivaceius*. Nippon Suisan Gakkaishi, 57: 845–850.

Taborsky, M. 1998. Sperm competition in fish: 'burgeois' males and parasitic spawning. Tree, 13: 222–226.

Takagi M, Sakai K, Taniguchi N. 2008. Direct evidence of multiple paternities in natural population of viviparous Japanese surfperch by allelic markers of microsatellite DNA loci. Fish Sci, 74: 976–982.

Takegaki, T, Nakazono, A. 1999. Reproductive behavior and mate fidelity in the monogamous goby, *Valenciennea longipinnis*. Ichthyol Res, 46: 115–123.

Taniguchi N., Han HS. Tsujimura, A. 1994. Variation in some quantitative traits of clones produced by chromosome manipulation in ayu, *Plecoglossus altivelis*. Aquaculture, 120: 53–60.

Takeuchi Y, Yoshizaki G, Takeuchi, T. 2001 Production of germ-line chimeras in rainbow trout by blastomere transplantation. Mol Reprod Dev, 59: 380–389.

Takeuchi, Y, Yoshizaki, G, Kobayashi, T, Takeuchi, T. 2002. Mass isolation of primordial germ cells from transgenic rainbow trout carrying the green fluorescent protein gene driven by the *vasa* gene promoter. Biol Reprod, 67: 1087–1092.

Takeuchi, Y, Yoshizaki, G, Kobayashi, T, Takeuchi, T. 2003. Generation of live fry from intraperitoneally transplanted primordial germ cells in rainbow trout. Biol Reprod, 69: 1142–1149.

Takeuchi, Y, Yoshizaki, G, Kobayashi, T, Takeuchi, T. 2004. Surrogate-broodstock produces salmonids. Nature, 430: 629–630.

Tatarenkov, A, Lina, SMQ, Taylor, DS, Avise, JC. 2009. Long term retention of self-fertilization in a fish clade. Proc Natl Acad Sci USA, 106: 14456–14459.

Taylor, CA, Burr, BM. 1997. Reproductive biology of the northern starhead top minnow *Fundulus dispar* (Osteichthyes, Fundulidae) with a review of data for freshwater members of the genus. Am Midl Nat, 137: 151–164.

Taylor, J, Mahon, R. 1977. Hybridization of *Cyprinus carpio* and *Carassius auratus*; the first two exotic species in the lower Laurentian GreatLakes. Env Biol Fish, 1: 205–208.

Thorgaard, GH. 1983. Chromosome set manipulation and sex control in fish. In: Fish Physiology, vol 9B. WS Hoar, DJ Randall, EM Donaldson (eds) Academic Press, Orlando, pp 405–434.

Thorgaard, GH. 1986. Ploidy manipulation and performance. Aquaculture, 57: 57–64.

Thorgaard, GH, Jazwin, ME, Stier, AR. 1981. Polyploidy induced by heat shock in rainbow trout. Trans Am Fish Soc, 110: 546–550.

Thorgaard, GH, Scheerer, PD, Hersberger, WK, Myers, JM. 1990. Androgenetic rainbow trout produced using sperm from tetraploid males show improved survival. Aquaculture, 85: 215–221.

Timmermans, LPM, Taverne, N. 1989. Segregation of primordial germ cells: their numbers and fate during early development of *Barbus conchonius* (Cyprinidae, Teleostei) as indicated by 3H-Thymidin incorporation. J Morphol, 202: 225–237.

Tiwari, BK, Kirubagaran, R, Ray, AK. 2000. Gonadal development in triploid *Heteropneustes fossilis*. J Fish Biol, 57: 1343–1348.

Toledo-Filho, SA, Almeida-Toledo, LF, Forestei, F, Bernardino, G, Calcagnotto, D. 1994. Monitoramento e conservacao em projecto de hildridacaoentre pacu e tambacu. In: *Cadernos de* Ictiogenetica, SA Toledo-Filho (ed) SaoPaulo, Brazil, pp 5–49.

Traut, W, Winking, H. 2001. Meiotic chromosomes and stages of sex chromosome evolution in fish: zebrafish, platyfish and guppy. Chromosome Res, 9: 659–672.

Trautman, MB. 1981. The Fishes of Ohio. Ohio State Univ Press, Ohio, USA.

Tripathy, N, Hoffman, M, Willing, EM, Lanz, C, Weigel, D, Dreyer, C. 2009. Genetic linkage map of the guppy *Poecilia reticulata* and quantitative loci analysis of male size and colour vision. Proc R Soc, London, 276B: 2195–2208.

Tsigenopoulos, CS, Rab, P, Naran, D, Berrebi, P. 2002. Multiple origins of polyploidy in the phylogeny of southern African barbs (Cyprinidae) as inferred from mtDNA markers. Heredity, 88: 466–473.

Tuan, PA, Mair, GC, Little, DC, Beardmore, JA. 1999. Sex determination and fertility of genetically male tilapia production in the Thai-Chitralada strain of *Oreochromis niloticus* (*L*.). Aquaculture, 173: 257–269.

Turner, B, Elder, JF Jr, Laughlin, TF, Davis, WP, Taylor, DS. 1992. The extreme clonal diversity and divergence in populations of a selfing hermaphroditic fish. Proc Natl Acad Sci, USA, 89: 10643–10647.

Tvelt, HB, Benfey, TJ, Martin-Robichaud, DJ, McGowan, C, Reith, M. 2006. Gynogenesis and sex determination in Atlantic halibut (*Hippoglossus hippoglossus*). Aquaculture, 252: 899–906.

Uchida, D, Yamashita, M, Kitano, T, Iguchi, T. 2002. Oocyte apoptosis during the transision from ovary like tissue to testis during sex determination of juvenile zebrafish. J Exp Biol, 205: 711–718.

Uedo, T. 1996. Chromosome aberrations in salmonid fish embryos using prolonged stored eggs in coelomic fluid. Chromosome Info. Ser, 61: 13–15.

Uedo, T, Kobayashi, M, Sato, O. 1986. Triploid rainbow trouts induced by polyethylene glycol. Proc Jap Acad, 62B: 161–164.

Uedo, T, Sawada, M, Kobayashi, J. 1987. Cytogenetic characteristics of embryos between diploid female and triploid male in rainbow trout. Jap J Genet, 62: 461–465.

Uedo, T, Sato, R, Iwata, M, Komaru, A, Kobayashi, J. 1991. The viable 3.5n trouts produced between diploid females and allotriploid males. Jap J Genet, 66: 71–75.

Umino, T, Arai, K, Maeda, K, Zhang, Q, Sakai, K, Niware, I, Nakagawa, H. 1997. Natural clones detected by multi locus DNA fingerprinting in gynogenetic triploid ginbuno *Carassius auratus langsdorfi* in Kurose River, Hiroshima. Fish Sci, 63: 147–148.

Underwood, JL, Hestand, RS II, Thompson, BZ. 1986. Gonadal regeneration in grass carp following bilateral gonadectomy. Prog Fish Cult, 48: 54–56.

Uneo, K, Arimoto, B. 1982. Induction of triploids in *Rhodeus ocellatus ocellatus* by cold shock treatment of fertilized eggs. Experientia, 38: 544–546.

Uyeno, T, Miller, RR. 1971. Multiple sex chromosomes in a Mexican cyprinodontid fish. Nature, 231: 452–453.

Valenzuela, N, LeClere, A, Shikano, T. 2006. Comparative gene expression of steroid oogenic factor 1 in *Chrysemys picta* and *Apalone mutica* turtles with temperature dependent and genotypic sex determination. Evol Dev, 8: 424–432.

van de Peer, Y, Taylor, JS, Meyer, A. 2003. Are all fishes ancient polyploids? J Struct Funct Genomics, 2: 65–73.

Vanderputte, M, Dupont-Nivet, M, Chavanne, H, Chatain, B. 2007. A polygenic hypothesis for sex determination in the European sea bass *Dicentrarchus labrax*. Genetics, 176: 1049–1057.

Van Doon, GS, Kirkpatrick, M. 2007. Turnover of sex chromosomes induced by sexual conflict. Nature, 231: 452–453.

Van Doornik, D, Parker, SJ, Millard, SR, Bernston, EA, Moran, P. 2008. Multiple paternity is prevalent in Pacific perch (*Sebastes alutus*) off the Oregon coast and is correlated with female size and age. Env Biol Fish, 83: 269–275.

Van Eenennaam, AL, Stocker, RK, Thiery, RG, Hagstrom, NT, Droshov, SF. 1990. Egg fertility, early development and survival from crosses of diploid female x triploid male grass carp (*Ctenopharyngondon idella*). Aquaculture, 86: 111–125.

Van Eenennaam, AL, van Eenennaam, JP, Medrano, JE, Droshov, SL. 1996. Rapid verification of meiotic gynogenesis and polyploidy in white sturgeon (*Acipenser transmontanus* Richardson). Aquaculture, 147: 177–189.

Van Eenennaam, AL, van Eenennaam, JP, Medrano, JE, Droshov, SL. 1999. Evidence of female heterogametic sex determination in white sturgeon. J Hered, 90: 231–233.

van Herwerden, AL, Choat, JH, Dudgeon, CL, Carlos, G, Newman, SL, Frisch, A, van Oppen, M. 2006. Contrasting patterns of genetic structure in two species of the coral trout *Plectropomus* (Serranidae) from east and west Australia: introgressive hybridization or ancestral polymorphism. Mol Phylogenet Evol, 41: 420–435.

Varadaraj, K. 1990. Genetic and endocrine manipulations and sex regulation in tilapias Ph.D Thesis, Madurai Kamaraj University, Madurai, India.

Varadaraj, K. 1993. Production of viable haploid *Oreochromis mossambicus* gynogens using Uv-irradiated sperm. J Exp Zool, 167: 460–467.

Varadaraj, K, Pandian, TJ. 1988. Induction of triploids in *Oreochromis mossambicus* by thermal hydrostatic pressure and chemical shocks. Proc Aquat Int Cong, Vancouver, Canada, 531–535.

Varadaraj, K, Pandian, TJ. 1989a. First report on production of supermale tilapia by integrating endocrine sex reversal with gynogenetic technique. Curr Sci, 58: 434–441.

Varadaraj, K, Pandian, TJ. 1989b. Induction of allotriploid in the hybrids of *Oreochromis mossambicus* female x red tilapia male. Proc Ind Acad Sci, (Anim Sci), 98: 351–358.

Varadaraj, K, Pandian, TJ. 1990. Production of all female sterile triploid *Oreochromis mossambicus*. Aquaculture, 84: 117–123.

Varadi, L, Benko, I, Varga, J, Horvath, L. 1999. Induction of diploid gynogenesis using interspecific sperm and production of tetraploid in African catfish *Clarias gariepinus* Burcell (1822). Aquaculture, 173: 401–411.

Varkonyi, E, Bercsenyi, M, Ozouf-Costaz, C, Billard, R. 1998. Chromosomal and morphological abnormalities caused by oocyte ageing in *Silurus glanis*. J Fish Biol, 52: 899–906.

Vasil'yev, VP. 1981. Chromosome numbers in fish-like vertebrates and fish. J Ichthyol, 20: 1–38.

Vasil'yev, VI, Vasil'eva, ED, Osinov, AG. 1990. The problem of reticulate species formation of vertebrates of the diploid-triploid-tetraploid complex in the genus *Cobitis* (Gobitidae). 4. tetraploid form. Veprosy iktiol, 30: 908–919.

Veith, AM, Froschauer, Korting, C, Nanda, I, Hanel, R, Schmid, M, Schartl, M, Volff, JN. 2003. Cloning of the *dmrt1* gene of *Xiphophorus maculatus*: dmY/dmrt1Y is not master sex-determining gene in the platyfish. Gene, 317: 59–66.

Venere, PC, Galetti, PM Jr. 1985. Natural triploidy and chromosome B in the fish *Carimata modesta* (Curimatidae, Characiformes). Rev Braz Genet, 8: 681–687.

Vicari, MR, Moreiro-Filho, O, Artoni, RF, Bertollo, LAC. 2006. ZZ/ZW sex chromosome system in an undescribed species of the genus *Apareiodon* (Characiformes, Parodontidae). Cytogenet Genome Res, 114: 163–168.

Vicari, MR, Artoni, RF, Moreira-Filho, O, Bertollo, LAC. 2008. Diversification of a ZZ/ZW sex chromosome system in *Characidium* fish (Crenuchidae, Characiformes) Genetica, 134: 311–317.

Vicari, MR, Nogaroto, V, Noleto, RB, Cestari, MM, Cioffi, MB, Almeida, MC, Moreira-Filho, O, Bertollo, LAC, Artoni, RF. 2010. Satellite DNA and chromosomes in Neotropical fishes methods, applications and perspectives. J Fish Biol, 76: 1094–1116.

Vilela, DAR, Silva, SGB, Peixoto, MTD, Godinho, HP, Franca, LR. 2003. Spermatogenesis in teleost: insights from Nile tilapia (*Oreochromis niloticus*) model. Fish Physiol Biochem, 28: 187–190.

Vitturi, R, Catalano, E, Lafargue, F. 1991. Evidence for heteromorphic sex chromosomes in *Zeus faber* (Pisces zeiformes): nucleolus organizer regions and C-banding pattern. Cytobios, 68: 37–44.

Voet, D, Voet, J. 1990. DNA replication, repair and recombination. In: Biochrmistry, D Voet, J Voet (eds) John Wiley, New York, pp 967–969.

Volckaert, FAM, Galbursa, PHA, Hellemant, BAS, van den Heute, C, Vanstaen, D, Olivier, F. 1994. Gynogenesis in African catfish (*Clarias gariepinus*). 1. Induction of meiogynogenesis with thermal and pressure shock. Aquaculture, 128: 221–233.

Volff, JN. 2005. Genome evolution and biodiversity in teleost fish. Heredity, 94: 280–294.

Volff, JN, Schartl, M. 2001. Variability of genetic sex determination in poeciliid fishes. Genetica, 111: 101–110.

Volff, JN, Kondo, M, Schartl, M. 2003a. Medaka dmY/dmrtY is not the universal primary sex-determining gene in fish. Trends Genet, 19: 196–199.

Volff, JN, Korting, C, Froschauer, A, Zhou, Q, Wilde, B, et al. 2003b. The *Xmrk* oncogene can escape non-functionalization in a highly unstable subtelomic region of the genome of the fish *Xiphophorus*. Genomics, 82: 470–479.

Volff, JN, Nanda, I, Sclunid, M, Schartl, M. 2007. Governing sex determination in fish: Regulatory putsches and ephemeral dictators. Sex Dev, 1: 85–99.

von Brandt, VI. 1979. Sozialle kontrolle des Wachtums und geschlechts differenzieung bei jungen anemonefischen (*Amphiprion bicinctus*). Universitat Muenchen, Germany.

von Hofsten, J, Olsson, PE. 2005. Zebrafish sex determination and differentiation: involvement of FTZ-F1 genes. Reprod Biol Endocrinol, 3: 63 doi: 10. 1186/ 1477-7827-3-63.

Vrijenhoek, RC, Schultz, RJ. 1974. Evolution of a trihybrid unisexual fish (*Poeciliopsis*, Poeciliidae). Evolution, 28: 305–319.

Waldbieser, GC, Bowsworth, BG. 2008. Utilization of a rapid DNA based assay for molecular verification of channel catfish, blue catfish F_1 hybrids and backcross offspring at several life stages. North Am J Aquacult, 70: 388–395.

Wallace, BMN, Wallace, H. 2003. Synaptonemal complex karyotype of zebrafish. Heredity, 90: 136–140.

Wallis, GP, Beardmore, JA. 1980. Genetic evidence for naturally occurring fertile hybrid between two goby species *Pomatoschistus minutus* and *P. lozanoi* (Pisces, Gobidae). Mar Ecol Prog Ser, 3: 309–315.

Walter, RB, Hazlewood, L, Kazianis, S. 2006. The *Xiphophorus* Genetic Stocking Centre Manual, KD Kallman, M Schartl (eds) Texas State University Press, Texas, USA.

Wang, X, Zhang, Q, Ren, J, Jiang, J, Wang, C, Zhuang, W, Zhai, T. 2009. The preparation of sex chromosome specific painting probes and construction of sex chromosome DNA library in half smooth tongue sole (*Cynoglossus semilaevis*). Aquaculture, 297: 78–84.

References 245

Wang, Y, Zhou, L, Yao, B, Li, Gui, JF. 2004. Differential expression of thyroid-stimulating hormone beta subunit in gonads during sex reversal of change-spotted and red spotted groupers. Mol Cell Endocrinol, 220: 77–88.

Webb, RO, Kingsford, MJ. 1992. Protogynous hermaphroditism in the half-banded sea perch *Hypoplectrodes maccullochi* (Serranidae). J Fish Biol, 40: 951–961.

Weidinger, G, Wolke, U, Koprunner, M, Thisse, C, Thisse, B, Raz, E. 2002. Regulation of zebrafish primordial germ cell migration by attraction towards an intermediate target. Development, 129: 25–36.

Wen, A, You, F, Tan, X, Sun, P, Ni, J, Zhang, Y, Xu, D, Wu, Z, Xu, Y, Zhang, P. 2009. Expression pattern of *dmrt* from olive flounder (*Paralichthys olivaceus*) in adult gonads and during embryogenesis. Fish Physiol Biochem. 35: 421–433.

Whitt, GS, Childers, WF, Tranquilli, J, Champion, M. 1973. Extensive heterozygosity in three enzyme loci in hybrid sunfish populations. Biochem Genet, 8: 55–72.

Wickler, W, Seibt, U. 1983. Monogamy. an ambiguous concept. In: Mate choice, PPG Bateson (ed) Cambridge University Press, New York, pp 35–50.

Wilde, GR, Echelle, AA. 1992. Genetic status of Pecos pubfish populations after establishment of a hybrid swarm involving an introduced congener. Trans Am Fish Soc, 121: 277–286.

Wilkins, NP, Courtney, HP, Curatolo, A. 1993. Recombinant genotypes in backcrosses of male Atlantic salmon x brook trout hybrids to female Atlantic salmon. J Fish Biol, 43: 393–399.

Wilson, C, Crim, L, Morgan, M. 1995. The effect of stress on spawning performance and larval development in the Atlantic cod *Gadus morhua* L. In: Proc. 5th Int Symp Reprod Physiol Fish, FW Goetz, P Thomas (eds) Austin, Texas, USA p 198.

Winge, O. 1930. On the occurrence of XX males in the *Lebistes* with some remarks on Aida's so called "Non-disfunctional" males in *Aplocheilus*. J Genet, 23: 69–76.

Winge, O. 1934. The experimental alteration of sex chromosomes and vice versa, as illustrated by *Lebistes*. CR Trans Lab Carbsburg. Ser Physiol, 21: 1–49.

Wittbrodt, J, Meyer, A, Schartl, M. 1998. More genes in fish ? Bio Essays, 20: 511–512.

Wong, AC, Van Eenennaam, AL. 2008. Transgenic approaches for the reproductive containment of genetically engineered fish. Aquaculture, 275: 1–25.

Wood, EM. 1986. Behaviour and social organization in anemonefish. Prog Underwater Sci, 11: 53–60.

Woolcock, B, Kazianis, S, Lucito, R, Walter, RB, Kallman, KD, et al. 2006. Allele specific marker generation and linkage mapping on the *Xiphophorus* sex chromosomes. Zebrafish, 3: 23–27.

Woram, RA, Gharbi, K, Sakamoto, T, Hoyheim, B, Holm, LE, et al. 2003. Comparative genome analysis of the primary sex determining locus in salmonid fishes. Genome Res, 13: 272–280.

Wu, C, Chen, YR, Liu, X. 1993. An artificial multiple triploid carp and its biological characteristics. Aquaculture, 111: 255–262.

Wu, Q, Ye, Y, Xinnhong, D. 2003. Two unisexual artificial polyploid clones constructed by genome addition of common carp (*Cyprinus carpio*) and crucian carp (*Carassius auratus*). Sci China, 460: 5950604.

Wyatt, PMW, Pitts, CS, Butlin, RK. 2006. A molecular approach to detect hybridization between bream *Abramis brama*, roach *Rutilus rutilus* and rudd *Scardinus erythrophthalmus*. J Fish Biol, 69A: 52–71.

Wylie, C. 2000. Germ cells. Cell, 96: 165–174.

Xia, W, Zhou, L, Yao, B, Li, CJ, Gui, JF. 2007. Differential and spermatogenic cell-specific expression of *dmrt* genes in embryos of the medaka fish. Mol Cell Endocrinol, 263: 156–172.

Yaakub, SM, Bellwood, DR, van Herdwerden, L, Walsh, FM. 2006. Hybridization in coral reef fishes: introgression and bidirectional gene exchange in *Thalassoma* (Family Labridae). Mol Phylogenet Evol, 40: 84–100.

Yamaha, E, Onozato, H. 1985. Histological investigation on the gonads of artificially induced triploid crucian carp *Carassius auratus*. Bull Fac Fish Hokkaido Uni, 36: 170–175.

Yamaha, E, Kazama-Wakabayashi, M, Otani, S, Fujimoto, T, Arai, K. 2001. Germ line chimera by lower-part blastoderm transplantation between diploid goldfish and triploid crucian carp. Genetica, 111: 227–236.

Yamaha, E, Murakami, M, Hada, K, Otani, S, et al. 2003. Recovery of fertility in male hybrids of a cross between goldfish and common carp by transplantation of PGC (Primordial germ cell) containing graft. Genetica, 119: 121–131.

Yamaha, E, Saito, T, Goto-Kazeto, R, Arai, K. 2007. Developmental biotechnology for aquaculture with special reference to surrogate production in the teleost fishes. J Sea Res, 58: 8–22.

Yamaki, M, Kawakami, K, Taniura, K, Arai, K. 1999. Live haploid mosaic charr *Salvelinus leucomaenis*. Fish Sci, 65: 736–741.

Yamamoto, T. 1955. Progeny of artificially induced reversal of male genotype (XY) in the medaka (*Oryzias latipes*) with special reference to YY male. Genetica, 40: 406–419.

Yamamoto, T. 1969. Sex differentiation. In: Fish Physiology, WS Hoar, DJ Randall (eds) Academic Press, NewYork, 3: 117–175.

Yamamoto, T. 1975. Medaka (Killifish) Biology and Strains. Keigaku Publishers, Tokyo, Japan.

Yamashita, M, Onozato, H, Nakanishi, T, Nagahama, Y. 1991. Breakdown of the sperm nuclear envelop is a prerequisite for male pronucleus formation: direct evidence from gynogenetic crucian carp *Carassius auratus langsdorfi*. Dev Biol, 137: 155–160.

Yan, J, Liu, L, Liu, S, Guo, X, Liu, Y. 2009. Comparative analysis of mitochondrial control region in polyploid hybrids of red crucian carp (*Carassius auratus*) and blunt snout bream *Megalobrama amblycephala*. Fish Physiol Biochem. DOI 10. 1007/s 10965-008-9251-0.

Yano, A, Suzuki, K, Yoshizaki, G. 2008. Flow-cytometric isolation of testicular germ cells from rainbow trout (*Oncorhynchus mykiss*) carrying the green fluorescent protein gene driven by trout *vasa* regulatory region. Biol Reprod, 78: 151–158.

Yao, B, Zhou, L, Wang, Y, Xia, W, Gui, JF. 2007. Differential expression and dynamic changes of *Sox3* during gametogenesis and sex inversion in protogynous hermaphroditic fish. J Exp Zool, 307A: 207–219.

Yao, HHC. 2005. The pathways of femaleness: current knowledge on embryonic development of the ovary. Mol Cell Endocrinol, 230: 87–93.

Yoon, C, Kawakami, K, Hopkins, N 1997. Zebrafish *vasa* homologue RNA is localized to the cleavage planes of 2-and 4-cell stage embryos and expressed in the primordial germ cells. Development, 124: 3157–3166.

Yoshida, K, Nagakawa, M, Wada, S. 2001. Pedigree tracing of a hatchery reared stock used for aquaculture and stock enhancement based on DNA markers. Fish Genet Breed Sci, 30: 24–35.

Yoshikawa, H, Morishma, K, Kusuda, S, Yamaha, E, Arai, K. 2007. Diploid sperm produced by artificially sex reversed clone loaches. J Exp Zool, 307A: 75–83.

Yoshikawa, H, Morishma, K, Fujimoto, T, et al. 2008. Ploidy manipulation using diploid sperm in the loach *Misgurnus anguillicaudatus*: a review. J Appl Ichthyol, 24: 210–214.

Yoshizaki, G, Oshiro, T, Takashima, F, Hirono, I, Aoki, T. 1991. Germ line transmission of carp α globin gene introduced in rainbow trout. Nippon Suisan Gakkaishi, 57: 2203–2209.

Yoshizaki, G, Sakatani, S, Tominaga, H, Takeuchi, I. 2000a. Cloning and characterization of a *vasa*- like gene in rainbow trout and its expression in the germ cell lineage. Mol Reprod Develop, 55: 364–371.

Yoshizaki, G, Takeuchi, Y, Sakatani, S, Takeuchi, T. 2000b. Germ line specific expression of green fluorescent protein in transgenic rainbow trout under control of the rainbow trout *vasa*- like promoter. Int J Dev Biol, 44: 323–326.

Yoshizaki, G, Takeuchi, Y, Kobayashi, T, Ihara, S, Takeuchi, T. 2002a. Primordial germ cells: the blueprint for a piscine life. Fish Physiol Biochem, 26: 3–12.

Yoshizaki, G. 2002b. Visualization and isolation of live primordial germ cells aimed at cell-mediated gene transfer in rainbow trout. In: Aquatic Genomics, Springer Verlag, Tokyo, pp 310–319.

You, C, Yu, X, Tong, J. 2007. Detection of hybridization between two loach species (*Paramisgurnus dabryanus* and *Misgurnus anguillicaudatus*) in wild populations. Env Biol Fish, DOI 10.1007/s 110641-0079282-X.

You, C, Yu, X, Tan, D, Tong, J. 2008. Gynogenesis and sex determination in large scale loach *Paramisgurnus dabryanus* (Sauvage). Aquacult Int, 16: 203–214.

Yu, X, Zhou, T, Li, K, Li, Y, Zhou, M. 1987. On the chromosomes of cyprinid fishes and a summary of fish chromosome studies in China. Genetica, 72: 225–236.

Yunhan, H. 1990. Tetraploidy induced by heat shock in big head carp *Aristichthys nobilis*. Acta Acad Sini, 36: 70–75.

Zarkower, D. 2001. Establishing sexual dimorphism: conservation amidst diversity? Nat Rev Genet, 2: 175–185.

Zhang, F, Oshiro, T, Takashima, F. 1992a. Chromosome synapsis and recombination during meiotic division in gynogenetic triploid ginbuna *Carassius auratus langsdorfi*. Jap J Ichthyol, 39: 151–155.

Zhang, F, Oshiro, T, Takashima, F. 1992b. Fertility of triploid backcross progeny (Gengoroubuna *Carassius auratus cuvier* ♀x carp *Cyprinus carpio* ♂) F_1 x carp or genogoroubuna ♂. Jap J Ichthyol, 39: 229–233.

Zhang, J. 2004 Evolution of *DMY*, a newly emergent male sex-determination gene of medaka fish. Genetics, 166: 1887–1895.

Zhang, Q, Arai, K. 1996. Flow cytometry for DNA contents of somatic cells and spermatozoa in the progeny of natural tetraploid loach. Fish Sci 62: 870–877.

Zhang Q., Arai, K. 1999a Aberrant meioses and viable aneuploid progeny of induced triploid loach (*Misgurnus anguillicaudatus*) when crossed to natural tetraploids. Aquaculture, 175: 63–76.

Zhang, Q, Arai, K. 1999b. Distribution and reproductive capacity of natural triploid individuals and occurrence of unreduced eggs as a cause of polyploidization in the loach, *Misgurnus anguillicaudatus*. Ichthyol Res. 46: 153–161.

Zhang, Q, Arai, K, Yamashita, M. 1998. Cytogenetic mechanisms for triploid and haploid egg formation in the triploid loach *Misgurnus anguillicaudatus*. J Exp Zool, 281: 608–619.

Zhang, Q, Nakayama, I, Fujwara, A, Kobayashi, T, Oohara, T, Masaoka, S, Kitamura, S, Devlin, RH. 2001. Sex identification by male specific growth hormone pseudogene (GH-ψ) in *Oncorhynchus masou* complex and a related hybrid. Genetica, 111: 111–118.

Zhao, G, Yu, QX, Zhang, WW, Zhang, YM, Chen, J, Long, H, Liu, JD. 2008. The SS rDNA related repetitive sequences in the sex chromosomes of the spiny eel (*Mastacembelus aculeatus*). Cytogenet Genome Res, 121: 143–148.

Zhou L, Gui JF. 2008. Molecular mechanisms underlying sex change in hermaphroditic groupers. Fish Physiol Biochem, DOI 10. 1007/s/10695.008 9219–0.

Zhou, R, Chang, H, Tiersch, TR. 2002. Differential genome duplication and fish diversity. Rev Fish Biol Fish, 11: 331–331.

Zou S, Li S, Cai W, Zhao J, Yang H. 2004. Establishment of fertile tetraploid population of blunt snout bream (*Megalobrama amblycephala*). Aquaculture, 238: 155–164.

Zou S, Li S, Cai W, Yang H, Jiang, X. 2008. Induction of interspecific allo-tetraploids of *Megalobrama amblycephala*♀× *Megalobrama terminalis*♂ by heat shock. Aquacult Res, 39: 1322–1327.

Zupanc, GKH. 1985. Fish and Behavior. How Fishes live. Tetra-Press, Melle, Germany.

Author Index

A

Abbas, K, 122
Aegerter, S, 92
Aida, S, 60, 61
Akhtar, N, 158
Alifaqih, MA, 195
Allsop, DJ, 166, 167, 170
Almeida-Toledo, LF, 19, 27, 32, 195
Alves, AL, 32
Alves, MJ, 126–128, 177, 184–189
Amer, MA, 148
Andreata, AA, 30
Anon, 70
Aparicio, SJ, 20
Arai, K, 22, 52, 55, 57–61, 70, 80, 90, 91, 93, 96, 98, 101, 102, 108, 109, 117, 118, 122–125, 127, 159
Arai, K, see Aida, S, Inamori, Y, Ishimoto, M, Mukaino, M, Zhang, Q
Araki, K, 80
Arimoto, B, see Uneo, K
Argue, BJ, 4, 36, 37, 43, 44, 47
Arkhipchuk, V, 7, 21
Arnold, ML, 37, 45
Artieri, CG, 194, 196
Artoni, RF, 21, 25, 29
Asoh, K, 192
Aspinwall, N, 41, 45
Avise, JC, 40, 44, 45, 46, 180
Avise, JC, see Mank, JE, Scribner, KT
Azevedo, MFC, 19

B

Balsano, JS, 179, 183
Balsano, JS, see Rasch, EM
Barlow, GW, see Francis, RC
Baroiller, JF, 191–193, 205
Barton, NH, 37
Basola, AL, 35
Beardmore, JA, 84
Beardmore, JA, see Wallis, GP
Bekkevold, D, 14

Beleyaeva, VN, see Romashov, DD
Bellafronte, E, 21, 25, 30
Bematchez, LH, 45
Benfey, TJ, 89, 102, 113
Benfey, TJ, see Schafhauser-Smith, D
Bercsenyi, M, 69, 71, 73
Berrebi, P, 46, 47, 113
Berrebi, P, see Phillipart, JC
Bertollo, LAC, 27, 30, 31, 33
Bertollo, LAC, see Artoni, RF
Beukeboom, LM, 180
Beulbens, K, 8
Bhowmick, RM, 48
Billard, R, see Linhart, O
Birstein, VJ, 113
Bishop, RD, 45
Black, DA, 18, 35, 192
Blanc, JM, 117, 118
Blom, JH, see Dabrowski, K
Bobyrev, A, 1
Bogart, JP, 178
Bongers, BJ, 22, 69, 81, 82
Borin, LA, 90, 113
Braat, AK, 145, 146
Brantley, RK, 12
Breden, F, see Lindholm, A
Brinster, RL, 147
Bromage, NR, 82
Brooks, S, 82
Brown, KH, 46
Brum, MJI, 23
Brusle, S, 130
Bull, JJ, see Charnov, EL
Burns, JR, see Flores, JA
Burr, BM, see Taylor, CA
Burton, MPM, see Manning, AJ

C

Cal, RM, 100
Campos-Ramos, R, 24
Carillo, M, 82
Carmona, JA, 184–186
Carrasco, LAP, 23

Carruth, LL, 165
Carvalho, RA, 181
Cassani, JR, 91, 92, 115
Castelli, M, 31, 46, 57, 60, 62, 63
Cavallaro, ZI, see Bertollo, LAC
Centofante, L, 25, 30
Chao, N, 139
Charlesworth, D, 20
Charnov, EL, 166, 171
Chaudhury, RC, 33
Chen, H, see Lu, R
Chen, J, 24
Chen, SL, 59, 60
Chen, TR, see Ebeling, AW
Chenuil, A, 47
Cherfas, NB, 51, 52, 54, 65, 92, 98, 99, 117, 126
Chevassus, B, 36, 39
Childers, WF, 40, 44
Choi, I, 202
Chopelet, J, 8, 167
Chourrout, D, 7, 57–59, 66, 98, 118
Cimino, MC, 109
Cloud, JG, see Krisfalusi, M, Nilsson, EE
Cole, KS, 164, 170, 174
Collares-Pereira, MJ, 90, 104, 186
Comai, L, 181
Conover, DO, 205
Coreley-Smith, GE, 22
Cotter, D, 89
Coughlan, T, 194
Crabtree, RE, 174
Csanyi, V, see Gervai, J
Cui, J-Z, 194
Cunado, N, 97, 100
Cunha, C, 186, 188, 189

D

Dabrowski, K, 52, 82
David, CJ, 5, 22, 69, 71, 73, 74, 79–83, 85–87, 91, 93, 96, 98, 101, 102, 128, 153, 194
Davies, M, 126
Dawley, RM, 42, 91, 96, 126, 179, 184
Dawley, RM, see Goddard, KA
DecCordier, I, 115
deGirolamo, M, 2
de Looze, EMA, see Kraak, SBM
Deng, Y, 98
deOliveira, RR, 30, 32
Desai, VR, 42, 44
Desperz, D, 50
Devlin, RH, 18, 20, 22, 23, 27, 29, 48, 59, 86, 194

DeWoody, JA, 14
Diniz, D, 24
Doitsidou, M, 134
Donaldson, EE, see Piferrer, F
Dowling, TE, 39
Dowling, TE, see Goddard, TE
Du, Q-Y, 194, 200, 201
Du, SJ, 194
Duan, W, 119
Dunham, RA, 22
Dunham, RA, see Argue, BJ

E

Ebeling, AW, 26
Echelle, AA, 126, 179
Echelle, AA, see Wilde, GR
Elder, JF, 179
Erisman, BE, 163
Ezaz, MT, 52, 64, 67, 86

F

Falco, JN, 29
Felip, A, 57
Ferguson-Smith, M, 192
Ferreira, IA, 24
Fernandes-Matioli, FMC, 90
Fineman, R, 82
Fischer, EA, 163
Fishelson, L, 8, 9, 11, 174
Flajshans, M, 92, 100, 105, 126
Flores, JA, 130
Flynn, SR, 31, 60, 62
Fopp-Bayat, D, 31, 52, 59
Foresti, F, 23
Foresti, F, see Galetti, PMJr
Foster, JW, 23
Franca, LR, 147, 149
Francis, RC, 165, 172
Fricke, HW, 167
Fricke, S, see Fricke, HW
Fridolfsson, AK, 23
Frisch, A, 168
Frolov, SV, 32, 34
Fu, P, 13
Fujiwara, A, 43, 109

G

Gaignon, JL, see Quillet, E
Galbreath, PF, 56, 126
Galbursera, P, 60
Galetti, PMJr, 29, 30, 32, 113, 193
Galetti, PMJr, see Venere, PC

Gante, HF, 45, 46
Gao, ZX, 195
Garcia-Vazquez, E, 40
Gehring, WJ, 115
George, T, 18, 22, 27, 39, 42, 43, 48, 62, 79, 85, 99, 192, 207
Gervai, J, 31, 60
Gillet, C, 82
Glamuzina, B, 41
Goddard, KA, 126, 180, 189
Goddard, KA, see Dawley, RM
Gold, JR, 19
Golubstov, AS, 113
Gomelsky, B, 52, 59, 60
Gordon, M, 35
Goudie, CA, 59, 61
Grant, BR, 36
Grant, PR, see Grant, BR
Graves, JA, 22
Graves, JA, see Foster, JW
Gray, AK, 96
Green, DM, 21
Gregory, TR, 112, 113
Gregory, TR, see Smith, EM
Grier, H, see Parenti, LR
Gromicho, M, 184, 189
Guan, G, 198
Guegan, JF, 113
Guerrero, GA, see Lo Nostro, F
Guest, WC, 96
Gui, JF, 63
Gui, JF, see Zhou, L
Guo, X, 99, 200
Guo, XH, 120
Guo, Y, 197, 198

H

Haaf, T, 7, 25
Haas, R, 47
Habicht, C, 94, 97
Haffray, P, 106
Hamaguchi, S, 47, 111
Han, Y, 106
Haniffa, MA, see Sridhar, S
Hardie, DC, 20
Harrington, RW, 161
Harrison, RG, 36
Hargreaves, JA, see Owusu-Frimpong, M
Harvey, SC, 19, 24
Hashimoto, Y, 3, 131, 156
Hawkins, MB, 163
Hayes, T, 1
He, C, 202

Heath, DD, 1, 12
Hebert, PDN, see Hardie, DC
Heins, SW, see Conover, DO
Hester, FE, see Smitherman, RO
Hett, AK, 200
Hewitt, GM, 36
Hewitt, GM, see Barton, NH
Hickling, CF, 49
Hilzerman, F, see Fishelson, L
Hinegardner, R, 20
Hobbs, J-PA, 171–173, 175
Hoese, DF, see Cole, KS
Hoffman, HA, 171
Horstgen-Schwark, G, 118
Horstgen-Schwark, G, see Muller-Belecke, A
Horvath, L, 53, 90, 105, 126, 128
Howell, WM, see Black, DA
Hosoya, K, see Kawamura, K
Hrbek, T, 76
Huang, X, 202–204
Hubbs, CL, 52, 126, 177, 188
Hubbs, LC, see Hubbs, CL
Hulak, M, 105, 128
Hussain, MG, 59, 64, 66, 67
Hutchings, JA, 12, 14

I

Ihssen, PE, see Phillips, RB
Imai, H, 40
Imamori, Y, see Arai, K
Immler, S, 11
Ishimoto, M, see Arai, K
Itono, M, 52, 55, 59
Iwama, GK, see Leggatt, RA

J

Jaillon, O, 20
Jalabert, B, see Aegerter, S
Jamieson, BGM, 4
Jensen, GL, 65
Jesus, CM, 32
Ji, X-S, 62, 67
Jia, ZY, 8, 63
Jianxun, C, 113
Johnson, KR, 42, 45, 56, 91, 94, 106, 126
Jones, AG, 11, 13, 14
Juchno, D, 115

K

Kallman, KD, 18, 35, 55, 192, 196
Kamler, E, 80, 81

Kanolh, Y, 12, 13
Karayucel, S, 61, 63, 67
Kato, K, 65
Kavumpurath, S, 22, 33, 40, 57, 61, 77, 84, 91, 93, 97, 98, 100, 105
Kavumpurath, S, see Pandian, TJ
Kawamura, K, 31, 42, 48, 51, 60, 62, 64, 104, 105, 128
Kazianis, S, 194
Kersten, CA, 158
Keyvanshokooh, S, 193
Khoo, L, 194
Khuda-Bukhsh, AR, 25
Kidwell, MG, 20
Kiflawi, M, 92
Kim, DS, 106
Kim, DS, see Nam, YK
King, J, 1, 15
Kingsford, MJ, see Webb, RO
Kirankumar, S, 22, 69, 70, 73–76, 82, 85–87, 128, 153, 172, 194
Kirankumar, S, see Pandian, TJ
Kirasnai, Z, 106
Kirkpatrick, M, see Van Doorn, GS
Kirpichnikov, VS, 84, 194
Kitamura, H, 98, 100, 105
Klinkhardt, MB, 19
Knapp, R, 12
Kobayashi, H, 104, 127
Kobayashi, K, 170
Kobayashi, M, 158
Kobayashi, S, 131
Kobayashi, T, 61, 64, 65, 100, 106, 107, 136–138, 141, 197, 207
Kojima, K, 126, 127
Kokokris, L, 49, 166, 171
Komen, J, 8, 51, 57, 59, 62, 64, 65, 69, 80
Kondo, M, 20, 24, 31, 192, 197
Kornfield, I, see McElroy, DM, Parker, A
Koteeswaran, R, 55, 73, 160, 166, 168, 171, 207
Koteeswaran, R, see Pandian, TJ
Kovacs, B, 194
Koya, Y, 8
Kraak, SBM, 27
Krisfalusi, M, 61, 65, 84, 100, 106
Krysanov, EY, see Golubstov, AS
Kumpf, K, see Noble, GK
Kurita, J, 38, 45, 96, 126
Kurokawa, H, 3, 131, 139, 142, 156, 201
Kuwamura, T, 11, 160, 171, 175
Kuwamura, T, see Nakazona, A
Kusunoki, T, 52, 59, 126, 127

L

Laarman, PW, 47
Larkin, JR, see Lowe, TP
Lacerda, SMSN, 147, 149, 150, 151, 154, 155
Lahn, BT, 199
Lajbner, Z, 46
Lakra, WS, see Pandey, N
Lam, T, 1
Lamatsch, DK, 8, 180, 181, 188
Lamatsch, DK, see Loewe, L
Lampert, KP, 10, 127, 180, 183, 189
Larkin, JR, see Lowe, TP
Lawrence, C, 174
Leary, RF, 39
Le Comber, SC, 113
LeGac, F, 149
Legendre, MG, 41, 42, 46
Leggatt, RA, 4, 112, 113
Lejeune, P, 171
Levin, DA, see Meyers, lA
Li, S-F, 96, 125
Libertini, A, 18, 25
Lincoln, RF, 98, 99, 104
Linhart, O, 52, 82, 101, 103–105, 128
Lindholm, A, 84, 194
Liu, D, 227
Liu, M, 172, 174, 175
Liu, M, see Sadavy, Y
Liu, Q, 51, 57, 128, 193
Liu, SJ, 44, 51, 55, 67, 119–122, 127, 129, 200
Liu, ZH, 198
Lodi, E, 1
Loewe, L, 8, 180
Loir, M, see LeGac, F
Lo Nostro, F, 174, 176
Love, MS, 15
Lovshin, LL, 49
Lowe, TP, 151, 158, 180
Lu, R, 127
Lucinda, PHF, 75
Luckenbach, JA, 60
Lutz, CG, 37
Lynch, M, 8, 180, 181

M

Mable, BK, 181
Mable, BK, see Gregory, TR
MacDonald,RI, 80
Madhu, K, 170
Madhu, R, see Madhu, K
Mahon, R, see Taylor, J
Mair, GC, 8, 18, 60, 63, 64, 66, 84, 207

Maistro, EL, 90
Mamcarz, A, 41
Mandal, RF, see Padhi, BK
Mani, I, 19
Mank, JE, 1, 3, 11, 12, 23, 24, 113
Manna, GK, 19
Manning, AJ, 106
Manning, MJ, 137, 146
Mantelman, II, 92
Marchand, U, 198
Marconato, A, 175
Margarido, VP, 34
Marian, IA, 64
Marie, AD, 46
Mariotto, T, 30
Martin, SB, 10
Martins, MJ, 186
Martinez, JL, 14
Mateos, M, 180
Matonda, N, 43, 44
Matsubara, K, 55, 108, 122–125
Matsuda, M, 23, 192, 195, 197, 198
McCombie, H, 115
McElroy, DM, 44
McKay, SJ, 141, 145
McMillan, M, 37
McPhail, JD, see Aspinwall, N
Mei, W, 3
Mestriner, A, 193
Meyer, A, 115
Meyers, LA, 112
Mia, MY, 39
Miller, RR, see Uyeno, T
Mims, SD, 60
Miranda, LA, 202
Mirza, JA, 60
Miura, C, 148
Miura, T, 147–149
Miura, T, see Schulz, RW
Mobley, KB, 11, 14
Moenhaus, WJ, 92, 96
Molley, PP, 168, 169
Momotani, S, 109, 126, 127
Moran, P, 25
Morand, S, see Guegan, JF
Moreira-Filho, O, 28, 29, 33, 34
Moreira-Filho, O, see Jesus, CM
Morescalchi, A, 30, 32
Mori, T, 102
Morishma, K, 55, 56, 126
Mukaino, M, see Arai, K
Muller-Belecke, A, 49–51, 64, 66, 84
Munday, PL, 161, 164, 175
Munday, PL, see Hobbs, J-PA

Munehara, H, 5
Muniyandi, N, 116
Munoz, RC, 170, 174
Myers, JM, 70
Myiazawa, CS, see Mariotto, 7

N

Naddafi, R, 174
Nagahama, Y, see Devlin, RH
Nagai, T, 131, 132
Nagasawa, K, 154
Nagler, JJ, 194
Nagoya, H, 80
Nagy, A, 56, 58
Nakamura, M, 2, 102, 106, 193
Nakanishi, T, see Manning, MJ
Nakayama, I, 194
Nakayama, I, see Chourrout, D
Nakazona, A, 9, 161
Nam, YK, 51, 60, 69, 98, 116, 128
NaNakorn, U, 41, 42
Nanda, I, 8, 18, 20, 24, 180–182, 192, 194, 195, 197
Naruse, K, 59
Neff, BD, 13
Nijhout, F, see Suzuki, Y
Nikoljukin, NI, 38
Nilsson, EE, 144
Noble, GK, 158
Noleto, RB, 25
Nomura, T, 60
Norberg, B, 82
Norbrega, RH, 148
Norma, MM, see Sandra, G-E
Nwadukwe, FO, 41

O

Ocalewicz, K, 66, 67, 84
Ohm, D, 165
Ohno, S, 4, 20, 26, 113, 115, 195
Ohta, K, 170
Ojima, Y, 31
Okada, YK, 158
Okudsu, T, 130, 134, 144, 146, 147, 151–153, 155
Oldfield, RG, 165
Oliveria, AV, 45
Oliveira, C, 19, 29, 31, 32
Olsen, LC, 131
Olsson, PE, see von Hofsten, J
Onozato, H, 76, 113
Onozato, H, see Yamaha, E

Orban, L, see Horvath, L
Orlando, EF, 161, 163
Oshiro, T, 65, 96
Oshima, K, 55, 104, 110, 111, 126
Ospina-Alvarez, N, 203
Otani, S, 7, 131, 133
Otto, SP, 102, 112
Owusu-Frimpong, M, 116

P

Padhi, BK, 47
Pala, I, 190, 196
Pajuelo, JC, 167
Pandey, N, 122
Pandian, TJ, 2, 4–6, 9, 11, 19, 42, 49–51, 54, 60,
      64, 66, 68–70, 72, 74, 75, 77–79, 89–91,
      98, 99, 101–103, 108, 113, 116, 122, 128,
      144, 146, 160, 161, 164, 166, 168, 174,
      177, 178, 181, 196, 199, 205, 207
Pandian, TJ, see David, CJ, George, T,
      Kavumpurath, S, Kirankumar,
      S, Koteeswaran, R, Marian, IA,
      Varadaraj, K
Parenti, LR, 3
Parise-Maltempi, PP, 32
Parker, A, 14
Parsons, JE, 97
Pavlidis, M, 48
Paull, GC, 174
Pedersen, RA, 20
Peichel, CL, 22
Pellegrini, E, 163
Penman, DJ, 61, 64, 99
Pendas, AM, 26
Perruzi, S, 59, 100, 102, 103, 128
Peters, HM, 160
Petersen, CW, 12, 164
Pfeifer, P, 56
Phillipart, JC, 41, 42
Phillips, RB, 25
Phillips, RB, see Reed, KM
Phillips, R, 23
Pifferer, F, 61, 65, 66, 89, 97, 106, 112, 115, 124,
      192, 193, 203
Piferrer, F, see Ospina-Alvarez, N
Poleo, GA, 7
Pongthana, N, 60, 61
Price, DJ, 7, 205
Pruginin, Y, 49
Purdom, CE, 89

Q

Quillet, E, 57, 59, 65, 66, 96
Quillet, E, see Chourrout, D
Quattro, JM, 180

R

Rab, P, 24, 113
Rab, P, see Berrebi, P, Phillips, R
Raicu, P, 113
Rao, KJ, see Desai, VR
Rasch, EM, 8, 91, 126, 181
Rasotta, MB, see Shapiro, DY
Raymond, CS, 197
Reed, KM, 26
Reis, RE, 19
Reinboth, R, 174
Ribeiro, F, 179, 184
Richter, CJJ, see Komen, J
Rico, C, 14
Rinchard, J, 31, 60, 62, 63
Rishi, KK, 19
Robalo, JI, 184
Robertson, DR, 2
Robertson, DR, see Cole, KS
Robertson, OH, 158
Rock, J, 113
Rodriquez-Mari, A, 205
Romashov, DD, 51
Rosa, R, 30, 31, 35
Ross, RA, 22
Ross, RM, 172
Ruiguang, Z, 113
Ruis-Carus, R, 176

S

Saber, MH, 51, 57, 59
Sadovy Y, 9, 16, 160, 161, 176
Sadovy, Y, see, Liu, M
Saito, T, 135, 136, 142–146, 156
Sakai, Y, 11
Sakaizumi, M, 41, 42, 96, 126
Sakaizumi, M, see Hamaguchi, S
Sandra, GE, 17, 192
Sarder, MR, 64, 66
Saunders, NC, see Avise, JC
Schaefer, SA, 19
Schartl, M, 180, 181, 189, 197
Schartl M, see Lampert, KP, Meyer, A, Volff,
      JN

Schaschl, H, 10
Schaufhauser—Smith, D, 106, 108
Scheerer, PD, 22, 70, 86
Schlosser, IJ, see Elder, JF
Schmid, M, see Haaf, T
Schulz, RW, 148
Schultheis, C, 24, 31
Schultz, RJ, 113, 126
Schultz, RJ, see Vrijenhoek, RC
Schwartz, FJ, 36, 38
Schwenk, K, 36
Scott, AG, 50
Scott, AP, see Lincoln, RF
Scribner, KT, 4, 36–39, 46
Seeb, JE, 94, 97
Seibt, U, see, Wickler, W
Shah, MS, 64
Shapiro, DY, 11, 12, 171, 174
Shapiro, DY, see Asoh, K
Sheela, SG, see, Pandian, TJ
Shelton, WL, 60
Shelton, WL, see Mirza, JA
Shen, J, 104
Shen, XY, 200
Shikina, S, 154, 155
Shimizu, Y, 113
Shimizu, YH, 111
Shinomiya, AI, 130, 155, 156
Shirak, A, 196, 198, 199, 205
Siegfried, KR, 193
Slanchev, K, 3, 142, 156
Smith, C, see Le Comber, SC
Smith, EM, 20, 113
Smitherman, RO, 96
Smitherman, RO, see Dunham, RA
Sofikitis, N, 147
Sogard, PS, 14
Sola, L, 25, 161, 188
Som, C, 180
Sousa-Santos, C, 184, 186
Sridhar, S, 40
Stanley, JG, 69, 79
Stein, J, 26, 194, 196
Steinemann, S, 20
Steinemann, M, see Steinemann, S
Steinke, D, 145
St Mary, CM, 164
Storchova, Z, 115
Sun, YD, 56, 105, 119, 128
Sutterlin, AM, 94
Suwa, M, 59
Suzuki, A, 113
Suzuki, K, see Kobayashi, K
Suzuki, R, 52, 59, 61

Suzuki, Y, 175
Swarup, H, 93

T

Tabata, K, 65
Taborsky, M, 11
Takagi, M, 14, 15
Takegaki, T, 11
Taki, Y, see Suzuki, A
Taniguchi, N, 59
Takeuchi, Y, 130, 132–138, 141, 146, 157
Tatarenkov, A, 9, 10, 161
Taverne, N, see Timmermans, LPM
Taylor, CA, 3
Taylor, J, 38
Thorgaard, GH, 57, 70, 80, 117, 118
Thorgaard, GH, see Galbreath, PF, Komen, J
Timmermans, LPM, 131
Tiwari, BK, 106
Toledo-Filho, SA, 41
Traut, W, 26
Trautman, MB, 96
Tripathy, N, 194
Tsigenopoulos, CS, 113
Tuan, PA, 82, 86
Turner, B, 10
Tvelt, HB, 60

U

Uchida, D, 31
Uedo, T, 92, 96, 105, 128
Umino, T, 52, 59, 90, 91, 126
Underwood, JL, 58
Uneo, K, 99
Uyeno, T, 27

V

Valenzuela, N, 196
van de Peer, Y, 115
Vanderputte, M, 7, 203, 205
Van Doon, GS, 22
Van Doornik, D, 14
Van Enenennam, AL, 31, 57, 60, 62, 97, 104, 128
Van Eenennam, AL, see Wong, AC
van Herwerden, AL, 46
Varadaraj, K, 7, 18, 22, 27, 50, 54, 60, 61, 64, 71, 91, 93, 95, 98
Varadaraj, K, see Pandian, TJ
Varadi, L, 116
Varkonyi, E, 92
Vasil'yev, VI, 113, 127, 179, 189

Veith, AM, 196
Venere, PC, 90
Vicari, MR, 25, 30
Vilela, DAR, 149
Vitturi, R, 33
Voet, D, 57
Voet, J, see, Voet, D
Volckaert, FAM, 61, 62
Volff, JN, 18, 20, 22, 23, 35, 84, 193, 195–197
von Brandt, VI, 172
von Hofsten, J, 200
Vrijenhoek, RC, 42

W

Waldbieser, GC, 41
Wallace, BMN, 24, 193
Wallace, H, see Wallace, BMN
Wallis, GP, 40
Walter, RB, 35
Wang, X, 20, 24, 25, 31
Wang, Y, 202
Warner, RR, see Munoz, RC
Webb, RO, 170, 171
Weidinger, G, 133
Wen, A, 198
West, SA, see, Allsop, DJ
Whithler, RE, see King, J
Whitt, GS, 47
Wickler, W, 11
Wilcove, D, see, McMillan, M
Wilde, GR, 45
Wilkins, NP, 45, 94
Wilson, C, 82
Winge, O, 7, 23, 84, 85
Winking, H, see Traut, W
Wittbrodt, J, 114
Wong, AC, 89
Wood, EM, 172
Woolcock, B, 192, 194
Woram, RA, 22, 194, 195
Wright, JE, see Johnson, KR

Wu, C, 92
Wu, Q, 179, 189
Wyatt, PMW, 47
Wylie, C, 131

X

Xia, W, 202

Y

Yaakub, SM, 46
Yamaha, E, 102, 127, 130, 135, 139, 140, 143, 151, 153, 156, 157
Yamaki, M, 54
Yamamoto, T, 23, 47, 50, 193
Yamashita, M, 139
Yan, J, 121
Yano, A, 154, 155
Yoon, C, 131
Yao, HHC, 202
Yao, B, 202
Yoshida, K, 14
Yoshikawa, H, 80, 128
Yoshizaki, G, 131, 134, 137, 141, 144, 146, 147, 149, 154, 159, 194
You, C, 31, 40, 42, 51, 56, 59, 60, 62
Yu, X, 113
Yunhan, H, 116

Z

Zarkower, D, 197
Zhang, F, 93, 98–100, 109, 127
Zhang, J, 196, 197
Zhang, Q, 55, 90, 93, 97, 98, 102, 108–110, 123, 125–127
Zhao, G, 24
Zhou, G, see Liu, Y
Zhou, L, 170, 171, 176, 193, 202
Zhou, R, 114
Zou, S, 117, 120
Zupanc, GKH, 160, 165, 167

Species Index

A

Abramis brama 6, 38, 41, 47
Acipenser spp 113
A. baeri 6, 31, 52, 57, 59, 90
A. brevirostrum 31, 60, 62, 113
A. mediarostrus 20
A. mikadoi 19, 113
A. ruthenus 6, 44, 52
A. stellatus 57, 59
A. strurio 200
A. transmontanus 31, 60, 62, 63, 65, 97, 98
Acanthopagrus schlegeli 6, 99, 202
Acanthurus leucosternon 46
A. nigricans 46
A. nigrofasciatus 92
Acheilognathus 48
A. cyanostigma 48
Acrossocheilus sumatransis 113
Alburnus alburnus 38
Alcichthys alcicornis 5
amago salmon, see *Oncorhynchus rhodorus* 63
Amazon molly, see *Poecilia formosa* 10
Amphioxus 115
Amphiprion 168
A. alkallopsis 167, 170, 171
A. bicinctus 172
A. percula 170
Anadonta woodiana 13
Anaecypris hispanica 179, 184, 186
Ancistrus sp 32
Ancistrus sp Balbina 32
Ancistrus sp Barcelos 32
A. dubius 30
Ancistrus sp Ragacu 30
A. ranunculus 30
anemonefish, see *Amphiprion bicinctus* 172
Anguilla anguilla 8
A. japonica 148
Anthias squamipinnis 11
Aparieodon sp 30
A. affinis 28, 31, 34
A. ibitiensis 21, 25, 30

A. pareiodon 21
Aphyosemion albrechti 38
A. australe 38, 44
A. christyi 38
A. cinnamomeum 38
A. cognatum 38
A. fasciolatus 38
A. gardneri 38, 44
A. scheeli 38
A. schoutedeni 38
Apolenichthys trimaculatus 171
Arctic charr, see *Salvelinus alpinus* 26
Aristichthys nobilis 39, 60
Astyanax scabripinnis 90
Atlantic cod, see *Gadus morhua* 82
Atlantic halibut, see *Hippoglossus hippoglossus* 82
Atlantic salmon, see *Salmo salar* 26
Aulopyge huegelii 113
Awous strigatus 31

B

bacu, see *Piaractus mesopotomicus* 41
Barbus 113, 122
B. barbus 6, 31, 41, 46, 47, 57, 60, 62, 63, 65, 68
B. carpenthicus 46, 47
B. conchonius 99
B. merionalis 41, 47
barbin flounder, see *Verasper moseri* 102
Betta splendens 22, 28, 51, 60, 61, 79, 91, 93, 98, 101, 158, 192
big head carp, see *Hypophthalmichthys nobilis* 116
black grouper, see *Mycteroperca bonaci* 168
black molly, see *Poecilia sphenops* var *melanosticta* 25
Blennius tentacularis 31
Blicca bjoerkna 38, 43, 44
blue catfish, see *Ictalurus furcatus* 39
blue gill sunfish, see *Lepomis macrochirus* 1
blue head wrasse, see *Thalossoma bifasciatum* 175

blunt snout bream see *Megalobrama*
    ambylocephala 120
Bodianus rufus 171
Brachyphopopomus pinnicaudatus 32
bream, see *Abramis brama* 47
Brevoortia aurea 40
brook trout, see *Salvelinus fontinalis* 40
brown trout, see *Salmo trutta* 39
Bryaninops 165
Buenos Aires tetra, see *Hemigrammus*
    caudovittatus 73
buck tooth parrotfish, see *Sparisoma radians*
    168
bull trout, see *Salvelinus confluentis* 39
Bynni 113

C

Calamus bajonado 102
Campostoma anomalum 38
Carassius 10, 116, 118
C. auratus 6, 22, 44, 65, 68, 71, 90, 101, 102,
    114, 117, 118, 120, 126, 127, 129, 131,
    139, 145, 158, 159, 178–180, 183, 189
C. auratus buergeri 114
C. auratus cuvieri 98, 99, 108, 109, 114, 119,
    127
C. auratus gibelio 52, 56, 63, 72, 91, 100, 101,
    104, 105, 114, 118, 119, 126, 128, 178
C. auratus grandoculis 114
C. auratus langsdorfi 52, 59, 90, 100, 114, 126,
    127, 139, 178
C. auratus sugu 178
C. carassius 114
C. catonensis 114
C. complex 10
Caribbean goby, see *Coryphopterus*
    personatus 170
cardinal shiner, see *Luxilus cardinalis* 45
Catasomus discolobus 36
Catla catla 42, 44
Centropyge 165
C. acanthrops 171
C. bicolor 171
C. fisheri 170, 171
C. vroliki 171
Cephalopholis boenak 172, 174, 175
Ceratophyres dorsata 122
Chaenodraco hamatus 31
C. myersi 31
C. wilsoni 31
Channel catfish, see *Ictalurus punctatus* 38
Channa punctatus 116
C. stewarti 114

Characidium alipionis 25, 30
C. fasciatum 30
C. gomesi 25, 30
C. lanei 25, 30
Chasmistes brevirostris 36
checkerboard cichlid, see *Crenicara*
    punctulata 165
Chinobathyscus dewitti 31
chimook salmon, see *Oncorhynchus*
    tshawytscha 96
Chondrostoma 45
chum salmon *Oncorhynchus keta* 96
Cichla 45
C. monoculus 172
Cichlassoma citrenellum 172
C. cyanoguttatum 44
C. nigrofasciatus 44
Cirrhitichthys 165
Cirrhinus mrigala 6
Cirrhilabrus temmencki 170
Clarias batrachus 122
C. gariepinus 40–42, 44, 60–62, 116, 194
C. macrocephalus 5, 6, 41, 42
Cobitis 10, 36, 90, 127
Cobitis biwae 6, 52, 59, 115, 126, 127
C. complex 114, 127
C. granoei 178, 179
C. granoei taenia 178–180, 183, 189
C. taenia 1, 178, 179
C. taenia bilineata 1
coho salmon, see *Oncorhynchus kisutch* 1
Colisa lalius 28
Colossa macroponum 41
Conger myriaster 31
Coregonus alpenae 28, 37
C. fera 44
C. lavaretus 114
C. pallasi 44
C. sardinella 28, 32, 34
Coris julis 130, 171
Corydoras 114
C. aeneus 114
Coryphopterus personatus 170, 171, 174
C. glaucofraenatus 171
Crenicara punctulata 160, 165, 167, 207
crucian carp see *Carassius auratus* 68
Ctenopharyngodon idella 4–6, 65, 69, 90–92,
    101, 114–116, 128, 158, 159
Curimata modesta 90
Cynoglossus semilaevis 24, 31, 59, 60, 62, 67
Cyprinodon alvarezi 47
C. nevadensis calidae 37
C. pocosensis 45

Cyprinus carpio 4, 6, 8, 22, 39, 44, 52, 54, 56, 58–60, 65, 68, 69, 71, 72, 90, 92, 98, 99, 101, 105, 108, 114, 117–119, 122, 126, 127, 129, 139, 178, 179, 191, 192, 200, 201
C. carpio chilia 114

D

Danio albolineatus 142, 156
D. frankei 40, 49
D. rerio 7, 22, 24, 26, 31, 40, 59, 72, 99, 101, 115, 130, 145, 156, 174, 193, 198, 200
Dascyllus 164
D. aruanus 162
Dicentrarchus labrax 5, 57, 82, 102, 193, 203
Diplodus 168
D. vulgaris 167
Diptychus dipogon 114
Displectrum 161
Ditrema temmencki 14
Drosoma capedianum 37
D. petenense 37
dusky grouper, see *Epinephelus marginatus* 41
dusky stripe shiner, see *Luxilus pilsbryi* 45

E

Eigenmannia sp 31
E. virescens 30
Epinephelus 202
E. aeneus 41
E. cocoides 202
E. guttatus 176
E. marginatus 41, 170, 171, 202
E. merra 202
E. tauvina 171, 202
Erythrinus erythrinus 30, 31
Esox masquinongy 31, 52, 60, 62, 63
Etheostoma olmstedi 14
European bitterling, see *Rhodeus amarus* 25
European catfish see *Silurus glanis* 92
European sea bass, see *Dicentrarchus labrax* 82

F

fancy carp, see *Cyprinus carpio* 54
Fengzheng crucian carp see *Carassius auratus gibelio* 104
Fundulus sp 178, 179
F. diaphanus 126, 178, 179
F. diaphanus heteroclitus 181

F. dispar 3
F. heteroclitus 6, 92, 126, 168, 179

G

Gadus morhua 12, 14, 82, 100, 102, 103, 128
Gambusia affinis 8, 18, 46, 192
G. affinis affinis 35
G. affinis holbrooki 35
G. amistadensis 37
G. holbrooki 46
Gasterosteus aculeatus 14, 93
Genicanthus watanabei 171
Gila ditaenia 7
G. eremica 7
gilthead sea bream, see *Sparus aurata* 48
Gnathopogon elongatus elongatus 6, 58
Gobiodon 164, 165
G. erythrospilus 172, 173, 175, 176
G. histrio 162, 164, 165
Gobiusculus flavescens 14
Gonostoma 168
G. bathyphilum 19
Grama loreto 192
grass carp, see *Ctenopharyngodon idella* 69
greater amberjack, see *Seriola dumerili* 25
green sturgeon, see *Acipenser mediostris* 20
Gymnotus carapo 90
G. pantanal 34
guppy, see *Poecilia reticulata* 26
Gymnocorymbus ternetzi 5, 22, 71, 72, 74, 82, 86, 91, 98, 99, 101

H

Halichoeres 2, 165
H. garnoti 2
H. maculipinna 2
H. melanurus 171
H. pictus 162
H. tenuispinnis 202
H. trimaculatus 171, 175
Haplochromis burtoni 44
H. nubilus 44
Haritta carvalhoi 31
Hemigrammus caudovittatus 5, 22, 69, 71–74, 80–83, 85, 86, 91, 98, 99, 101, 102, 128, 172, 193, 194
Hemiancistrus spilomma 30
Herichthys cyanoguttatus 6
Heterobranchus longifili 40–42, 44
Heteropneustes fossilis 40, 54, 73, 91, 102–104, 106, 114, 122, 128
H. microps 40

Hisonotus sp 30
H. leucofrenatus 30
Hippoglossus hippoglossus 60, 82
Hoplestethus atlanticus 148
Hoplias malabaricus 24, 28, 31–33
Hucho perryi 148
Huso huso 6, 14, 193
Hypoplectrodes maccullochi 170, 171
Hypoplectrus 161
Hypophthalmichthys molitrix 4, 39, 60, 92
H. nobilis 5, 6, 91, 116
Hypopomus sp 32
Hypostomus sp G 30

I

Iberian minnow, see *Squalius alburnoides*
       184
Ictalurus furcatus 6, 39, 44, 61
I. punctatus 6, 22, 38, 44, 59, 61, 101
Imparfinis mirini 30
Indian catfish see *Heteropneustes fossilis* 73

J

Japanese crucian carp, see *Carassius auratus
       cuvieri* 119
Japanese goby, see *Valenciannea longipinnis*
       11
Japanese huch, see *Hucho perryi* 148
Japanese wrasse, see *Pseudolabrus sieboldi*
       170

K

Kryptolebias marmoratus 10, 161, 163, 207
K. ocellatus 10

L

Large scale loach, see *Paramisgurnus
       dabryanus* 40
Labeo calbasu 48
L. rohita 42, 44, 48
Labroides 165
L. dimidiatus 10, 171
Lates 168
L. calcarifer 107
Lepidosiren paradoxa 20
Lepomis auritus 14, 38, 40, 44, 47, 96
L. cyanellus 38, 40, 42, 47, 96, 126
L. gibbosus 38, 42, 96, 126
L. gulosus 38, 40
L. humulis 38

L. macrochirus 1, 38, 44, 47, 96, 195
L. macrolophus 40, 47, 96, 126
L. megalotis 96
L. punctatus 14
Lepornis conirostris 29
L. aff. Brunnus 29
L. elongatus 31, 193
L. lacustris 193
L. macrocephalus 29
L. obtusidens 29
L. reinhardti 29
L. trifasciatus 29
Leptoscarus vaigiensis 167, 168, 207
Limanda yokohamae 60, 61
Liobagrus marginatus 24
L. styani 24
Liza auratus 130
Luxilus albeslus 36
L. cardinalis 45
L. pilsbryi 45
Lythrypnus 164, 165, 168
L. dalli 162
L. zebra 171

M

Macropodus chinensis 50
M. concolor 50
M. opercularis 31, 50, 60, 165, 166, 168, 207
masu salmon, see *Oncorhynchus masou* 96
Mastacembelus aculeatus 24
medaka, see *Oryzias latipes* 7
Mediterranean porgy, see *Pagrus pagrus* 48
Megalobrama 116
M. ambycephala 116–118, 120, 125, 127, 129
M. terminalis 116
Megupsilon aporus 47
Menidia 16
M. beryllina 178, 179
M. clarkhubbsi 55, 126, 178–182
M. menidia 205
M. notata 6, 92
M. penninsulae 178, 179
Mereone chrysops 44, 45
M. dolomeiu 96
M. punctatus 96
M. salmoides 96
M. saxatilis 44, 45
Microlepidogaster leucofrenatus 30
Micropterus dolomieui 7
M. salmoides 7, 14
Misgurnus 116

M. anguillicaudatus 6, 22, 40, 42, 52, 55–61, 69, 72, 80, 90, 91, 93, 97, 98, 101, 102, 104, 108, 110, 114, 116, 117, 122–129, 145, 191
M. fossilis 114
M. mizolepis 6, 101, 116, 128
Monopterus albus 3, 203, 204
mosaic charr, see *Salvelinus leucomaensis* 54
Mycteroperca bonaci 163, 168, 171
M. rosacea 163
Mylocheilus caurinus 41, 44, 45

N

Nematolosa come 40
N. japonica 40
Neoditrema ransonneti 14
Nigerian cichlid, see *Pelvicachromis pulcher* 1
Noechilus barbatulus 90, 104
Notropis chrysocephalus 40
N. cornutus 40

O

Odontesthes bonaiensis 202
Odontostible heterodon 29
O. microcephala 29
O. paranaensis 29
Oncorhynchus 114, 192, 194, 206
O. apache 45, 94
O. clarki 45, 94
O. gorbuscha 42, 94, 95
O.keta 42, 94, 95
O. kisutch 1, 6, 42, 61, 65, 95, 97, 192, 206
O. masou 6, 43, 44, 95, 101, 102, 141, 145, 153, 157, 192, 206
O. masou rhodurus 190
O. mykiss 4, 6, 22, 42–45, 57–59, 61, 65–67, 70–72, 80, 82, 84, 86, 94, 98, 101, 105, 107–109, 116–118, 128, 129, 141, 145, 153, 157–159, 195, 198
O. nerka 158, 159
O. rhodurus 61, 65, 76
O. tshawytscha 1, 4, 6, 22, 26, 86, 94, 95, 194
Opsodoras sp 30
Ophiiodon elongatus 1
Orange roughy, see *Hoplostethus atlanticus* 46
Oreochromis 7, 18, 35, 49, 196
O. aureus 6–8, 18, 23, 24, 49, 60, 63, 64, 96, 116
O. aureus honorum 7
O. honorum 6, 7, 49, 50
O. leucostictus 7, 49
O. macrochir 7, 49

O.mossambicus 6, 7, 18, 22, 27, 44, 49, 50, 54, 60, 61, 64, 66, 68, 71, 86, 91, 93, 96, 98, 99, 101, 116
O. niloticus 6, 7, 22–24, 49, 50, 52, 59, 61, 65, 66, 70, 72, 84, 86, 96, 97, 99, 105, 116, 126, 150, 151, 159, 192, 198, 200, 202, 205
O. nigra 49, 50
O. spilurus niger 7
O. veriabilis 7, 49
O. vulcani 7, 49
O. urolepis-honorum 50
Oryzias 206
O. celebensis 41, 96
O. curvinotus 17, 38, 41, 42, 45, 47, 96, 192, 198, 205
O. hubbsi 31
O. javenicus 31
O. latipes 17, 38, 41, 42, 45, 47, 59, 82, 96, 115, 126, 130, 156, 158, 192, 193, 195–198, 205
O. luzonensis 38, 41, 96, 126
Osteochilus hosselti 6
Otocinclus vestitus 30

P

Pacific tuna, see *Thunnus orientalis* 154
Pagellus 161
Pagetopsis macropterus 31
Pagrus pagrus 99, 106, 167, 171
P. major 6, 65, 99, 101
Pangasius schwanenfeldii 6
P. sutchi 5, 6, 41, 42
Paradise fish, see *Macropodus opercularis* 165
Paralichthys lethostigma 60
P. olivaceus 6, 65, 198, 200
Paragobiodon 165
P. echinocephalus 162
Paramisgurnus dabryanus 31, 40, 42, 51, 56, 59, 60, 62, 63
Parodon hilarii 21, 33, 34
P. ibitiensis 21, 30
P. moreirai 30
P. vladi 30
peacock wrasse, see *Cirrhilabrus temmencki* 170
Pearl danio, see *Danio albolineatus* 142
Pelvicachromis pulcher 1
Perca flavescens 101
Percocypris 113
Perinotus paralatus 7
P. alatus 7
Phoxinus 16

P. eos 178, 179
P. eos neogaeus 55, 126, 178–180, 183, 184
P. neogaeus 178, 179, 189
P. oreas 6
Piaractus mesopotomicus 41
pink salmon, see *Oncorhynchus gorbuscha* 96
plaice, see *Pleuronectes platessa* 98
platyfish, see *Xiphophorus helleri, X.*
     maculatus 1
Plecoglossus altivelis 59
Plectropomus 46
P. caudofasciata 47
P. nigrofasciata 47
Pleuronectes platessa 99, 104
Poecilia 16, 18, 35, 36, 114
P. formosa 10, 17, 52, 55, 90, 91, 126, 127,
     177–183, 188, 189
P. latipinna 4, 178, 179, 188
P. mexicana 178, 181, 188
P. mexicana limantouri 181, 188
P. reticulata 7, 18, 22, 77, 84, 85, 192–194, 200
P. sphenops 6, 18, 22, 25, 27, 28, 39, 42, 43, 48,
     62, 85, 192
P. shenops-melanosticta 25
P. velifera 6, 39, 42, 43, 48
Poeciliopsis 16, 36, 55, 109, 114, 178, 179, 188
P. infans 178
P. lucida 178, 179, 188
P. monacha 42, 179, 188
P. monacha lucida 126, 180, 183
P. occidentalis 178, 188
P. viriosa 42
Pomatochistus minutus 13, 14, 40
Polyodon spathula 60, 113
Polypterus aethiopicus 20
P. palmas 115
Pond loach, see *Misgurnus anguillicaudatus*
     40
Premnas biaculeatus 172
Prinotus paralatus 7
P. alatus 7
Protopterus dollie 113
Prussian carp, see *Carassius gibelio* 105
Psetta maxima 102
Pseudochromis 165
Pseudoanthias squamipinnis 174
Pseudogrammus 161
Pseudolabrus 113, 165
P. sieboldi 170
Pseudotropheus zebra 14
Pufferfish, see *Tetradon fluviatilis* 20
Puntius conchonius 5, 6, 22, 69–71, 73–76,
     78–80, 82, 86, 87, 128, 153, 172, 193, 194

P. gonionotus 6, 60, 61, 191
P. tetrazona 5, 6, 22, 69, 71, 72, 74, 82, 86

R

rainbow trout, see *Oncorhynchus mykiss* 96
red crucian carp, see *Carassius auratus* 119
Rhodeus 48
R. amarus 25
R. ocellatus 13
R. ocellatus ocellatus 31, 60, 62, 63, 65, 105,
     128
rice field eel, see *Monopterus albus* 3
Richardsonius balteatus 41, 44, 45
rivulus, see *Kryptolebias marmoralus* 161
roach, see *Rutilus rutilus* 43
rockcod, see *Cephalopholis boenak* 175
rose bitterling, see *Rhodeus ocellatus* 13
rosy barb, see *Puntius conchonius* 73
rudd, see *Scardinus erythrophthalmus* 47
Rutilus rutilus 37, 38, 43, 44, 47, 174

S

Salmo gaidneri 116
S. salar 6, 14, 26, 40, 42, 45, 56, 57, 91, 94, 96,
     102, 116, 126, 194–196
S. salmo 79
S. trutta 6, 42, 44, 45, 56, 91, 94, 96, 126, 195
sailfin molly, see *Poecilia latipinna* 10
Salvelinus alpinus 26, 37, 82, 102, 193–195
S. confluentis 39, 45
S. fontinalis 6, 36, 39, 42, 44, 45, 72, 96, 108,
     114
S. leucomaensis 54
S. namycush 25, 44, 45, 194, 196
Saratherodon galilaeus 11, 44
Sarpa salpa 162
Scardinius erythrophthalmus 37, 38, 47
Schizothorax 114
shell brooding cichlid see *Telematochromis*
     vittatus 1
Scophthalmus maximus 31; 100
Sebastes alutus 14
S. altrovitens 14
S. schlegeli 14
sea bream, see *Calamus bajonado* 102
Semaprochilodus taeniatus 30
Semotilus atromaculatus 6
Seriola dumerili 25
Serraniculus 161
Serranus 161
Serranus baldwini 162–164, 207
S. hepatus 130
S. fasciatus 163, 207

S. tortugarum 162
Serrapinus notomelas 29
Silurus asotus 60, 61, 98
S. glanis 82, 92, 106
silver bream, see *Blicca bjoerkna* 43
silver carp, see *Hypophthalmichthys molitrix*
    39
Sinocyclocheilus 113
Sparus 161
S. aurata 6, 48, 106
Sparisoma cretense 2
S. radians 168, 172
S. viride 162
Spinachia spinachia 14
spinous loach, see *Cobitis biwae* 52
Squalias 116
S. alburnoides 8, 10, 155, 126–128, 178–181,
    184–189
S. carolittertii 178, 184–186
S. pynaicus 178, 179, 186
Stephanolepis cirrhifer 31
stickleback, see *Gasterosteus aculeatus* 93
stone loach, see *Noemacheilus barbatulus* 104
St. Peter's tilapia, see *Saratherodon galilaeus*
    11
Synbranchus marmoratus 176
Syngnathus floridae 14
S. typhle 14

T

Takifugu rubripes 20, 115, 193, 194
tampaqu, see *Colossoma macroponum* 41
Tanakia lanceolata 48
T. limbata 48
T. koreensis 25
T. signifier 25
Telematochromis vittatus 1
tench, see *Tinca tinca* 41
Tetraodon fluviatilis 20
T. nigroviridis 20, 193
tiger barb, see *Punctius tetrazona* 75
Tinca tinca 6, 52, 90, 92, 101, 103–105, 126,
    127

Thalassoma 46
T. bifasciatum 170, 171, 174–176
Thoracocharax stellatus 30
Thunnus orientalis 154
Thymallus thymallus 6
Tomeurus gracilis 76
tongue role, see *Cynoglossus semilaevis* 24
Tor 113
Tricanthus brevirostris 28, 33
Trichomycterus davisi 90
Trimma 165
Tripotheus 21, 23, 25
T. albus 21, 29
T. angulatus 29
T. elongatus 21, 29
T. flavus 29
T. guentheri 21, 29, 33
T. paranensis 21, 29
T. signatus 29
turbot see *Psetta maxima* 102

V

Valenciennea longipinnis 11
Varicorhinus 113
Verasper moseri 102

W

Wallago attu 114
write grouper, see *Epenephelus aneus* 41
widow tetra, see *Gymnocorynbus ternetzi* 72

X

Xiphophorus sp 35, 130
X. alvarezi 31
X. helleri 1, 35, 44, 192
X. maculatus 18, 23, 26, 35, 44, 192, 194, 196
X. milleri 196
X. nezahualcotyl 196

Z

Zeus faber 33

Subject Index

A

Acrosome 4, 37
Alu elements 20
Androdiocious 9, 10
Anthropogenic influence 37
Apopostis 8, 149
Aromatase 161, 163
Asymmetrical hybridization 46
Autosomal factor 50

B

Band sharing index (BSI) 57, 108
Biomarkers 170
Bisexual potency 16, 130, 152, 159, 206, 207

C

Chemo attractant 3, 133, 142
Chemokine SDF1 133
Chimera 135, 142, 143
Chromosomes
    B or micro 8, 16, 19, 179–182
    Bacterial artificial (BAC) 24
    Balanced 124
    banding 19
    Heteromorphic sex 7, 21, 23, 24, 26–28
    Homomorphic 21, 23, 161
    Heterologous 100
    Homologous 100, 105
    Maternal 79
    Neo-sex 195
    Paternal 43, 79
    Sex 7, 17, 19–21, 23–27, 34, 50, 63, 84, 176, 193–197
Clonal
    diversity 10, 184
    egg 91
    sperm 186, 189
Colonizing efficiency 136, 139, 141, 145, 146, 153–155
Comparative genome hybridization (CGH) 26

Cross
    Back 40, 44–47, 62
    Hybrid 37, 39
    Interfamily 41
    Intergeneric 37, 40, 41
    Interploid 97, 125
    Interspecific 49
    Reciprocal 39, 40, 41, 43
Cytonuclear linkage disequilibria 46
Cytogenetic
    characterization 19, 23
    evidence 29, 31
    features 25
    transformation 20
Cytoplasmic incompatibility 43
Cytotypes 30, 32

D

Differentiation
    Chromosome 20, 24–26
    Gonadal 8, 48, 142, 197
    Molecular 18
    Ovarian 3
    Sex 3, 16, 18, 62, 175
    Spermatogonial 197
    Testicular 3, 175, 197
    Trans 3
Dimensionless theory 170

E

Endemic species 27, 46
Estrogen (E$_2$) 144, 148
Estrogen receptor 148, 161
Estrogen treated 76, 142, 144, 145, 158

F

Fertilization
    external 2, 37
    internal 2
Fluorescent *in situ* hybridization (FISH) 23–25

Fluorescence microscopy 150
Frankenstein monster 144

G

Genome
    addition 10, 129, 180, 183
    Divergent 36
    DNA 74, 78, 85, 172
    doubling 114, 115
    mapping 171
    Maternal 10, 57, 65, 69, 70, 75
    Mitochondreal 46, 187
    Novel 36
    Paternal 10, 57, 66, 69, 180
    sequencing 193
    size 20, 94, 112, 125, 203
Genital ridge (GR) 132, 133, 137, 139, 141,
    143
Genetic
    contamination 57, 62, 66, 75
    pollution 45
    retrogression 4, 44, 45, 89
Germ cell
    identity 3
    lineage 131, 134
    transplantation 135
Germ line transmission 137
Global warming 16
Gonadal
    anlage 3, 139, 145
    sterility 100
Gonado somatic index (GSI) 82, 83, 104, 168,
    169

H

Haploid syndrome 53, 79
Harem 2, 9, 11, 164, 167, 168, 172, 174, 175,
    207
Hereditary transmission 19
Hermaphrodite
    Cyclical 162, 164, 175
    Diandric 18, 162, 168, 174
    Dichromatic 162, 164, 168
    Digynic 18, 162, 164, 168
        '*Lates*' model 168
        '*Gonostoma*' model 168
    Gamete exchanging 2, 9, 161–164, 207
    Marian 9, 18, 162, 164, 168
    Monandric 18, 162, 164, 168
        '*Amphiprion*' model 168
        '*Diplodus*' model 168
    Monochromatic 18, 162, 164, 168
    Monogynic 18, 162, 164, 168
    Okinawan 9, 18, 162, 164
    Protandric 9, 15, 18, 48, 50, 99, 164, 166,
        168, 176, 202
    Protogynic 9, 15, 18, 48, 50, 106, 162, 164,
        166, 168, 170, 175, 202, 207
    Self-fertilizing 1, 2, 9, 10, 15, 161, 162,
        207
    Serial 1, 9, 16, 18, 160, 161, 162, 164
    Sequential 1, 11, 18, 161, 164, 201
    Unidirectional 161, 164
    Simultaneous 160–162
Heterochromatization 21, 25
Heterochronic transoplantation 135, 142
Heterologous activation 4, 51
Heterologous egg 79
Heterologous fertilization 4
Heterospecific insemination 4, 6, 36, 94
Hormonal profile 100
Hox gene clusters 115
Hybrid
    fertility 43
    gonad 141, 146
    line 56
    triploid 96
    tetraploid 120, 121
    zone 37
Hybridogenesis 2, 8, 55, 178–180, 183–186

I

Inbreeding 10, 11, 47
Incidence of natural hybrid 37
Interploid 52, 91, 98, 123, 125
Intracytoplasmic sperm injection (ICSI) 6
Introgressive hybridization 45–47
Isochromic transplantation 135
Isoforms 163, 203
Isogenic clone 51, 69

K

Karyology 19
Karyotype 19, 24, 26, 28, 33–35, 40, 42, 75

L

Laser microdissection (LMD) 24
Leydig cells 3, 149, 150
Lophopodial movement 145

M

Marker
    Allozyme 13, 40, 57, 59, 72
    Barbel 122
    Chromosome 93, 122
    Diagnostic DNA fragment 40, 122
    DNA fingerprinting 10, 57–59, 62, 108, 153, 188
    Erythrocytes 93, 120–122, 185
    Flow cytometry 59, 94, 137, 147, 185
    Green Flurescent Protein (*Gfp*) gene 75, 134–139, 142, 147, 152–155
    Isozyme 57, 72
    Microsatellite 10, 40, 57, 63, 185
    Minisatellite 10
    Molecular 17, 44, 57, 72, 78, 154, 193
    mtDNA 10, 40, 45, 100, 153
    Multi locus nDNA 46, 57
    Multiple 39, 40
    Phenotype 37, 40, 57, 93
    Protein 39, 57, 93
    Tissue grafting 59
Mating system 1, 9
Meiogynogenic 56, 59, 61, 62, 64, 65, 177
Mesonephric blastema 3
Micropyle 4, 5, 7, 73, 78, 89, 104, 115, 118
Microsatellite linkage 194
Mitogynogenic 56, 59, 61, 62, 64, 65
'Montecarlo' simulation 180
Molecular pedigree analysis 105
Morphotypes: male
    α or territorial 11, 13, 165, 175, 176
    Cuckolder 1, 11–13
    Heramic 1, 2
    Hooknose 1
    Jack 1
    Parental 1, 13
    Pairing 1, 12
    Piracy 1, 11, 12
    Primary 1, 12, 164, 174–176
    Secondary 1, 164, 174
    Satellite 11, 12
    Sneaker 1, 11, 12
    Streaker 1, 11
    Sub-male 165
Mosaics 54, 116, 184
mtDNA clock calibration 10
Muller's ratchet 8, 10, 16, 178, 180
Mutational
    analysis 132
    meltown 8, 16, 178, 180
    load 180

Multiple maternity 13, 15
Multiple patermity 13, 15
Multiple sex chromosome system 26–32, 35
Multiploid eggs 126, 127
Multiploid sperm 128

N

Neo female 76, 77, 84, 85, 94
Neo male 23, 60, 64, 94
Nuclear Organization Region (NOR) 23, 25
nuage 131
nanos 131

O

Oocyte 4, 61, 65, 105, 106
Oogenesis 48
Oogenic rescue mechanism 177

P

Paternal chromosomal fragments 66, 67, 184
Paternal genomic leakage 8, 179–181
Paternal triploid 79, 90, 92, 96–99, 102
PEG incubation 73, 80, 81
Premeiotic endomitosis 42, 52, 55, 56, 90, 91, 112, 127
Primordial Germ Cells (PGCs) 2, 16, 130–142, 144–147, 153–157
Plolygenic sex determination 7, 203, 205
Proliferative mitotic phase 48, 153
Post-ovulatory aging 90, 92, 105

R

Recombination frequency 58
Repetitive elements 24
Robertsonian fussion 34

S

Sandwich 139, 140, 156
Sandwich derived goldfish 142
Second generation tetraploid 118
Sertoli cells 3, 148–150
Sex
    Genetic 17, 172, 176, 193
    Inter 48
    Ratio 11, 13, 35, 47–50, 62, 64–66, 97, 99, 118–120, 136, 155, 157, 168, 175, 185, 186
Sex determining genes 17, 24, 191, 193, 195
Sex determination locus SEX 194
Sex determining master gene 195, 196, 205, 206

Sex determining mechanism (SMD) 7, 18, 47–49, 97, 159, 190, 191, 195, 206
Sex determining system 23, 35, 48, 59, 191, 195
Sex specific organization 26
Sex linked
    Bep, Rdt 194
    Dmrt1bY 197
    GH pseudogene *GH-ψ* 194
    RAPD marker *CgaY1 CgaY2* 194
Sex specific probes 193
Size advantage model 27
Sexual dimorphism 3, 131, 139, 155, 156
Somatic cell lineage 3
Sperm
    Aneuploid 53, 73, 101–103, 105, 119, 188
    donor 58
    count 58, 73, 82, 83, 103, 104
    Heterologous 54, 56, 70, 79, 91, 96
    Homologous 54, 56, 91, 96
    Irradiated 56, 64
    Multiploid 128
    Multiple tailed 42, 102
    Unreduced 128
Spermatocyte 42, 48, 101, 150, 151
Spermatid 101, 103, 149, 150
Spermatogenesis 42, 48, 103, 140, 148, 149
Spermatogonium 100, 103, 148–151, 153, 154, 202
Spermatogonial cell line 154
Spermatogonial stem cells (SSCs) 16, 130, 146–155, 157–159
Spermatozoa 4, 37, 101, 109, 135, 137, 138, 153, 200
Splake 44, 50

Splicing site 203, 204
Superfemale 22, 58, 82, 85
Supermale 22, 50, 77–79, 81–87
Sub metacentric marker chromosomes 120
Surrogate 69, 130, 147, 153
Synaptonemal complex (SC) 23, 24

T

Transposable elements (TE) 20, 197
Trigenomic hybrid triploid 45, 94, 95, 119
Trigenomic triploid 45, 94, 96, 98, 119
Transmission frequency 136, 137, 139, 146, 153, 155

U

Unexpected male 61, 66
Unexpected female 84, 87
Unisexual lineage 128
Unitary origin 3
Unreduced eggs 42, 52, 54, 55
Unreduced sperm 128

V

Vasa 131, 134, 151, 154
vitellogenesis 49

X

Xenogenic goldfish 145

Z

Zygoparity 2

Colour Plate Section

Chapter 3

Fig. 8 External appearance of *Poecilia velifera* (a) male, (b) female, *P.sphenops* (c) male, (d) female, and their mottled hybrid (e) male and (f) female and striped hybrid (g) male and (h) female (from George and Pandian, 1997).

Chapter 4

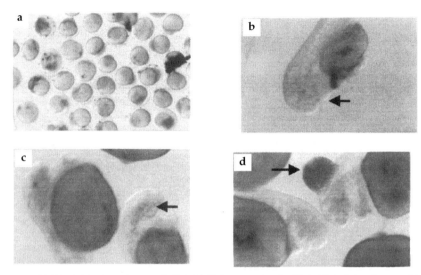

Fig. 11 (a) Haploid eggs of normal diploid *Heteropneustes fossilis* initiated development by haploid sperm. Note the failure of (b) jaw formation, (c) eye formation, and (d) formation of tumor (from Koteeswaran and Pandian, 2011).

Chapter 5

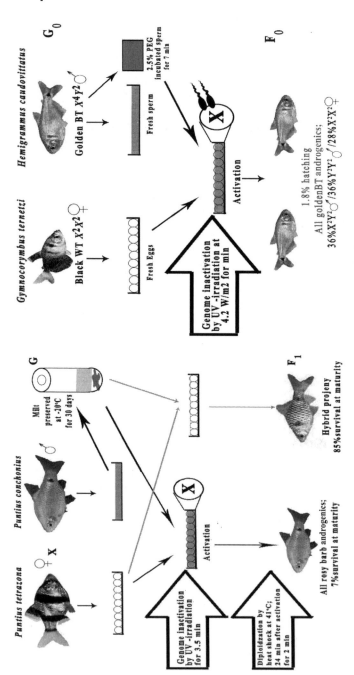

Fig. 15 Left panel = Induction of allo-androgenics of *Puntius conchonius* using its preserved sperm for activation of genome-inactivated (surrogate) eggs of *P.tetrazona* (from Pandian and Kirankumar, 2003). Right panel = Allo-androgenics of *Hemigrammus caudovittatus* using its preserved sperm for activation of genome-inactivated (surrogate) eggs of *Gymnocorymbus ternetzi* (from David 2004).

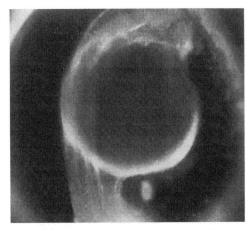

Fig. 16 Green Fluorescent Protein (*Gfp*) gene expression in 16 hr-old haploid androgenic blond *P.conchonius* (from Pandian and Kirankumar, 2003).

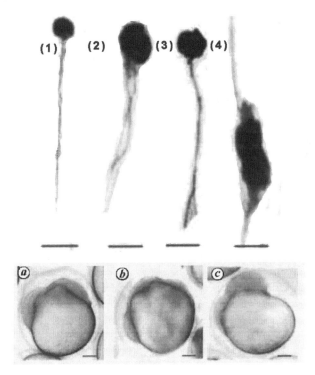

Fig. 22 Effect of PEG incubation of milt on fusion of sperm of BT tetra *Hemigrammus caudovittatus*. Upper panel 1 = normal sperm, 2 = incompletely fused sperm with to independently moving tails, 3 = completely fused double sperm and 4 = head to head wrongly fused sperm. Lower panel : Photograph showing the first division after entry of (a) double, (b) triple BT sperm, (c) abnormal first mitotic division after entry of double sperm (from David and Pandian, 2008).

Chapter 7

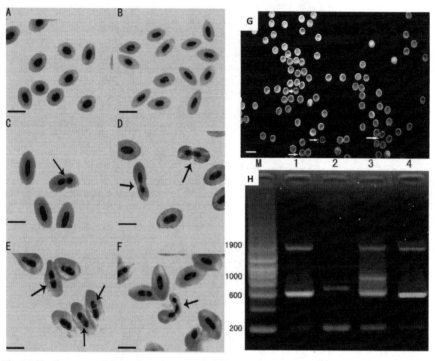

Fig. 34 Erythrocytes of the red crucian carp (RCC) and blunt snout bream (BSB) and polyploid hybrids. (A) normal erythrocytes with one nucleus in RCC (B) normal erythrocytes with one nucleus in BSB (C) normal erythrocytes with one nucleus and unusual erythrocyte with two nyclei (arrows) in a 3n RB hybrid (D) normal erythrocytes with one nucleus and unusual erythrocyte with two nuclei (arrows) in a 4n RB hybrid (E) normal erythrocytes with one nucleus and unusual erythrocyte with two nuclei (arrows) in a 5n RB hybrid (F) normal erythrocytes with one nucleus and unusual erythrocyte with two or three nuclei (arrows) in a 5n RB hybrid. Bar in A-F = 0.01 mm. (G) Simultaneous production of two sized eggs by 4n RB hybrids. Smaller eggs are shown by arrows. Bar = 0.4 cm. (H) Amplified DNA fragments resulting from PCR based on the primers of HMG of *Sox* genes in BSB, RCC, 3n RB and 4n RB hybrids. M = DNA ladder marker. Lane 1 = 3DNA fragments in RCC, Lane 2 = 2 DNA fragments in BSB, Lane 3 = 4 DNA fragments in 4n RB hybrids. Lane 4 = 3 DNA fragments in 3n RB hybrids (from Liu et al., 2007a).

Chapter 8

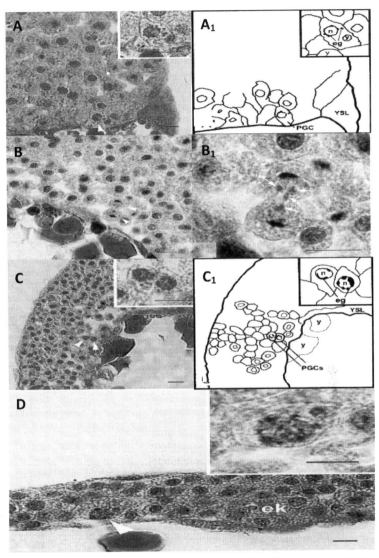

Fig. 35 Transverse sections of zebrafish embryos showing the development of PGC. A and A_1 at sphere stage (4 hpf), B and B_1 epiboly stage (6 hpf) C and C_1 shield stage (8 hpf), and D at bud stage (10 hpf). All inserts are higher magnifications. Arrow heads indicate PGCs. n = nucleus, y = yolk spherule, eg = eosinophilic granules, ek = embryonic keel. Scale bars indicate 10 μm (reproduced from Fig. 1, 2 of Nagai et al., 2001, *Zoological Science*, 18: 215–223).

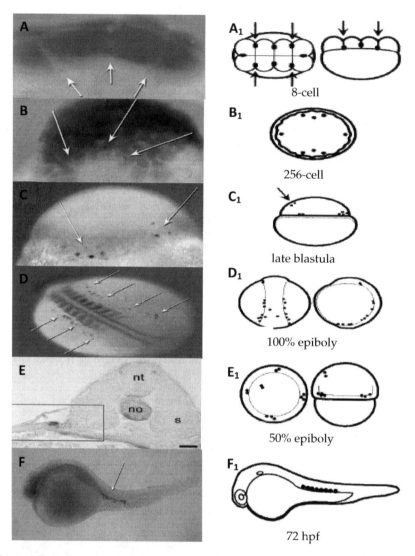

Fig. 36 Whole-mount *in situ* hybridization of cleavage (A,B,C) and segmentation (D, E, F) stages of goldfish embryos A_1, B_1, C_1, D_1, E_1 and F_1 are schematic figures of the corresponding developmental stages. Arrows indicate *vasa* signals. nt = neural tube, no = notochord, s = somite (reproduced from Fig. 1, 3 and 4 of Otani et al., 2002, Zoological Science, 19: 519–526).

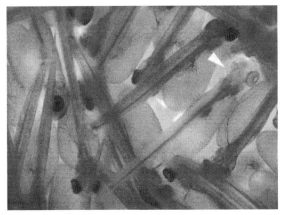

Fig. 38 Germ-line transmission of donor-derived orange coloured progenies of rainbow trout (from Takeuchi et al., 2003).

278   *Sex Determination in Fish*

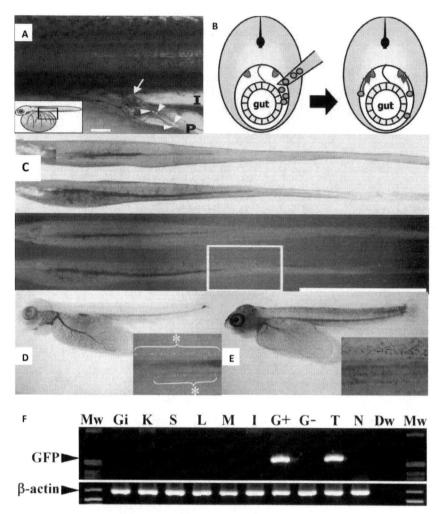

Fig. 39 Transplantation of *Gfp*-labelled PGCs into rainbow trout embryo. (A) Donor PGCs injected into the peritoneal cavity of an alevin. Arrow indicates the position at which the PGCs were injected. Arrow heads represent the PGCs. I = intestine; p = micropipette. (B) Schematic representation of the migration of donor PGCs on the peritoneal wall. (C) Bright view (top) and fluorescent view (bottom) of a pair of testes excised from a male parent (D) Orange coloured progenies with *Gfp* expression, which were sired by donor–derived spermatozoa and (E) the wild control without *Gfp* expression (right). (F) PCR analysis of the allogenic DNA extracted from 180 days old fry : Gi = gill, K = kidney, s = spleen, L = liver, M = muscle, I = intestine, G+ = donor-derived gonad; G- = gonad control, T = adipose tissue of an allogenic, N = control adipose (from Takeuchi et al., 2003).

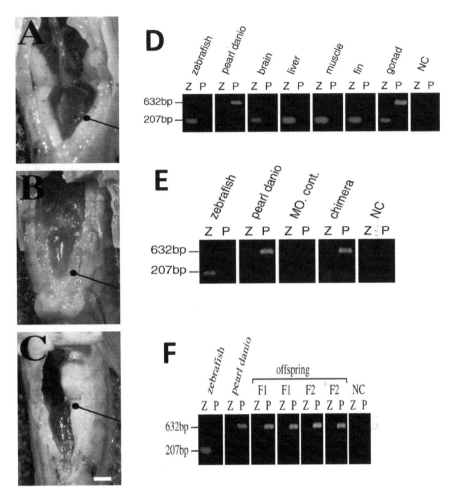

Fig. 41 Xenogenic pearl danio. A, B and C external appearance of gonads in males. A. Wild control, B.Sterile control injected with antigens *dnd* morpholino and C. Xenogenic pearl danio. (G) PCR genotyping of recipient, donor and chimeric fishes. The results of PCR with zebrafish specific primers are in the colum Z and those of pearl danio in the coloumn P. The zebrafish PCR product was expected to be 207 bp and the pearl danio 632 bp. (H) RT-PCR genotyping of gonads of the recipient, donor and their progenies. The gonad of the chimera had only pearl danio specific PCR product. (F) Genotyping of F_1 and F_2 progenies produced by the chimera. Z and P represent amplification of zebrafish and pearl danio specific primers, respectively (from Saito et al., 2008).

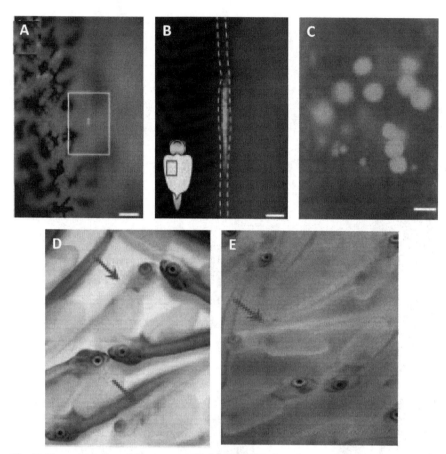

Fig. 44 Transplantation of donor *pvasa-Gfp* labelled SSCs into the recipient gonad of rainbow trout. (A) Fluorescent view of incorporation of donor SSCs into the recipient at 20 dpf, (B) Fluorescent view of proliferation of donor SSCs germ cells in a recipient testis at 7 months post-transplantation (pt) (C) Fluorescent view of donor SSCs–derived oocytes in a recipient ovary at 7 months pt (D) Donor derived F_1 albino progenies of a male recipient and (E) F_1 albino progenies from SSCs-derived female recipient (from Okutsu, T, Takeyuchi, Y, Takeyuchi, T, Yoshizaki, G, 2006b. Testicular germ cells can colonize sexually undifferentiated embryonic gonade and produce functional egg in fish. Proceedings of the National Academy of Sciences, USA, 103: 2725–2729).

Chapter 9

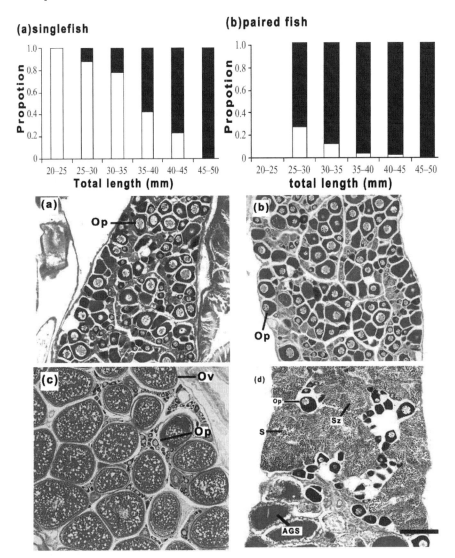

Fig. 49 Upper panel: The proportion of immature (open square) and mature (closed square) individuals in each size class for (a) single fish and (b) paired *Gobiodon erythrospilus*. Lower panels: Longitudinal section from (a) a typical juvenile (immature female) prior to experimental manipulation, (b) a juvenile (immature female) after remaining as a single for 42 days, (c) a juvenile paired with adult male for 42 days and (d) a juvenile paired with an adult female for 42 days. Op = previtellogenic oocytes, Ov = vitellogenic oocytes, S = spermatocytes, Sz = spermatozoa, AGS = accessory gonad structure. Scale bar = 0.1 mm (With kind permission by The Royal Society, London; from Hobbs, J-PA, Munday, PL, Jones, GP, 2004. Social induction of maturation and sex determination in a coral reef fish. Proc R Soc, London, 271B: 2109–2114).

Chapter 11

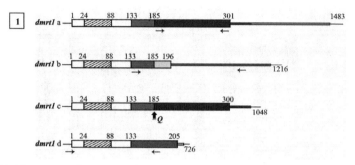

Fig. 56(1) *Monopterus albus*: Upper figure: Splicing of *dmrt* forms four kinds of mRNAs: *dmrt1a, dmrt1b, dmrt1c* and *dmrt1d*, which encode 4 proteins with 301, 196, 300 and 205 amino acids, respectively. *dmrt* has deleted an amino acid Q compared with that of *dmrt1a* because of alternative splicing. DM domains are indicated by shaded boxes from amino acids 24–88; sequences from amino acids 1–133 are common among the four transcripts. Alternatively spliced region in the 3′ are shown by different colours and same colour represents the same DNA sequence. The numbers in the end under the lines indicate nucleotide numbers of these cDNAs. Middle panel: (a) RT-PCR and (b) Northern blot analysis of *dmrt1* of the rice field eel. (a) Expression of *dmrt1* in three kinds of gonads. (b) The expression of *dmrt1* is high on testis, with a slightly lower expression in ovary and ovotestis. Lower panel: The relative expression of *dmrt1* isoforms to *beta-actin* in ovary, ovotestis I (pre-intersex stage), ovotestis II (medium intersex stage), ovotestis III (post intersex stage) and testis measured by Real time PCR (from Huang et al., 2005).